COLLAPSE OF BURNING BUILDINGS

COLLAPSE OF BURNING BUILDINGS

A GUIDE TO FIREGROUND SAFETY — SECOND EDITION

Vincent Dunn

> **Disclaimer:** The recommendations, advice, descriptions, and the methods in this book are presented solely for educational purposes. The author and publisher assume no liability whatsoever for any loss or damage that results from the use of any of the material in this book. Use of the material in this book is solely at the risk of the user.

Copyright © 2010 by
PennWell Corporation
1421 South Sheridan Road
Tulsa, Oklahoma 74112-6600 USA

800.752.9764
+1.918.831.9421
sales@pennwell.com
www.FireEngineeringBooks.com
www.pennwellbooks.com
www.pennwell.com

Marketing Coordinator: Jane Green
National Account Executive: Barbara McGee
Director: Mary McGee
Managing Editor: Jerry Naylis
Production Manager: Sheila Brock
Production Editor: Tony Quinn
Book Designer: Susan E. Ormston
Cover Designer: Alan McCuller

Library of Congress Cataloging-in-Publication Data

Dunn, Vincent, 1935-
 Collapse of burning buildings : a guide to fireground safety / Vincent Dunn. -- 2nd ed.
    p. cm.
 Includes index.
 ISBN 978-1-59370-233-5
 1. Fire extinction--Safety measures. 2. Building failures--Safety measures. I. Title.
 TH9182.D86 2010
 628.9'25--dc22
                                        2010013871

All rights reserved. No part of this book may be reproduced,
stored in a retrieval system, or transcribed in any form or by any means,
electronic or mechanical, including photocopying and recording,
without the prior written permission of the publisher.

Printed in the United States of America

8   9   10   18   17

DEDICATION

This book is dedicated to the 58 New York City Fire Department (FDNY) chiefs, company officers, and firefighters who have been killed by burning buildings that collapsed in the 53 years from 1956 to 2009, as well as the 343 chiefs, company officers, and firefighters killed in the terrorist attack on World Trade Center on 9/11:

1. Deputy Chief Thomas Reilly: floor collapse
2. Battalion Chief Walter Higgins: floor collapse
3. Battalion Chief Frank Tuttlemondo: wood-frame building collapse
4. Captain Walter Bersig: floor collapse
5. Captain Scott Lapiedra: floor collapse
6. Captain William Russell: wall collapse
7. Captain John Stelmack: parapet wall coping stone
8. Lieutenant Joseph Beetle: ceiling collapse
9. Lieutenant James Blackmore: floor collapse
10. Lieutenant Howard Carpluk: floor collapse
11. Lieutenant John Clancy: floor collapse
12. Lieutenant James Cutillo: roof collapse
13. Lieutenant Robert Dolney: wood-frame building collapse
14. Lieutenant John Finley: floor collapse
15. Lieutenant Charles Hunt: pier collapse
16. Lieutenant John Molloy: parapet wall collapse
17. Lieutenant Joseph Priore: floor collapse
18. Firefighter Richard Andrew: wall collapse
19. Firefighter John Berry: wall collapse
20. Firefighter Bernard Blumenthal: floor collapse
21. Firefighter Charles Bouton: roof collapse
22. Firefighter Edward Carroll: parapet wall collapse
23. Firefighter John Downing: wall collapse
24. Firefighter Thomas Earl: ceiling collapse
25. Firefighter Thomas Egan: parapet wall collapse
26. Firefighter Brian Fahey: floor collapse
27. Firefighter John Farragher: floor collapse
28. Firefighter Harry Ford: floor collapse
29. Firefighter James Galanaugh: floor collapse
30. Firefighter Richard Gifford: wall collapse
31. Firefighter Arthur Hanson: parapet wall collapse
32. Firefighter Harold Hastings: roof collapse
33. Firefighter Fred Hellaver: parapet wall collapse
34. Firefighter William Hollan: parapet wall collapse
35. Firefighter Charles Infosino: parapet wall collapse
36. Firefighter Rudolph Kaminsky: floor collapse
37. Firefighter Kevin Kane: ceiling collapse
38. Firefighter Joseph Kelly: floor collapse
39. Firefighter Carl Lee: floor collapse
40. Firefighter Ernest Marquart: ceiling collapse
41. Firefighter James Marino: wall collapse
42. Firefighter Robert Meill: roof collapse
43. Firefighter William McCarron: floor collapse
44. Firefighter John McKenna: roof collapse
45. Firefighter Peter McLaughlin: ceiling collapse
46. Firefighter James McManus: roof collapse
47. Firefighter William O'Connor: roof collapse
48. Firefighter Daniel Rey: floor collapse
49. Firefighter Mike Reilly: floor collapse
50. Firefighter George Rice: roof collapse
51. Firefighter Alfred Ronaldson floor collapse
52. Firefighter Charles Sanchez: floor collapse
53. Firefighter William Schmid: floor collapse
54. Firefighter Stanley Skinner: ceiling collapse
55. Firefighter Bernard Tepper: floor collapse

56. Firefighter Edward Tuite: roof collapse
57. Firefighter Louis Valentino: roof collapse
58. Firefighter George Zahn: wall collapse

343 who died on 9/11

1. First Deputy Commissioner William M. Feehan
2. Chief of Department Peter J. Ganci, Jr., COD
3. Assistant Chief Gerard A. Barbara, Operations
4. Assistant Chief Donald J. Burns, Operations
5. Deputy Chief Dennis A. Cross, Battalion 57
6. Deputy Chief Raymond M. Downey, SOC
7. Deputy Chief Edward F. Geraghty, Battalion 9
8. Department Chaplain Mychal F. Judge, OFM
9. Deputy Chief Charles L. Kasper, SOC
10. Deputy Chief Joseph R. Marchbanks, Jr., Battalion 12
11. Deputy Chief Orio J. Palmer, Battalion 7
12. Deputy Chief John M. Paolillo, SOC
13. Battalion Chief James M. Amato, Squad Co. 1
14. Battalion Chief Thomas P. DeAngelis, Battalion 8
15. Battalion Chief Dennis L. Devlin, Division 3
16. Battalion Chief John J. Fanning, Haz-Mat Operations
17. Battalion Chief Thomas J. Farino, Engine Co. 26
18. Battalion Chief Joseph D. Farrelly, Engine Co. 4
19. Battalion Chief Joseph Grzelak, Battalion 48
20. Battalion Chief Thomas T. Haskell, Jr., Ladder Co. 132
21. Battalion Chief Brian C. Hickey, Rescue Co. 4
22. Battalion Chief William J. McGovern, Battalion 2
23. Battalion Chief Louis J. Modafferi, Rescue Co. 5
24. Battalion Chief John M. Moran, Soc
25. Battalion Chief Richard A. Prunty, Battalion 2
26. Battalion Chief Matthew L. Ryan, Battalion 4
27. Battalion Chief Fred C. Scheffold, Battalion 12
28. Battalion Chief Lawrence T. Stack, Safety Battalion 1
29. Battalion Chief John P. Williamson, Battalion 6
30. Captain Daniel J. Brethel, Ladder Co. 24
31. Captain Patrick J. Brown, Ladder Co. 3
32. Captain Vincent E. Brunton, Ladder Co. 105
33. Captain William F. Burke, Jr., Engine Co. 21
34. Captain Frank J. Callahan, Ladder Co. 35
35. Captain Martin J. Egan, Jr., Ladder Co. 118
36. Captain Michael A. Esposito, Squad Co. 1
37. Captain John R. Fischer, Ladder Co. 20
38. Captain Vincent F. Giammona, Ladder Co. 5
39. Captain Terence S. Hatton, Rescue Co. 1
40. Captain Walter G. Hynes, Ladder Co. 13
41. Captain Frederick J. Ill, Jr., Ladder Co. 2
42. Captain William E. McGinn, Squad Co. 18
43. Captain Thomas C. Moody, Engine Co. 310
44. Captain Daniel O'Callaghan, Ladder Co. 4
45. Captain William S. O'Keefe, Engine Co. 154
46. Captain Vernon A. Richard, Ladder Co. 7
47. Captain Timothy M. Stackpole, Ladder Co. 103
48. Captain Patrick J. Waters, Haz-Mat Co. 1
49. Captain David T. Wooley, Ladder Co. 4
50. Lieutenant Joseph Agnello, Ladder Co. 118
51. Lieutenant Brian G. Ahearn, Engine Co. 230
52. Lieutenant Gregg Atlas, Engine Co. 10
53. Lieutenant Steven J. Bates, Engine Co. 235
54. Lieutenant Carl J. Bedigian, Engine Co. 214
55. Lieutenant John A. Crisci, Haz-Mat Co. 1
56. Lieutenant Edward A. D'Atri, Squad Co. 1
57. Lieutenant Manuel Del Valle, Jr., Engine Co. 5
58. Lieutenant Andrew J. Desperito, Engine Co. 1
59. Lieutenant Kevin W. Donnelly, Ladder Co. 3
60. Lieutenant Kevin C. Dowdell, Rescue Co. 4
61. Lieutenant Michael N. Fodor, Ladder Co. 21
62. Lieutenant David J. Fontana, Squad Co. 1
63. Lieutenant Andrew A. Fredericks, Squad Co. 18
64. Lieutenant Peter L. Freund, Engine Co. 55
65. Lieutenant Charles W. Garbarini, Ladder Co. 61

66. Lieutenant Ronnie E. Gies, Squad Co. 288
67. Lieutenant John F. Ginley, Engine Co. 40
68. Lieutenant Geoffrey E. Guja, Engine Co. 82
69. Lieutenant Joseph P. Gullickson, Ladder Co. 101
70. Lieutenant David Halderman, Squad Co. 18
71. Lieutenant Vincent G. Halloran, Ladder Co. 8
72. Lieutenant Harvey L. Harrell, Rescue Co. 5
73. Lieutenant Stephen G. Harrell, Ladder Co. 157
74. Lieutenant Michael K. Healey, Squad Co. 41
75. Lieutenant Timothy B. Higgins, Squad Co. 252
76. Lieutenant Anthony M. Jovic, Ladder Co. 34
77. Lieutenant Thomas R. Kelly, Ladder Co. 105
78. Lieutenant Ronald T. Kerwin, Squad Co. 288
79. Lieutenant Joseph G. Leavey, Ladder Co. 15
80. Lieutenant Michael F. Lynch, Ladder Co. 4
81. Lieutenant Patrick J. Lyons, Squad Co. 252
82. Lieutenant Charles J. Margiotta, Ladder Co. 85
83. Lieutenant Peter C. Martin, Rescue Co. 2
84. Lieutenant Paul R. Martini, Engine Co. 201
85. Lieutenant Paul T. Mitchell, Ladder Co. 110
86. Lieutenant Dennis Mojica, Rescue Co. 1
87. Lieutenant Raymond E. Murphy, Ladder Co. 16
88. Lieutenant Robert B. Nagel, Engine Co. 58
89. Lieutenant John P. Napolitano, Rescue Co. 2
90. Lieutenant Thomas G. O'Hagan, Engine Co. 52
91. Lieutenant Glenn C. Perry, Ladder Co. 34
92. Lieutenant Philip S. Petti, Ladder Co. 148
93. Lieutenant Kevin J. Pfeifer, Engine Co. 33
94. Lieutenant Kenneth J. Phelan, Engine Co. 217
95. Lieutenant Michael T. Quilty, Ladder Co. 11
96. Lieutenant Robert M. Regan, Ladder Co. 118
97. Lieutenant Michael T. Russo, Squad Co. 1
98. Lieutenant Christopher P. Sullivan, Ladder Co. 111
99. Lieutenant Robert F. Wallace, Engine Co. 205
100. Lieutenant Jeffrey P. Walz, Ladder Co. 9
101. Lieutenant Michael P. Warchola, Ladder Co. 5
102. Lieutenant Glenn E. Wilkinson, Engine Co. 238
103. Fire Marshal Ronald P. Bucca, Manhattan Base
104. Fire Marshal Andre G. Fletcher, Rescue Co. 5
105. Fire Marshal Vincent D. Kane, Engine Co. 22
106. Fire Marshal Kenneth B. Kumpel, Ladder Co. 25
107. Fire Marshal Paul J. Pansini, Engine Co. 10
108. Firefighter Eric T. Allen, Squad Co. 18
109. Firefighter Richard D. Allen, Ladder Co. 15
110. Firefighter Calixto Anaya, Jr., Engine Co. 4
111. Firefighter Joseph J. Angelini, Sr., Rescue Co. 1
112. Firefighter Joseph J. Angelini, Jr., Ladder Co. 4
113. Firefighter Faustino Apostol, Jr., Battalion 2
114. Firefighter David G. Arce, Engine Co. 33
115. Firefighter Louis Arena, Ladder Co. 5
116. Firefighter Carl F. Asaro, Battalion 9
117. Firefighter Gerald T. Atwood, Ladder Co. 21
118. Firefighter Gerard Baptiste, Ladder Co. 9
119. Firefighter Matthew E. Barnes, Ladder Co. 25
120. Firefighter Arthur T. Barry, Ladder Co. 15
121. Firefighter Stephen E. Belson, Ladder Co. 24
122. Firefighter John P. Bergin, Rescue Co. 5
123. Firefighter Paul M. Beyer, Engine Co. 6
124. Firefighter Peter A. Bielfeld, Ladder Co. 42
125. Firefighter Brian E. Bilcher, Engine Co. 33
126. Firefighter Carl V. Bini, Rescue Co. 5
127. Firefighter Christopher J. Blackwell, Rescue Co. 3
128. Firefighter Michael L. Bocchino, Battalion 48
129. Firefighter Frank J. Bonomo, Engine Co. 230
130. Firefighter Gary R. Box, Squad Co. 1
131. Firefighter Michael Boyle, Engine Co. 33
132. Firefighter Kevin H. Bracken, Engine Co. 40
133. Firefighter Michael E. Brennan, Ladder Co. 4
134. Firefighter Peter Brennan, Squad Co. 288
135. Firefighter Andrew C. Brunn, Ladder Co. 5
136. Firefighter Gregory J. Buck, Engine Co. 201
137. Firefighter John P. Burnside, Ladder Co. 20

138. Firefighter Thomas M. Butler, Squad Co. 1
139. Firefighter Patrick D. Byrne, Ladder Co. 101
140. Firefighter George C. Cain, Ladder Co. 7
141. Firefighter Salvatore B. Calabro, Ladder Co. 101
142. Firefighter Michael F. Cammarata, Ladder Co. 11
143. Firefighter Brian Cannizzaro, Ladder Co. 101
144. Firefighter Dennis M. Carey, Haz-Mat Co. 1
145. Firefighter Michael S. Carlo, Engine Co. 230
146. Firefighter Michael T. Carroll, Ladder Co. 3
147. Firefighter Peter J. Carroll, Squad Co. 1
148. Firefighter Thomas A. Casoria, Engine Co. 22
149. Firefighter Michael J. Cawley, Ladder Co. 136
150. Firefighter Vernon P. Cherry, Ladder Co. 118
151. Firefighter Nicholas P. Chiofalo, Engine Co. 235
152. Firefighter John G. Chipura, Engine Co. 219
153. Firefighter Michael J. Clarke, Ladder Co. 2
154. Firefighter Steven Coakley, Engine Co. 217
155. Firefighter Tarel Coleman, Squad Co. 252
156. Firefighter John M. Collins, Ladder Co. 25
157. Firefighter Robert J. Cordice, Engine Co. 152
158. Firefighter Ruben D. Correa, Engine Co. 74
159. Firefighter James R. Coyle, Ladder Co. 3
160. Firefighter Robert J. Crawford, Safety Battalion 1
161. Firefighter Thomas P. Cullen, III, Squad Co. 41
162. Firefighter Robert Curatolo, Ladder Co. 16
163. Firefighter Michael D. D'Auria, Engine Co. 40
164. Firefighter Scott M. Davidson, Ladder Co. 118
165. Firefighter Edward J. Day, Ladder Co. 11
166. Firefighter Martin N. DeMeo, Haz-Mat Co. 1
167. Firefighter David P. DeRubbio, Engine Co. 226
168. Firefighter Gerard P. Dewan, Ladder Co. 3
169. Firefighter George DiPasquale, Ladder Co. 2
170. Firefighter Gerard J. Duffy, Ladder Co. 21
171. Firefighter Michael J. Elferis, Engine Co. 22
172. Firefighter Francis Esposito, Engine Co. 235
173. Firefighter Robert E. Evans, Engine Co. 33
174. Firefighter Terrence P. Farrell, Rescue Co. 4
175. Firefighter Lee S. Fehling, Engine Co. 235
176. Firefighter Alan D. Feinberg, Battalion 9
177. Firefighter Michael C. Fiore, Rescue Co. 5
178. Firefighter John J. Florio, Engine Co. 214
179. Firefighter Thomas J. Foley, Rescue Co. 3
180. Firefighter Robert J. Foti, Ladder Co. 7
181. Firefighter Thomas Gambino, Jr., Rescue Co. 3
182. Firefighter Thomas A. Gardner, Haz-Mat Co. 1
183. Firefighter Matthew D. Garvey, Squad Co. 1
184. Firefighter Bruce H. Gary, Engine Co. 40
185. Firefighter Gary P. Geidel, Rescue Co. 1
186. Firefighter Denis P. Germain, Ladder Co. 2
187. Firefighter James A. Giberson, Ladder Co. 35
188. Firefighter Paul J. Gill, Engine Co. 54
189. Firefighter Jeffrey J. Giordano, Ladder Co. 3
190. Firefighter John J. Giordano, Engine Co. 37
191. Firefighter Keith A. Glascoe, Ladder Co. 21
192. Firefighter James M. Gray, Ladder Co. 20
193. Firefighter Jose A. Guadalupe, Engine Co. 54
194. Firefighter Robert W. Hamilton, Squad Co. 41
195. Firefighter Sean S. Hanley, Ladder Co. 20
196. Firefighter Thomas P. Hannafin, Ladder Co. 5
197. Firefighter Dana R. Hannon, Engine Co. 26
198. Firefighter Daniel E. Harlin, Ladder Co. 2
199. Firefighter Timothy S. Haskell, Squad Co. 18
200. Firefighter Michael H. Haub, Ladder Co. 4
201. Firefighter John F. Heffernan, Ladder Co. 11
202. Firefighter Ronnie L. Henderson, Engine Co. 279
203. Firefighter Joseph P. Henry, Ladder Co. 21
204. Firefighter William L. Henry, Rescue Co. 1
205. Firefighter Thomas J. Hetzel, Ladder Co. 13
206. Firefighter Jonathan R. Hohmann, Haz-Mat Co. 1
207. Firefighter Thomas P. Holohan, Engine Co. 6
208. Firefighter Joseph G. Hunter, Squad Co. 288
209. Firefighter Jonathan L. Ielpi, Squad Co. 288

210. Firefighter William R. Johnston, Engine Co. 6
211. Firefighter Andrew B. Jordan, Ladder Co. 132
212. Firefighter Karl H. Joseph, Engine Co. 207
213. Firefighter Angel L. Juarbe, Jr., Ladder Co. 12
214. Firefighter Paul H. Keating, Ladder Co. 5
215. Firefighter Richard J. Kelly, Jr., Ladder Co. 11
216. Firefighter Thomas W. Kelly, Ladder Co. 15
217. Firefighter Thomas J. Kennedy, Ladder Co. 101
218. Firefighter Michael V. Kiefer, Ladder Co. 132
219. Firefighter Robert C. King, Jr., Engine Co. 33
220. Firefighter Scott M. Kopytko, Ladder Co. 15
221. Firefighter William E. Krukowski, Ladder Co. 21
222. Firefighter Thomas J. Kuveikis, Squad Co. 252
223. Firefighter David J. LaForge, Ladder Co. 20
224. Firefighter William D. Lake, Rescue Co. 2
225. Firefighter Robert T. Lane, Engine Co. 55
226. Firefighter Peter J. Langone, Squad Co. 252
227. Firefighter Scott A. Larsen, Ladder Co. 15
228. Firefighter Neil J. Leavy, Engine Co. 217
229. Firefighter Daniel F. Libretti, Rescue Co. 2
230. Firefighter Robert T. Linnane, Ladder Co. 20
231. Firefighter Michael F. Lynch, Engine Co. 40
232. Firefighter Michael J. Lyons, Squad Co. 41
233. Firefighter Joseph Maffeo, Ladder Co. 101
234. Firefighter William J. Mahoney, Rescue Co. 4
235. Firefighter Joseph E. Maloney, Ladder Co. 3
236. Firefighter Kenneth J. Marino, Rescue Co. 1
237. Firefighter John D. Marshall, Engine Co. 23
238. Firefighter Joseph A. Mascali, Rescue Co. 5
239. Firefighter Keithroy M. Maynard, Engine Co. 33
240. Firefighter Brian G. McAleese, Engine Co. 226
241. Firefighter John K. McAvoy, Ladder Co. 3
242. Firefighter Thomas J. McCann, Engine Co. 65
243. Firefighter Dennis P. McHugh, Ladder Co. 13
244. Firefighter Robert D. McMahon, Ladder Co. 20
245. Firefighter Robert W. McPadden, Engine Co. 23
246. Firefighter Terence A. McShane, Ladder Co. 101
247. Firefighter Timothy P. McSweeney, Ladder Co. 3
248. Firefighter Martin E. McWilliams, Engine Co. 22
249. Firefighter Raymond M. Meisenheimer, Rescue Co. 3
250. Firefighter Charles R. Mendez, Ladder Co. 7
251. Firefighter Steve J. Mercado, Engine Co. 40
252. Firefighter Douglas C. Miller, Rescue Co. 5
253. Firefighter Henry A. Miller, Jr., Ladder Co. 105
254. Firefighter Robert J. Minara, Ladder Co. 25
255. Firefighter Thomas Mingione, Ladder Co. 132
256. Firefighter Manuel Mojica, Squad Co. 18
257. Firefighter Carl E. Molinaro, Ladder Co. 2
258. Firefighter Michael G. Montesi, Rescue Co. 1
259. Firefighter Vincent S. Morello, Ladder Co. 35
260. Firefighter Christopher M. Mozzillo, Engine Co. 55
261. Firefighter Richard T. Muldowney, Jr., Ladder Co. 7
262. Firefighter Michael D. Mullan, Ladder Co. 12
263. Firefighter Dennis M. Mulligan, Ladder Co. 2
264. Firefighter Peter A. Nelson, Rescue Co. 4
265. Firefighter Gerard T. Nevins, Rescue Co. 1
266. Firefighter Dennis P. O'Berg, Ladder Co. 105
267. Firefighter Douglas E. Oelschlager, Ladder Co. 7
268. Firefighter Joseph J. Ogren, Ladder Co. 3
269. Firefighter Samuel P. Oitice, Ladder Co. 4
270. Firefighter Patrick J. O'Keefe, Rescue Co. 1
271. Firefighter Eric T. Olsen, Ladder Co. 15
272. Firefighter Jeffrey J. Olsen, Engine Co. 10
273. Firefighter Steven J. Olson, Ladder Co. 3
274. Firefighter Kevin M. O'Rourke, Rescue Co. 2
275. Firefighter Michael J. Otten, Ladder Co. 35
276. Firefighter Jeffrey A. Palazzo, Rescue Co. 5
277. Firefighter Frank Palombo, Ladder Co. 105

278. Firefighter James N. Pappageorge, Engine Co. 23
279. Firefighter Robert E. Parro, Engine Co. 8
280. Firefighter Durrell V. Pearsall, Rescue Co. 4
281. Firefighter Christopher J. Pickford, Engine Co. 201
282. Firefighter Shawn E. Powell, Engine Co. 207
283. Firefighter Vincent A. Princiotta, Ladder Co. 7
284. Firefighter Kevin M. Prior, Squad Co. 252
285. Firefighter Lincoln Quappe, Rescue Co. 2
286. Firefighter Leonard J. Ragaglia, Engine Co. 54
287. Firefighter Michael P. Ragusa, Engine Co. 279
288. Firefighter Edward J. Rall, Rescue Co. 2
289. Firefighter Adam D. Rand, Squad Co. 288
290. Firefighter Donald J. Regan, Rescue Co. 3
291. Firefighter Christian Regenhard, Ladder Co. 131
292. Firefighter Kevin O. Reilly, Engine Co. 207
293. Firefighter James C. Riches, Engine Co. 4
294. Firefighter Joseph R. Rivelli, Jr., Ladder Co. 25
295. Firefighter Michael E. Roberts, Engine Co. 214
296. Firefighter Michael E. Roberts, Ladder Co. 35
297. Firefighter Anthony Rodriguez, Engine Co. 279
298. Firefighter Matthew S. Rogan, Ladder Co. 11
299. Firefighter Nicholas P. Rossomando, Rescue Co. 5
300. Firefighter Paul G. Ruback, Ladder Co. 25
301. Firefighter Stephen Russell, Engine Co. 55
302. Firefighter Thomas E. Sabella, Ladder Co. 13
303. Firefighter Christopher A. Santora, Engine Co. 54
304. Firefighter John A. Santore, Ladder Co. 5
305. Firefighter Gregory T. Saucedo, Ladder Co. 5
306. Firefighter Dennis Scauso, Haz-Mat Co. 1
307. Firefighter John A. Schardt, Engine Co. 201
308. Firefighter Thomas G. Schoales, Engine Co. 4
309. Firefighter Gerard P. Schrang, Rescue Co. 3
310. Firefighter Gregory R. Sikorsky, Squad Co. 41
311. Firefighter Stephen G. Siller, Squad Co. 1
312. Firefighter Stanley S. Smagala, Jr., Engine Co. 226
313. Firefighter Kevin J. Smith, Haz-Mat Co. 1
314. Firefighter Leon Smith, Jr., Ladder Co. 118
315. Firefighter Robert W. Spear, Jr., Engine Co. 26
316. Firefighter Joseph P. Spor, Rescue Co. 3
317. Firefighter Gregory M. Stajk, Ladder Co. 13
318. Firefighter Jeffrey Stark, Engine Co. 230
319. Firefighter Benjamin Suarez, Ladder Co. 21
320. Firefighter Daniel T. Suhr, Engine Co. 216
321. Firefighter Brian E. Sweeney, Rescue Co. 1
322. Firefighter Sean P. Tallon, Ladder Co. 10
323. Firefighter Allan Tarasiewicz, Rescue Co. 5
324. Firefighter Paul A. Tegtmeier, Engine Co. 4
325. Firefighter John P. Tierney, Ladder Co. 9
326. Firefighter John J. Tipping, II, Ladder Co. 4
327. Firefighter Hector L. Tirado, Jr., Engine Co. 23
328. Firefighter Richard B. Van Hine, Squad Co. 41
329. Firefighter Peter A. Vega, Ladder Co. 118
330. Firefighter Lawrence G. Veling, Engine Co. 235
331. Firefighter John T. Vigiano, II, Ladder Co. 132
332. Firefighter Sergio G. Villanueva, Ladder Co. 132
333. Firefighter Lawrence J. Virgilio, Squad Co. 18
334. Firefighter Kenneth T. Watson, Engine Co. 214
335. Firefighter Michael T. Weinberg, Engine Co. 1
336. Firefighter David M. Weiss, Rescue Co. 1
337. Firefighter Timothy M. Welty, Squad Co. 288
338. Firefighter Eugene M. Whelan, Engine Co. 230
339. Firefighter Edward J. White, Engine Co. 230
340. Firefighter Mark P. Whitford, Engine Co. 23
341. Firefighter Raymond R. York, Engine Co. 285
342. EMS Lieutenant Ricardo J. Quinn, Ems Battalion 57
343. Paramedic Carlos R. Lillo, Ems Battalion 49

CONTENTS

    Foreword xiii
    Acknowledgments xv
1. General Collapse Information 1
2. Terms of Construction and Building Design 11
3. Building Construction: Firefighting Problems and Structural Hazards 39
4. Masonry Wall Collapse 53
5. Collapse Dangers of Parapet Walls 65
6. Wood Floor Collapse 75
7. Sloping Peak Roof Collapse 91
8. Timber Truss Roof Collapse 103
9. Flat Roof Collapse 127
10. Lightweight Steel Roof and Floor Collapse 141
11. Lightweight Wood Truss Collapse 155
12. Ceiling Collapse 163
13. Stairway Collapse 179
14. Fire Escape Dangers 189
15. Wood-Frame Building Collapse 199
16. Collapse Hazards of Buildings Under Construction 211
17. Collapse Caused by Master Stream Operations 225
18. Search-and-Rescue at a Building Collapse 237
19. Safety Precautions Prior to Collapse 255
20. Why the World Trade Center Towers Collapsed 271
21. High-Rise Building Collapse 293
22. Post-Fire Analysis 303
23. Early Floor Collapse 319
    Epilogue: Are Architects, Engineers, and Code-Writing Officials Friends of the Firefighter? 327
    Index 331

Foreword

The burning building collapse experienced by firefighters of this nation over the past 200 years has not yet been identified or documented. This book is an attempt to examine the dangers of structural failure caused by fire. From 1956 to 2009, 58 FDNY firefighters died when burning buildings suddenly collapsed. And 343 firefighters were killed when the World Trade Center towers burned and collapsed after the impact of the terrorists' jet planes. These men were burned to death when floors or roofs caved in below them, crushed to death by falling brick, wood, and steel, or slowly asphyxiated, trapped beneath heavy ceilings. Many others narrowly escaped death but were crippled, seriously burned, and forced to leave the fire service.

Throughout this nation, structural collapse is one of the leading causes of firefighter deaths, yet, except for this book, there is a complete lack of useful, accurate information about the danger. An abundance of safety information is written about other causes of firefighter deaths, such as stress, falls, and products of combustion, but not the fireground danger of sudden burning building collapse. This is the second edition of the first textbook written to warn firefighters, company officers, and fire chiefs about exactly how structures collapse when destroyed by fire.

There are books on the subject of building construction and firefighting strategy which briefly mention the important subject of collapse. This book is different: it does more than give a surface treatment to this critical life-and-death subject. This book examines the subject of burning building collapse in great detail. It is based on 40 years of firefighting experience in the New York City Fire Department, analysis of hundreds of collapse sites, and interviews with firefighters rescued from beneath tons of collapse rubble.

This book shows exactly how brick walls fall down, explains exactly how floors cave in, and describes exactly how stairs collapse, fire escapes crumble, and truss roofs crash down. More important, this textbook, unlike any other publication, instructs firefighters and fire officers in how to survive burning building collapse.

This book is also written for the building code official who wonders why the fire service resists changes in building codes that reduce structural fire resistance. This book will also assist architects and engineers who are required to examine and analyze fire scenes where a structure has burned and collapsed.

This second edition contains new information. Collapse zones have been increased to one and one half or two times the height of a wall and are determined by the officer in command. The reasons that the World Trade Center collapsed are examined, including a summary of the National Institute of Standards and Technology's (NIST) findings on why the towers fell. This new edition also looks at the future of high-rise building collapse, the collapse problems created by illegal renovations, the shift to the performance code from the specification code, and the increasing "through the door and through the floor" collapse dangers created by lightweight wood trusses and wood I-beams. The text examines how Type II noncombustible steel construction is replacing ordinary construction and distorting the fire size-up experience; and how firefighters can manage collapse dangers such as floor collapse, peaked roof collapse, and lightweight construction collapse. There is a timber truss trilogy case study; a section covering buildings under demolition included with the chapter on buildings under construction. There are lessons learned from the terrorist attacks in Oklahoma City and New York City that have changed the collapse search-and-rescue plan. NIOSH case study references are included in related chapters, and finally there is a firefighting exit strategy for use when there is a collapse danger.

ACKNOWLEDGMENTS

I would like to thank the following people who greatly assisted in the writing of this book: Patricia Dunn, my wonderful wife, for her inspiration and love; Faith Deutsch, for cover design, Carl Dunn and Julie Manso for graphics; Jeff Barrington and Janet Kimmerly for great editing (the secret to my best writing), and most important, the photographers for the outstanding building collapse photos: Steven Spak, Wea Township Fire Department, Tod Conner, John Badman, Ian Stronach, Richie Kubler , Steve White, Joe Berry, Joe Hoffman, Doug Boudrow, Chuck Wehrli, Pat Grace, David Novak, Mat Daly, Warren Fuchs, Harvey Isner, Alan Simmons, and Bill Thompson.

1

GENERAL COLLAPSE INFORMATION

All firefighters fear the thought of sudden building collapse. Veteran firefighters will tell you, "I know how flames spread. I know the dangers of fire-flashover and backdraft, and I know ways to protect myself from these hazards. The only thing I cannot predict is sudden building collapse. The floors, walls, and roof can cave in all of a sudden, and there is nothing I can do about it." Being buried alive beneath tons of bricks, smoldering timbers, and plaster is the dread of all firefighters.

The sudden collapse of burning buildings can kill large numbers of firefighters at one time. The largest number of firefighters were killed in a single building collapse in Chicago, Illinois: 21 firefighters died in the collapse of a stockyard building in 1910. The wall of the building collapsed on top of the shed over a loading platform. During that same year, Philadelphia lost 14 firefighters in a single collapse when the floors and walls of a leather factory building suddenly caved in. First the floors collapsed and trapped three firefighters and then the walls collapsed on rescuers and killed 11 more firefighters. In Brockton, Massachusetts in 1941, 13 firefighters were killed when a steel truss roof in a movie theater collapsed. Firefighters pulling ceiling up in the balcony of the Strand Theater were crushed to death when the truss roof and wire lath ceiling collapsed. In 1966, 12 firefighters were lost when the floor of a drug store collapsed into a burning cellar.

In 1967, five mutual aid firefighters from Ridgefield, New Jersey, were killed when a timber truss roof of a bowling alley collapsed in Cliffside Park, New Jersey. The falling roof pushed out a masonry wall on top of firefighters outside the burning building. In Boston, Massachusetts, in 1972, nine firefighters were killed when the floors and walls of the Vendome Hotel collapsed. In 1988 in Hackensack, New Jersey, five firefighters operating a hose line died when a wood timber truss roof collapsed during a fire in an auto dealership. In Brackenridge, Pennsylvania, in 1991, four firefighters died when a floor in a wood refinishing building collapsed (fig 1–1). In Seattle, Washington, in 1995, four firefighters died when a fire in a warehouse caused a floor collapse, and in New York City, 343 firefighters died when the World Trade Center collapsed on September 11, 2001 after a terrorist attack and fire. Burning building collapse is one of the leading causes of fireground death. The National Fire Protection Association states that leading causes of firefighting deaths are:

- Stress
- Responding and returning to fires and emergencies
- Falls
- Falling objects

- Coming in contact with objects
- Firefighters caught and trapped
- Burning building collapse

Fig. 1–1. Column failure was the cause of this floor collapse that killed four Brackenridge, PA, firefighters.

Collapse of Burning Buildings

Structural collapse during firefighting can be expected to increase in the twenty-first century. Four factors that will increase burning building collapse are:

- Age of buildings
- Abandonment
- Lightweight construction materials
- Faulty renovations

Age of a building

A building, like a person, has an expected life span. A structure that is 75 or 100 years old, like a person, is near the end of its life expectancy. Beyond this age, a structure becomes badly deteriorated. Many structures in this nation are over a century old and are weakened by age. Wood shrinks and rots, mortar loses its adhesive

Chapter 1 | **General Collapse Information**

qualities, and steel rusts. Unless there is a much greater effort toward building preservation, rehabilitation, and legal renovation, the older structures of this country will collapse at an increasing rate and kill more firefighters (fig. 1–2).

Fig. 1–2. The age of a building increases its collapse danger.

Abandoned derelict buildings

During the 1960s, 1970s, and 1980s, large numbers of the United States population moved to the sunbelt from the Northeast and Midwest urban areas. A large number of abandoned buildings were left behind in the wake of this migration. These dangerous unoccupied buildings increase the collapse danger to firefighters. However, in the 21st century some cities have experienced a rebirth; people moved to the urban areas and the number of vacant buildings has been reduced. But there are still many dangerous abandoned and neglected, unoccupied structures in cities. These are so-called *target hazards*. Buildings such as these must be targeted by fire departments for frequent inspections, board-up orders, and fire preplanning. They present a major collapse danger during a fire. A building, vacant for several years and exposed to rain, snow, summer heat, and freezing temperatures, will collapse more quickly during a fire than a building that is well maintained and occupied by residents or a business. In a vacant building, rain and snow penetrating broken windows quickly rots a wood floor, roof, or foundation. Wind and rain erode mortar between bricks. Water seepage into an unheated building's masonry walls will freeze, expand, and crack the bricks. This freeze–thaw cycle cracks masonry walls. In an unoccupied building, fire-retarding coverings are quickly stripped from structural supports by moisture in an unheated building. Homeless people and drug addicts take shelter in these structures and start deadly fires. In 1999, a fire started by a homeless person in a vacant storage warehouse killed six Worcester, Massachusetts, firefighters.

Lightweight materials and methods of construction

The widespread use of lightweight construction materials presents another serious danger to the firefighters of America. Today's buildings are constructed of building material such as lightweight wood trusses, sheet metal C-beams, wooden I-beams, open bar steel joists. Firefighting inside a burning building constructed of lightweight materials is much more dangerous than firefighting inside a burning building constructed of traditional materials and by traditional methods. For example, underwriters testing laboratories has documented lightweight wood trusses, fastened together by thin pieces of sheet metal, collapse more readily than do solid 2×8-inch continuous wood beams.

The National Fire Protection Association states: "Unprotected lightweight steel bar joists fail when exposed to 5 or 10 minutes of fire exposure." We know bar joist collapse more readily than solid steel I-beams. Today, high-rise buildings are increasingly using lightweight materials. The World Trade Center towers had 60-foot unsupported steel bar joist floors. This was a factor in the rapid collapse of the towers after the terrorist jet plane attack and fire. One tower totally collapsed in 8 seconds and the other in 10 seconds.

The National Institute of Standards and Technology (NIST) investigated the collapse of the towers and discovered that the floors were the first part of the building to collapse. Fire protection engineers know fire resistance and collapse resistance are directly related to the mass of a building. A wood bearing wall with studs 24 inches or 30 inches on center will fail more quickly than a wall with studs 16 inches on center. A ¼-inch thick floor or roof deck will fail before a deck ¾-inch thick one. Lightweight construction materials and methods are one of the answers given by the building industry as the solution to affordable housing, however, lightweight construction will kill and injure more firefighter in the future. Lightweight building construction includes small unseen changes also. For example: wood bracing between floor and roof beam are eliminated; wood stud wall and floor construction are spaced 36 inches on center instead of 16 inches; wood sheathing for roof and floor is ¼-inch thick instead of ¾- or 1-inch thick; roofs are supported by 2×4-inch wood beams instead of 2×8- or 2×10-inch beams. This trend will not stop, so the fire service must change its firefighting strategy and tactics when combating fires in lightweight structures.

Illegal renovations

Another cause of burning building collapse is illegal and improper building construction and renovation methods. Buildings newly constructed, and those renovated or enlarged must comply with existing building codes. Often they do not because an illegal or improper renovation is undertaken. The Vendome Hotel floor and wall collapse in Boston and the New York City 23rd Street Wonder Drug Store floor collapse had improper and illegal renovations as contributing causes. The cause of several recent burning building collapses in New Youk City that killed and injured firefighters were determined to be due to poor illegal or improper renovations. For example: fire walls that supported a floor were removed to increase rentable space; firefighters operating hose lines on this floor above a fire were killed when the unsupported floor collapsed (Captain Scott LaPedria and Lieutenant James Blackmore); the roof of a building constructed without building department notification or approval collapsed killing a firefighter advancing with a hose team (Firefighter Louis Valentino); illegal renovation was a factor in a store fire floor collapse that took the lives of Lieutenant Howard Carpluk and firefighter Mike Reilly (fig. 1–3).

Fig. 1–3. Floor collapse in this 99-cent store killed two FDNY firefighters.

Data on Building Collapse

Despite the potential danger to firefighters of sudden building collapse, local fire departments have compiled little information about the subject. Simple questions such as how a building falls down or what part of a burning structure collapses first are rarely answered after a structural failure. In all the studies and research on firefighting, little, if any, attention is paid to collapse causes: age of a building, abandonment, lightweight construction and illegal and/or shoddy building renovation methods. One reason for this is that any research into the subject of burning building collapse offers small benefit to anyone except firefighters. The general public will not receive any spillover benefit from information discovered about burning building collapse. Occupants have usually been safely evacuated from burning buildings before the collapse danger becomes great. Very few persons, other than firefighters, are killed by burning building collapse. Only firefighters are close to a burning building when it has been weakened by flames to the point of collapse danger. On the other hand, research and study into the other causes of firefighter death, such as physical stress, toxic smoke, and burns have been great, and information gained from this research has contributed to society and the general public.

Another reason there is little information about such an important subject is the fire service itself. After a burning building collapses and kills or seriously injures a member, the chiefs, company officers, and firefighters are usually unable to analyze objectively why the building collapsed or even how the structure collapsed. Emotions such as sorrow, guilt, and anger distort the investigation. Outside impartial investigators, who were not operating at the fire scene at the time of the collapse and who can objectively evaluate the collapse rubble, are needed to conduct an analysis of the incident. These investigators must be trained in the techniques and practices of post-fire investigation. Since 1990, objective investigations of burning building collapses are conducted by the National Institute for Occupational Safety and Health (NIOSH) and the National Institute of Standards and Technology (NIST), and can be found on the Internet.

A further reason why there has been so little accurate information obtained during an investigation of a collapse is the attitude of those conducting the investigation. Because of the legal considerations involved when someone is killed or injured at a fire, often the officials in charge of the investigation are concerned only with placing blame or avoiding legal problems. Valuable information about collapse danger and safety lessons which could be given to firefighters, company officers, and chiefs is often overlooked and lost during the investigation.

Another reason there is very little information about the danger of burning building collapse and firefighter survival during collapse is the absence of a standard definition of the term *collapse*. There is no fire service definition of the term collapse. Unless several tons of brick and mortar suddenly crash down into the street at a fire, any small part of a building that collapses may be called a *falling object*. The absence of an accurate, standard definition of the term "structural collapse" has led to an underestimation of the collapse problem.

The term *structural collapse* is defined in this book as "any portion of a structure that fails as a result of fire." If a burned section of plaster ceiling falls on top of a firefighter, it is a structural collapse—not a falling object. If a heated stone stair landing collapses beneath a firefighter, it is an injury caused by structural collapse—not a fall. If a small part of a lightweight truss floor collapses and a firefighter falls into a fire and dies, it is a death caused by collapse—not exposure to fire products.

The definition of a collapse in this book is when any part of a building falls and does not matter how small. If the falling piece is part of the structure, it is defined as a structure collapse. If one brick falls from a parapet and strikes a firefighter, the cause of injury should be structural collapse not falling object.

The definition of falling object is any object other than the structure that falls, is thrown, or knocked loose from burning buildings and strikes firefighters operating below. Firefighters are killed and injured by things falling near and around a burning building. These things are not part of the structure like a wall, floor, roof, or ceiling. Falling objects may be small and large objects but deadly to firefighters operating below. Some objects that have fallen from buildings during fires that killed and injured firefighters are: broken glass, tools, window air conditioners, smoldering mattresses, stuffed chairs, TV antennas, flower boxes, and even people jumping to escape fire. These deadly airborne missiles are a leading cause of firefighter death and injury.

The perimeter of a burning building is a dangerous area on the fireground because of building collapse and falling objects. There are more incidents of firefighters being injured by falling objects than by structure collapse each year. The front sidewalk, side alley, and rear yard are danger zones for falling objects. Firefighters have to work around the outside of a burning building near the front sides and rear walls where things fall. Because they must raise ladders, operate hose streams, vent windows, and conduct forcible entry around the perimeter of a burning building, they are frequently injured by falling objects, so they must be aware of dangers from above. As soon as the firefighting task is complete at the perimeter of the burning building, firefighters should either enter the building or withdraw from the danger zone.

Broken glass from windows being vented is the most common falling object at a structure fire. If glass falls from a window because the window frame was heated by fire and distorts allowing the glass to fall out of the frame or the glass cracks from the heat and falls, that is considered a burning building structural collapse. If the glass is broken by a firefighter venting, it can be considered a falling object. When firefighters are searching for trapped victims in a smoke-filled building that limits visibility, windows must be vented to clear smoke and increase visibility for the search. In older buildings that have many coats of paint preventing opening the windows, the glass must be broken. Several firefighters performing a primary search in a burning building can create a rainstorm of broken glass around the front, side, or rear of a building. Firefighters below can be cut by this falling glass. Glass in a residential building window is $1/16$- or $1/8$-inch thick and it is subdivided into small windowpanes. Commercial building window glass is larger, thicker, and heavier.

A case study

A firefighter was connecting a supply hose line to a standpipe inlet. Firefighters above were venting windows at a high-rise residence building at a particularly smoky fire. Glass shards were falling near the firefighter connecting the hose to the standpipe. The firefighter was wearing protective equipment. When bending over to connect the supply line to a Siamese inlet, a 3-inch sliver of glass went through the pump operator's turnout coat and severed part of his spinal column. The firefighter fell to the ground and could not get up. He was placed on a stretcher and rushed to the hospital where a three-hour operation was performed to remove the glass and prevent permanent paralysis.

Glass in a commercial building is much more dangerous because it is thicker and heavier, and can be ¼ or ½ inch thick and weigh 2½ or 5 pounds per square foot. This means an 8×4-foot display window of ½-inch thick glass when broken by a firefighter to vent smoke can create four, 40-pound razor-sharp glass shards. A firefighter could be decapitated by one of these falling objects. So, firefighters operating around the perimeter of a building where glass windows are being vented should beware.

Who is responsible when a firefighter inside a burning building performing search and rescue breaks a window in a smoke- and heat-filled room and the falling glass injures a firefighter operating on the ground around the perimeter of the burning building? Is it the firefighter who broke the window or the firefighter on the ground?

The firefighter inside, who broke the glass, is not responsible. The firefighter outside the burning building is responsible when injured from falling glass, because the firefighter inside is operating in a superheated, smoke-filled fire environment searching for life. However, if we change the question and the stage of the fire operation, then who is responsible for an injury that occurs after the fire has been controlled when during overhauling operations a smoldering object is thrown out of a window, or a glass is knocked out of a window frame and it strikes a firefighter below? Is it the firefighter inside or the firefighter outside? The answer, now, is the firefighter inside. Why? It's because the firefighter operating inside the building is not working in a life-threatening environment. The fire is out, the rescue operations have ended, and his actions must be more controlled. Firefighters inside a burned out structure performing overhaul and salvage should never throw any smoldering object out a window or trim glass shards by knocking them outside unless the area below has been cleared and a firefighter is standing guard outside at ground level. It is not sufficient to yell, "Watch out below" and then throw a smoldering chair or mattress out a window. Such a deadly, irresponsible act has killed and injured firefighters.

The safe procedure taken by firefighters when an object must be thrown out of a window during overhauling is:

1. Obtain permission from the officer in command of the fire.
2. Notify or assign a firefighter outside the building to clear the area of civilians and act as a safety guard.
3. After the area is clear, the firefighter acting as guard signals when to throw the smoldering objects out or breaks off the jagged glass shards.
4. When all objects have been discarded out the window, notify the firefighter below who's been assigned as a safety guard.

Several years ago at a fire, a company extinguished a small blaze in a stuffed chair. They removed a badly burned man who had started the fire by falling asleep in the chair with a cigarette. Firefighters dragged the smoldering chair to a window, pushed it out on to a fire escape and threw it over the rail into a back yard. Unfortunately, a firefighter assigned to the outside vent position was in the rear yard about to climb the fire escape. He was struck with the smoldering chair, knocked unconscious, suffered a disabling head injury, and was forced to leave the fire service.

1. Do not throw objects from a window during overhaul unless the area is clear, and you have been signaled to do so by another firefighter acting as a safety guard below.
2. When trimming broken glass from windows, knock the glass shard inside, not outside.
3. When assigned to operate around the perimeter of a burning building, be aware of the danger of falling objects and wear proper protective clothing. A well-fitted helmet, gloves, and an eyeshield in the down position can protect you.
4. When venting windows from inside, attempt to open the window before breaking glass. Double-paned windows in new and renovated buildings can be more quickly and fully opened manually then by breaking glass.

5. If a window is vented by breaking glass from the inside, first break a small section to warn firefighters inside and then take the entire window out.
6. A stuffed chair left inside a building often reignites. And, when dragging a stuffed chair outside, fresh air in the hallway can cause the chair to burst into flame, so have a portable extinguisher or hoseline in the hall near the chair ready to quench a flash fire.
7. Realize that commercial glass is more dangerous when broken then residential glass. The thickness and weight of falling glass pieces can cause serious lacerations and cut hoselines.
8. The perimeter of a burning building is a dangerous place. After completing your assignment there, go inside the building or withdraw outside the collapse danger zone.
9. When anything must be thrown out a window during salvage or overhaul, notify the incident commander and ask for a clearance below.

The final and most important reason for the absence of collapse information for firefighters after 200 years of firefighting experience in this country is the lack of fire department documentation and record of collapse. There is almost never a record or written report of a collapse unless several firefighters have been killed and the collapse is of national interest. Most firefighters who die in burning building collapses do so one at a time. These facts are rarely recorded in a written document which can be used as a source of training. The fire service records all types of information that is of little importance to a firefighter. For example, some fire reports include the community political district where the fire took place. The name and make of the electric product which overheats is sometimes required to be listed on the fire record also. However, to my knowledge there is no fire report that requires information about structural collapse to be recorded.

Recently the federal government and the National Fire Protection Association (NFPA) have provided useful information about burning building collapse. Starting in 1996, the National Institute of Occupational Safety (NIOSH) was mandated by Congress to investigate every firefighter death in this nation. Before this, many burning building collapses that killed firefighters were never investigated and causes were never determined. In 1999, the NFPA published a 10-year study of firefighters killed by building collapse. This study documented the exact parts of a building that collapse during fire. This landmark study found that 56 firefighters died in burning building collapses during the period from 1990 to 1999; 21 died by floor collapse, 19 by roof collapse, 14 by wall collapse, and two by ceiling collapse. This study instructs the fire service in which part of a building is more likely to collapse and kill firefighters. The floors collapsing kill most firefighters, the next most dangerous structural failure is roof collapse, then wall collapse, and finally ceiling collapse.

A study of burning building collapse was conducted by the National Institute of Standards and Technology (NIST) after the World Trade Center towers terrorist attacks. (See chapters 20 and 21.) This government agency conducted a comprehensive study of the disaster, and issued a report on the collapse of the World Trade Center Towers and issued a final report on the collapse of Building 7 in 2008.

Post-Fire Analysis

When a burning building collapse kills or seriously injures a firefighter, a post-fire investigation and analysis should be conducted by the fire department. A fact sheet recording building information should be completed, a fireground diagram should be drawn up showing the area of collapse, and a photographic color slide documentary of the area should be prepared. This information should be used for firefighter safety training sessions and to improve firefighting strategy and tactics.

2
TERMS OF CONSTRUCTION AND BUILDING DESIGN

After knocking down a fire in a three-story brick residence, firefighters are working on the first floor of the fire-damaged, oven-like house, opening up walls and ceilings and checking for concealed fire (fig. 2–1). Hot water drips from the soot-covered ceiling, and pieces of plaster drop to the floor. Smoke drifts up from the charred lumps of wood and plastic that once were furniture. Every square inch of the room is scorched black. Heat, smoke, and steam hang in the humid atmosphere.

Fig. 2–1. Fuel load consists of content fuel and fuel in the structure.

Firefighters shine flashlights around the room, searching for the charred remains of a body. For a moment everyone is silent, and only the distant sound of the pumper in the street can be heard. The captain of Ladder 6 turns to his crew. "First, let's get some portable lights in here and get rid of that smoldering upholstered chair," he says,

"I'll get the lights," says one firefighter, leaving the room. Two others start to overhaul the chair before removal. One takes what is left of the smoldering chair cushions and places them in a water-filled basin in the bathroom. The other examines the wall behind the chair, then pulls smoking handfuls of stuffing from the chair frame and places them in a small puddle of water and soggy plaster on the floor. Together they carry the smoldering chair outside.

The captain turns to a rookie firefighter and tells him, "Open up the ceiling over where the chair was." Holding a pike pole, the young firefighter checks the forward position of the metal hook, glances up at the point on the ceiling, and, putting his head down, forcefully drives the hook point up through the charred plaster and wood lath ceiling, As he makes several short, sharp, downward pulls, a part of the ceiling breaks open, and plaster, wood strips, and dust roll out. The officer shines his light up into the opening, looking for concealed fire. "Pull more of the ceiling, over toward the stairway," he says.

The rookie pulls down the ceiling near the stairway, exposing a double-header beam and the floor stair opening. Sparks fall from the crack between the double beam. One end of the double beam is deeply charred and almost burned though. Turning toward another firefighter, the captain orders, "Spread the header beam apart with your Halligan tool just enough so that the hose stream be driven up between. Be careful—don't pull the header beam down or it will collapse the floor on us."

As the rookie firefighter regains his breath after opening the plaster ceiling, the captain again calls to him. "Open up the ceiling where I shine my light. Just put small observation holes in the plaster." He then directs his light beam first to a corner where a radiator steam pipe riser penetrates ceiling and then to the center of the room where the scorched remains of a recessed ceiling light fixture are located. The young firefighter opens the ceiling at both points. He sees no fire extension.

The officer and firefighters walk into the adjoining burned-out room. "Move those pictures from the top of the table," orders the captain, "and put them into the top drawer. Cover the table with the curtain on the floor and move the piece of furniture away from that window." As two firefighters struggle with the table, the captain tells the rookie, "Take off the top piece of trim around the window." The firefighter taps the hook end of the pike pole into the plaster wall and gets a bite with it behind the charred piece of decorative wood trim at the center. Pulling the hook, he snaps the burned top section of wood off the window. Smoke drifts out of the concealed space now exposed.

"Okay remove the vertical trim on each side of the window," directs the officer. The firefighter silently follows the orders, pulling the vertical strips out of their nailed surfaces. Once the concealed spaces on either side of the charred window is opened, two fallen metal window sash weighs are revealed, with burning pieces of rope still tied to them, causing the smoke. Using the end his pike pole, the firefighter pulls the metal weights out of the window cavities and on to the wet floor. He steps on the rope ends and extinguishes the flames.

"Say, captain," says the rookie, "should I remove the 2×4-inch piece of wood over the window opening? It looks like smoke is coming from it."

"No," says the officer. "That's just steam coming from the fire-heated bricks around the wood. Besides that's a lintel beam and if you remove it you could collapse the brick wall above it." The captain continues, "Just as there are parts of your body that are more important than others when it comes to staying alive, there are parts of a building that are more important than others when it comes to staying upright and not collapsing. A lintel is one of them."

Many concepts and terms of structural design are known and taken for granted by veteran firefighters and officers but are unknown to young recruit firefighters. To effectively combat and survive fires in structures, all firefighters should be familiar with the following concepts of building design, construction, and collapse:

ACTIVE FIRE PROTECTION. Fire protection provided by automatic sprinklers and firefighters hose streams. See *passive fire protection*.

ARCH. A curved masonry structure used as a support over an open space. The removal or destruction of any part of an arch will cause the entire arch to collapse.

BALLOON CONSTRUCTION. One of the three basic methods of constructing wood-frame residential buildings (braced-frame and platform are the other two types). Balloon frame buildings' exterior walls have studs extending continuously from the structure's foundation sill to the top plate near the attic. The concealed space between these studs can spread fire, smoke, and heat from the cellar area or the intermediate floors to the attic space. If a nonbearing wall collapses during a fire, the continuous studs will cause the wall to fall straight outward, in one section, at a 90-degree angle. If a bearing wall of a balloon-constructed building fails, it can cause a second collapse of the floors it supports (fig. 2–2).

BEAMS. A beam is a horizontal structural member, subject to compression, tension, and shear, supported by one of three methods:

- *Cantilever Beam Support*. A beam supported or anchored at only one end, which is considered a collapse hazard during fire exposure. Examples of cantilever structures are an ornamental stone cornice, a marquee, a canopy, a fire escape, and an advertising sign attached perpendicularly to a wall. Of the types of beam supports, it has the least amount of structural stability during a fire.

- *Continuous Beam Support*. A beam supported at both ends and at the center. During a fire, it has the greatest structural stability of the three types beam supports.

- *Simple Supported Beam*. A beam supported at both ends. If the deflection at the center of such a beam becomes excessive, a collapse may occur. A simple supported beam is more stable under fire conditions than a cantilever beam but less stable than a continuous supported beam.

Fig. 2–2. Balloon frame construction has exterior walls that have studs extending continuously from the foundation to the top plate near the attic. This concealed space allows rapid fire spread from the cellar to the attic.

BRACED-FRAME CONSTRUCTION. One of three basic methods of constructing wood-frame residential buildings (balloon and platform are the other methods); sometimes called "post-and-girt" construction. Vertical timbers called posts reinforce each of the four corners of the structure, and horizontal timbers called girts reinforce each floor level. Posts and girts are connected by fastenings called mortise-and-tenon joints. During a fire, a braced-frame building wall often fails in an inward/outward collapse. The wall breaks apart with the top part collapsing inward on top of the pancaked floors, and the bottom part collapsing outward on the street (fig. 2–3).

Fig. 2–3. Brace-framed construction has posts and girts fastened by mortise-and-tenon joints.

BRIDGE TRUSS. A perpendicular truss used is steel bar truss floor systems to provide lateral stability. *See purlin*.

BUTTRESS. A wall reinforcement of a brace built on the outside of a structure, sometimes called a "wall column" (fig. 2–4). On a masonry wall, a buttress is a column of bricks built into the wall. When separated from the wall of a buttress on an exterior wall can indicate the point where roof trusses or girders are supported by a bearing wall. A buttress constructed on the inside of a wall is called a pilaster.

COLD-FORMED STEEL BEAM. A sheet steel C-beam used in lightweight steel floor construction.

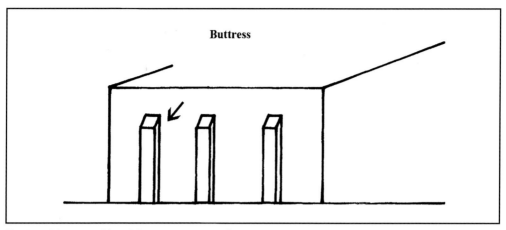

Fig. 2–4. A buttress adds stability to a masonry wall

COLLAPSE. The failure of any portion of a structure during a fire. A section of falling plaster ceiling, a broken fire escape step, a falling coping stone, and the collapse of several tons of brick wall are all structural failures and should be classified as structural collapse.

- *Curtain-Fall Wall Collapse.* One of the three types of masonry wall collapse, it occurs when an exterior masonry wall drops like a falling curtain cut loose at the top. The collapse of a brick veneer, brick cavity, or masonry-backed stone wall often occurs in a curtain-fall manner. The impact of an aerial platform master stream striking a veneer wall at close range can cause a curtain-fall collapse of bricks.
- *Inward/Outward Collapse.* The collapse of an exterior wall that breaks apart horizontally. The top collapses inward, back on top of the structure; the bottom collapses outward on to the street. Wood-braced-frame-constructed buildings collapse in this manner, and a timber truss roof collapse can cause a secondary collapse of a front wall in this manner.
- *Lean-Over Collapse.* A type of wood-frame building collapse indicated by the burning structure slowly starting to tilt or lean over to one side.
- *Lean-To Floor Collapse.* A floor collapse in which one end of the floor beams remains partially supported by the bearing wall, and the other end of the floor beams collapses on to the floor below or collapses but remains unsupported. A lean-to collapse can be classified as supported or unsupported, depending upon the position of the collapsed beam ends (fig. 2–5).
- *Ninety-Degree-Angle Wall Collapse.* A type of burning building wall collapse. The wall falls straight out as a monolithic piece at a 90-degree angle, similar to a falling tree. The top of the falling wall strikes the ground a distance from the base of the wall that is equal to the height of the falling section. Bricks or steel lintels may bounce or roll out beyond this distance.

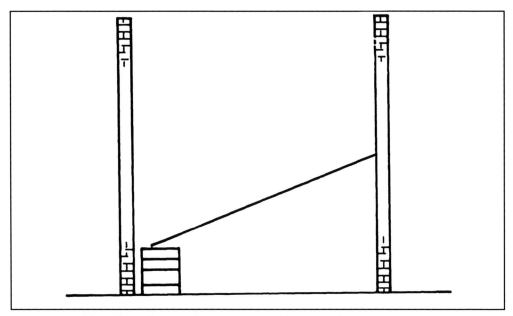

Fig. 2–5. A wood floor often collapses and creates a supported lean-to configuration.

- *Pancake Floor Collapse.* The collapse of one floor section down upon the floor below in a flat, pancake-like configuration. When floor beams pull loose or collapse at both ends, a pancake collapse occurs (fig. 2-6).

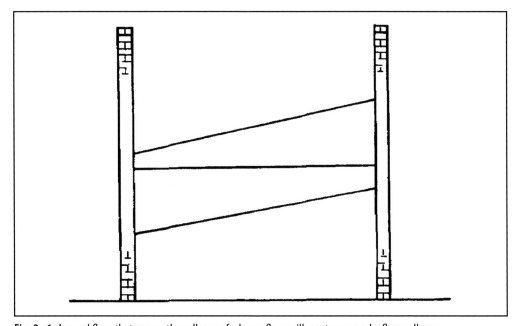

Fig. 2–6. A wood floor that causes the collapse of a lower floor will create a pancake floor collapse.

- *Secondary Collapse*. The collapse of portions of burning taller structures on to smaller structures, causing the collapse of the smaller building. On December 12, 1946, in New York City, borough of the Bronx, a 65-foot brick wall of a burning vacant building collapsed against an adjoining smaller tenement and crushed the entire rear, killing 37 people inside. Large buildings collapsing on top of smaller buildings cause the collapse of the smaller buildings.
- *Tent Floor Collapse*. A floor collapse in the shape of a tent. When a floor collapses and an interior partition or wall holds up the center of the fallen floor, a tent floor collapse occurs (fig. 2–7). If the interior partition were not present, the result would be a pancake collapse.

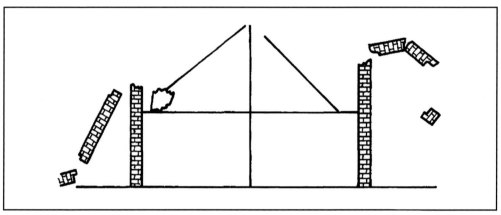

Fig. 2–7. A wood floor may collapse in the shape of a tent. This process is called a tent floor collapse.

- *V-Shape Floor Collapse*. The collapse of a floor at the center of the floor beams. The broken center of the floor section collapses down upon the floor below, and both ends of the floor section remain partially supported or rest up against the outer bearing walls (fig. 2–8).

COLUMN. A vertical structural member subject to compressive forces (fig. 2–9). The structural framework of a building consists of vertical and horizontal elements. Columns and bearing walls are parts of the vertical framework; girders and beams are parts of the horizontal framework. Bearing walls, columns, and girders can be classified as primary structural members. They support other parts of the structure, and their collapse can trigger a secondary collapse of other structural members. Other primary structural members are ridgepoles, hip rafters, headers, and trimmer floor beams.

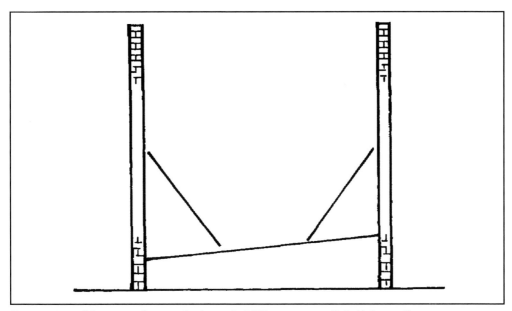

Fig. 2–8. A wood floor may collapse in the shape of a "V." This process is called a V-shape collapse.

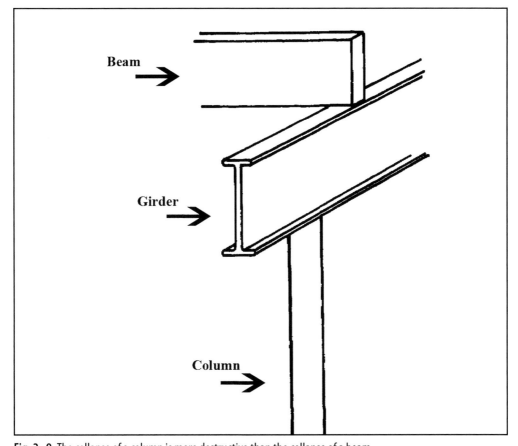

Fig. 2–9. The collapse of a column is more destructive than the collapse of a beam.

CONVENIENCE STAIR. An opening in a floor slab for a stair between floors. Sometimes called an access stair.

COPING STONE. The top masonry tile or stone of a parapet wall, designed to carry off rainwater. Sometimes called a "capstone," it weighs between 5 and 50 pounds. A coping stone can be dislodged and fall from a parapet under the impact of a high-pressure master stream or when struck by a retracting aerial ladder or aerial platform. Firefighters have been killed or crippled by falling coping stones.

CORBEL. A bracket or extension of masonry that projects from a masonry wall (fig. 2–10). It can be a decorative ornament on the top of a parapet front wall, or it can be used on the inside of a brick wall as a support for a roof beam end. A corbel used on the inside of a masonry wall to support a beam is called a "corbel ledge" or "corbel shelf." Under the weight of a firefighter, a roof beam end that is resting on a corbel ledge can rotate off its support if the center of the beam has been burned away. A flat roof that has roof beam ends supported only by gravity—that is, resting on a corbel ledge—is not as stable as one that has roof beam ends rigidly encased in cavities of a brick bearing wall.

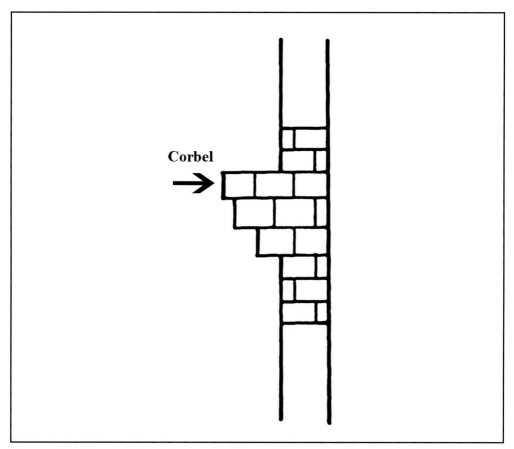

Fig. 2–10. A corbel shelf can be used to support roof beams. A beam supported on a corbel shelf is an unrestrained beam; the beam end can rotate off the shelf if the center is burned away.

CORNER SAFE AREAS. Four flanking zones around a burning building. When you look at a four-sided building from a bird's eye view, and if you imagine the four walls collapsing and covering the ground with bricks, you will find there are four areas at the corner of the collapse building that have fewer brick.

DECK. A horizontal surface covering supported by a floor or a roof beam. When an arsonist spills an accelerant on a floor or roof to start a fire, the deck area inside the spill is charred and weakened. Firefighters searching or advancing hoselines in such a building often plunge through collapsing decks. Floor beams prevent the firefighter from falling into the fire or the area below; however, if the firefighter cannot extricate himself quickly from the collapsing deck and fire is burning below, he will suffer severe and disabling foot, leg, and groin burns. It is very difficult for him to remove his foot from a hole in a plywood deck because the splintered edges of the hole trap him.

DEFLECTION. A bend, twist, or curve of a structural element under a load. All structures deflect slightly when supporting a load, but a structural element is designed to withstand a load without showing signs of deflection. When a firefighter notices the deflection of a column, beam, or wall, he or she knows that this condition indicates structural overload or failure and should be reported to the officer in command.

FACADE. The front or face of a building (fig. 2–11). A facade includes four collapse dangers—a marquee, a cornice, a canopy, and a parapet wall. The portion of the facade wall that extends above the roof level is called a *parapet wall*. The facade parapet is often a freestanding, decorative wall which frequently collapses during a fire. An ornamental facade parapet wall with a decorative stone corbel coping is one of the most unstable walls a firefighter will encounter. The slightest shift of the supports below a parapet wall caused by the effects of a fire will create a collapse danger. Even under non-fire conditions, these walls are collapse prone. In areas where earthquakes occur frequently, the height and ornamentation of parapet walls are limited by law.

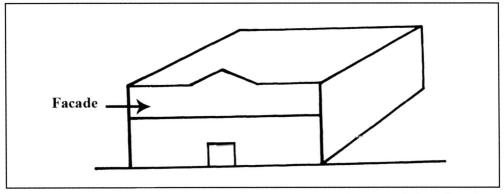

Fig. 2–11. The facade of a building often contains an ornamental stone or brick parapet wall. This is often a freestanding wall and collapses during a fire.

FIRE CUT BEAM. A gravity-support beam end designed to release itself from the masonry wall during a collapse. The end of the wood beam, where it rests within the cavity of a masonry wall, is cut at an angle. This self-releasing beam is designed to save the expensive masonry wall during a fire and resulting collapse. An unintended indirect advantage of the fire cut beam end is the safety of firefighters operating outside the building near the enclosing walls—the floor collapse will not topple the bearing walls outward on top of them. A disadvantage of the fire cut beam to firefighters operating inside a burning building is early floor collapse.

FIRE LOAD. The measure of maximum heat release when all combustible material in a given fire area is burned. The content and structure of a building contribute to fire load; structural collapse during a fire is directly proportional to the fire load. The greater the fire load, the greater the possibility of structural collapse during a fire. Weakening of unprotected structural steel, spalling of concrete, or fire destruction of wood structural members are processes that require the large quantities of heat given off by a large fire load.

FIRE RESISTANCE RATING. A relative rating to indicate in hours how long a wall, floor, ceiling, beam, or column will sustain performance during a fire. It is a relative rating.

FLUTED METAL (STEEL) DECK. A wavy piece of sheet steel deck used to support concrete floor. Sometimes called a corrugated metal deck.

FORCE. The cause of a motion, a change in motion, or a stoppage of motion. Force acting on a structure can be external or internal. An external force is a *load*, such as a dead load, a live load, a wind load, or an impact load. An internal force is a *stress*, such as a compressive stress, a tension stress, or a shear stress. External or internal force can collapse a structure.

FRAME TUBE CONSTRUCTION. The World Trade Center towers were frame tube construction. The building structure had tubular hollow exterior (perimeter columns) bearing walls. Frame tube exterior hollow bearing wall varied from 4-inch thickness near the bottom to only ¼-inch thickness near the top.

GIRDER. A structural element that supports a floor or roof beam. It can be considered a primary structural member, along with a column and a bearing wall. The collapse of a primary structural member can cause the collapse of another structural member.

GLOBAL COLLAPSE. A total collapse of a building. The World Trade Center towers were an example of a global collapse. *See progressive collapse.*

GRAVITY LOAD. A combination of dead load and live load.

GUSSET PLATE. A metal fastener in the form of a flat plate used to connect structural members. A steel plate with steel bolts and nuts is an example of a gusset plate connection. An inferior type of gusset plate, called a "sheet metal surface fastener," is used on lightweight wood trusses. It is a ¼ inch-thick piece of sheet metal with many small, triangular holes punched through it by a stamping machine. The V-shaped points caused by the whole punches substitute for the steel bolts and nuts of the old gusset plates. These points are only ½ inch long and act as nails. The nailing points penetrate the wood trusses only a fraction of an inch. During a fire, the sheet metal surface fasteners quickly fall off the structure—heat warps and bends the sheet metal connectors, and surface charring weakens the nailing surface. During shipping and unloading at the construction site, these connectors may be knocked loose from the wood trusses. From a fire protection point of view, a sheet metal surface fastener is an inferior, dangerous connector.

HAT TRUSS. A means of load distribution connecting core columns and perimeter columns. A hat truss was used at the top floor in the World Trade Center towers.

HEADER BEAM. A support used to reinforce an opening in the floor of a wood-frame, ordinary, or heavy timber building. A header beam (sometimes doubled for increased strength) is placed perpendicularly between two trimmer beams and supports the shorter, cut-off beams called "tail beams" (fig. 2-12) An opening in the floor is encircled by the header and trimmer beams. If a firefighter cuts through or pulls down a header beam, the tail beams and floor deck can subsequently collapse.

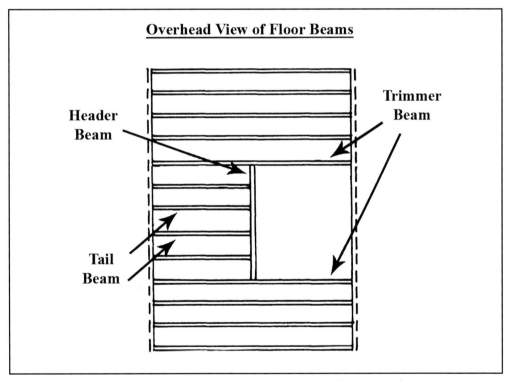

Fig. 2–12. The header and trimmer beams are primary structural supports. The collapse of a primary support can cause the collapse of other structural members.

HIERARCHY OF BUILDING ELEMENTS. Horizontal and vertical structural elements of a building arranged in a collapse hierarchy. The collapse of certain structural elements is more dangerous than the collapse of other structural elements, depending upon where they fit in the hierarchy.

Structural element	Seriousness of Collapse
Deck	Less
Beams, floor, and roof	
Girder	
Column	
Bearing Wall	Great

I-BEAM. A wood or steel I-beam consists of a center called a web section and top and bottom flanges. Wood I-beam has a web section of glue and wood shavings connecting the top and bottom flanges made of 2×2-inch wood sections.

JOIST. A piece of lumber used as a floor or roof beam. The terms joist, beam, and rafter are used interchangeably. A joist supports a roof or floor deck and is often supported by a girder.

INTERSTITIAL SPACE. A concealed space between floors used to contain large mechanical and electrical equipment. The space can be up to 8 feet high and contain a walkway for access for maintenance, repairs, and renovations (fig. 2–13). This is also called a plenum space.

KIP. One kip equals 1,000 pounds. When measurements expressed in pounds become large and unwieldy, kips are used to simplify the figures. The term kips is used most often when stating the strength of steel; KSI is the abbreviation for kips per square inch.

LAMINATED BEAM. A glued or layered composition beam. Lamination wood chips are used in lightweight I-beam, and layered bent wood is used in timber truss chord construction (fig. 2–14).

LINTEL. A horizontal piece of timber, stone, or steel placed over an opening in a wall. The lintel is a load-bearing structural element that supports and redistributes the load above the opening. The failure of a lintel during a fire can cause the collapse of the wall above the opening. Flame and heat issuing from a window of a burning building weaken the lintel spanning the top of the window opening.

Chapter 2 | **Terms of Construction and Building Design**

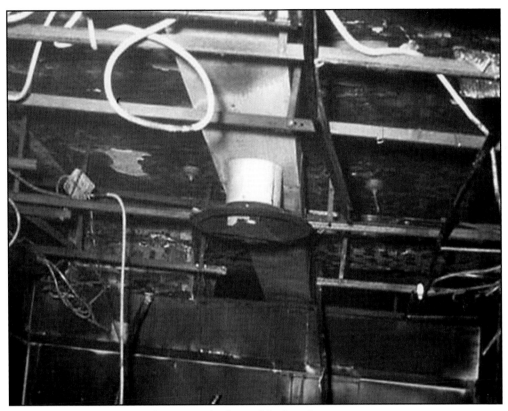

Fig. 2–13. The intervening space between and ceiling and the floor above

Fig. 2–14. Wood I-beams have a web section connected with top and bottom wood flanges.

LOADS. Forces acting upon a structure. The loads which can cause a collapse during a fire are dead loads, live loads, impact loads, and fire loads:

- *Axial Load.* One of the three ways a load can be imposed upon a supporting structural element (eccentric and torsional loads are the other two methods). An axial load passes through the center of a structure and is the most efficient manner by which a load can be transmitted through a structural support like a column or a bearing wall. A structural element can withstand its greatest load—and is less likely to collapse—when the load imposed is axial (fig. 2–15).

  When flames are destroying a building, structural elements become deformed and shift slightly because of heat and flame destruction. As a result, loads designed to be transmitted as axial loads become eccentric (off-centered) or torsional (twisting) loads. Structural collapse can occur during a fire when a dead load transmitted through a column or bearing wall changes from an axial load to an eccentric or torsional load.

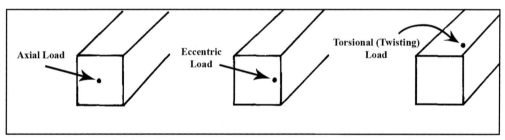

Fig. 2–15. A load transmitted axially through a structural member is least likely to cause a collapse.

- *Concentrated Load.* A load applied at one point or within a limited area of a structure. The concentration of heavy cast-iron fixtures inside a small bathroom can be considered a concentrated load. When weakened by fire, the bathroom floor can collapse under the weight of the heavy bathroom fixtures confined in the small floor area. A load distributed over a large area creates less strain on a supporting structure during a fire than does a load in a small area.

- *Dead Load.* One of the five major loads that must be considered in the design of a building (live, wind, impact, and seismic loads are the others). A dead load is a static or fixed load created by the structure itself and all permanent equipment within. Walls, floors, columns, and girders are part of the structural dead load. Air-conditioning machinery, fire escapes, suspended ceilings, roof water-storage tanks, and advertising signs are part of the equipment dead load. Firefighters have been killed by the collapse of dead loads.

- *Eccentric Load.* A load transmitted off-center or unevenly through a structural member. During a fire, a load designed to be transmitted axially can slowly become an eccentric load when steel columns or girders expand, timber surfaces char, or concrete spalls and exposes reinforcement steel. When load transmission slowly changes during a fire from axial to eccentric, the strength of the structure is seriously diminished. When floors collapse inside a burning brick-and-joist building, the load on the bearing walls shifts violently from a vertical axial load to a lateral eccentric load. This shift often causes the walls to collapse.

- *Impact Load.* A load applied to a structure suddenly, such as a shock wave or a vibrating load. It can cause a structure to collapse more readily than a slow, steady, evenly applied load. An explosion and a master stream directed at a structure in a whipping or pulsating manner are examples of impact loads.
- *Lateral Load.* Any type of load applied to an upright structure from a direction parallel to the ground. Examples of lateral loads are wind loads, tips of ladders placed against structures, horizontal explosion shock waves, and hose streams. Most upright structures are not designed to withstand lateral loads. A structure capable of withstanding great vertical loads may collapse under slight lateral loads.
- *Live Load.* A transient or movable load, such as a building's content, the occupants, the weight of firefighters, the weight of fire equipment, and the water discharge from hose streams.
- *Static Load.* A load that remains constant, applied slowly. A structure may support a greater amount of static load than impact load. An example of a static load is the content of a floor used for storage. A firefighter who slowly and softly applies his weight along a roof deck above a fire or on a fire escape step is demonstrating the principle of static loading.
- *Torsional Load.* A load that creates a twisting stress on a structural member. When a large steel girder collapses at one end, the other end experiences a torsional or twisting stress.
- *Wind Load.* A lateral load imposed on a structure by wind. A wind load may tear a roof off a structure or collapse a freestanding parapet wall or chimney.

LOAD STRESS. An internal stress created by a load in a structural element, including compression stress, tension stress, and shear stress. A collapse can occur when a stress created by a load exceeds the load-carrying capability of the structural element. An example of collapse by compression stress is the failure of a steel column weakened by the heat of combustion; one of collapse by tension stress is the failure of fire-weakened steel or wood hanger straps holding up a suspended ceiling; one of collapse by shear stress is the collapse of a brick veneer wall breaking away from the cement bonding and falling in a curtain-fall collapse.

MORTISE. A structural connection often used in braced wood-frame construction, it is a hole cut into a timber that receives a tenon (fig. 2–16). The mortise opening reduces the thickness of the timber and reduces the load-carrying capability of the timber at this point. When the timber-braced framework fails during a fire, it often breaks apart at the mortise-and-tenon connection, which has become weakened by flames and decomposition.

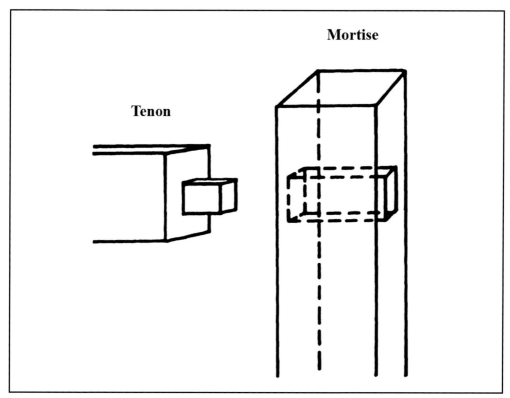

Fig. 2–16. Cutting a mortise and tenon into a timber reduces its load-bearing capacity.

MULTILEVEL FLOOR COLLAPSE. A floor failure that causes one or more floors below and one or more enclosing walls to collapse.

OPEN-WEB STEEL BAR JOISTS. A lightweight steel truss used as a floor or roof beam. It is made from a steel bar, bent at 90-degree angles, welded between angle irons at the top and bottom bar bends. This open-web bar joist is used for floor and roof beams in noncombustible buildings. An unprotected open-web steel bar joist can collapse after 5 to 10 minutes of fire exposure.

PASSIVE FIRE PROTECTION. The fire containment provided by a structure is considered "passive" fire resistance. Partition walls of spray-on fire-retarding and fire-resistive construction are examples of passive fire resistance. Passive fire resistant in modern buildings often fails and can not be depended upon to stop fire spread.

PILASTER. A masonry column bonded to and built as an integral part of the inside of a masonry wall. Sometimes called a "wall column," a pilaster can carry the load of a girder or timber, or it can be designed to provide lateral support to a wall.

PLATFORM WOOD-FRAME CONSTRUCTION. One of the three methods of wood-frame residential building construction (the others are balloon and braced-frame). A building of this construction has one complete level of 2×4-inch wood enclosing walls raised and nailed together; the floor beams and deck for the next level are constructed on top of these walls. The next level of 2×4-inch wood enclosing walls is then constructed on top of the first completed level. From a fire protection standpoint, platform construction is superior to balloon and braced-frame construction because there are no concealed wall voids extending for more than one floor level. Lightweight wood construction, using small wood trusses for floors and roofs, may replace platform construction in the future, but from a fire protection and a collapse point of view, this type of construction is inferior and dangerous to firefighters.

PRIMARY STRUCTURAL MEMBER. A structure that supports another structural member in the same building, such as a bearing wall, a column, or a girder. The collapse of a primary structural member will often cause the collapse of the structural member it supports.

PROGRESSIVE COLLAPSE. This is when the initial structural failure spreads from structural element to structural element resulting in the collapse of an entire structure or a disproportionately large part of it.

PURLIN. A timber laid horizontally perpendicular to support the common rafters of a roof.

RESTRAINED BEAM END. A welded, nailed, bolted, or cemented end of a floor or roof beam. It is one of two methods of supporting a beam end, the other being an unrestrained beam (fig. 2–17). During a fire, a restrained beam end will not collapse as readily as an unrestrained beam end. A wood beam cemented into a cavity of a brick wall is an example of a restrained beam end.

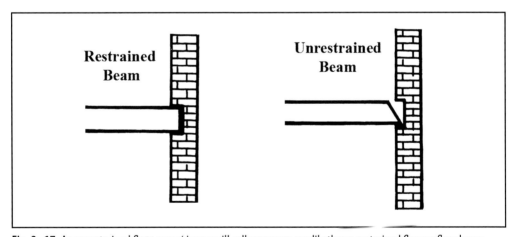

Fig. 2–17. An unrestrained floor or roof beam will collapse more readily than a restrained floor or floor beam.

RIDGEPOLE. A horizontal timber that frames the highest point of a peak roof. Roof rafters are fastened to the ridgepole.

ROOF. The sheltering structure of a building that protects the interior spaces from natural elements. It is designed to support dead loads such as the roof deck, roof shingles, suspended ceilings, and suspended lights, and live loads such as snow. A roof is not designed to support the weight of firefighters and their equipment.

SAFETY FACTOR. The quotient of the load that will cause a structure to collapse divided by the load a structure is designed to support. If a floor beam is designed to support a load of 100 pounds per square foot and the floor has been tested and found to collapse at 200 pounds per square foot, the safety factor of the floor is 2. Most structural elements are designed with a safety factor of 2 or more. A safety factor provides the structural engineer and the designer with a cushion or a margin for error in case there is an unknown factor in the load-bearing capability of a structural element. Years ago, little was known about failure points and accurate strength testing of structural materials; to compensate, built-in safety factors were large. Today, with improved testing techniques and computers to calculate precise collapse points, the built-in safety factor is being reduced and, in some instances, eliminated. Lightweight building construction is one result of the reduction of safety factors.

STRESS. A force exerted upon a structural member that strains or deforms its shape. The terms *stress* and *load* are often used interchangeably. The three common types of stress are compression, shear, and tension (fig. 2–18):

- *Compression.* A force pressing or squeezing a structure together. A steel column is subject to compression, but, when heated by a fire, it loses its compressive strength. When the heated steel column buckles and distorts, it collapses under the forces of compression from the load it is supporting. Some materials are strong in compressive strength and weak in tensile strength. Steel is strong in compressive and tensile strength; concrete is strong in compressive strength and weak in tensile strength; rope is strong in tensile strength and has no compressive strength.

- *Shear.* A stress causing a structure to collapse when contacting parts or layers of the structure slide past one another. A brick veneer wall breaking away from the cement bonding to the back wall and collapsing to the sidewalk is a structural failure caused by shearing. Also, a steel bolt connecting a fire-escape step to a stringer will shear apart when it has rusted over the years and is forced to respond suddenly to the impact of a firefighter's foot.

- *Tension.* Stress placed on a structural member by the pull of forces causing extension. Tension is the opposite of compression. For example, the hanger straps supporting a suspended ceiling are under the stress of tension.

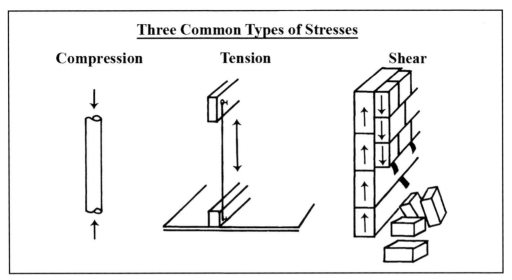

Fig. 2–18. Excessive stress on a structure can cause it to collapse more quickly during a fire.

SPRAYED FIRE-RESISTIVE MATERIAL. Spray-on fire-retarding material must cover the steel entirely, must be of proper thickness and or density, and must adhere to the steel and not flake off when exposed to air movement in a HVAC plenum area.

S.F.R.M. Initials used for "sprayed fire resistive material" referred as insulation or fluffy spray-on-fire-retarding used to protect steel from heat of fire.

SUSPENDED CEILING. A ceiling built several inches or feet below the supporting roof or floor beams above; sometimes called a "hanging" or "dropped" ceiling. The concealed space above the ceiling is sometimes called a "cockloft." The suspended ceiling is attached to beams above by means of vertical wood, wire, or steel straps. The ceiling is held up by the collective strength of all the hanger straps. If several ceiling hanger straps are destroyed by fire or removed during overhauling, the remaining straps may not be able to support the ceiling. A progressive total collapse of an entire stucco or cement ceiling can occur when the support of several hanger straps is eliminated.

TENON. A projecting, reduced portion of a timber designed to be inserted into the mortise hole of another timber. The tenon used with a mortise is a connection employed in braced wood-frame residential construction. Because of its reduced size, the tenon connection is the weakest portion of the timber; it can be destroyed by fire and decomposed by exposure to moisture.

TERRAZZO. A polished floor covering made of small marble chips set in several inches of cement. A terrazzo floor is a collapse hazard—it adds weight to floor beams, conceals the heat of a serious fire below, and, because it is watertight, allows water to accumulate and build up to dangerous proportions. In New York City in 1966, a fire burning in a cellar below a terrazzo floor burned away the floor beams, although little heat and smoke penetrated the floor itself. The floor suddenly collapsed, killing 12 firefighters (fig. 2–19).

Fig. 2–19. Terrazzo floors have polished marble chips in cement.

TIMBER. Wood larger than 2×4 inches.

TRIMMER BEAM. A wood beam constructed around the perimeter of a floor opening. A trimmer beam supports the header beam, which in turn supports the tail beams. When a floor area is designed to create an opening for a stairway, the edge of the opening is surrounded by header and trimmer beams. Header and trimmer beams are primary structural members that, if they fail, will cause the collapse of other sections of the floor.

TRUSS. A braced arrangement of steel or wood-frame work made with triangular connecting members. The truss presents several dangers to firefighters. It suffers early collapse during a fire because its exposed surface area is greater than the exposed surface area of a solid beam spanning the same distance. Also, there are a greater number of connections in a truss and, if any one fails during a fire, it can trigger the entire truss to collapse. Truss roof beams are spaced farther apart than solid beams, creating larger areas of unsupported roof deck. When the truss collapses, large areas of roof deck collapse. The failure of one timber truss of a number of trusses spaced 20 feet apart, supporting a 100×100-foot roof, can collapse 4,000 square feet of roof deck. A bowstring or peak timber truss roof creates a concave space on the under side of the roof, where great quantities of heat and flame can accumulate. A firefighter walking upright in a fire area where heat and flame have accumulated high above his head can miscalculate the amount of fire inside an occupancy.

UNRESTRAINED BEAM END. A beam end resting on a support, held in place only by gravity. During a fire, an unrestrained beam end will collapse more readily than a restrained or fixed beam end. An example of an unrestrained beam end is a fire cut beam or a beam resting on a corbel ledge or on a girder.

WALLS. Firefighters should be familiar with the following types of walls:

- *Area Wall.* A freestanding masonry wall surrounding or partly surrounding an area (for example, a masonry fence).
- *Bearing Wall.* An interior or exterior wall that supports a load in addition to its own weight. Part of the skeletal framework of a structure, it most often supports the floors and roof of a building. The collapse of a bearing wall is more serious than the collapse of a column, a floor or roof beam, a floor or a roof deck, or a nonbearing wall.
- *Demising Wall.* A partition wall that extends from floor slab to floor slab above.
- *Fire Wall.* A nonbearing, self-supporting wall designed to prevent the passage of fire from one side to another. Any door or window built into a fire wall must be equipped with an opening protective closure designed to prevent fire spread. The fire wall must be independent of the roof structures on either side and be designed to withstand complete collapse of a structure on either side. Party walls with parapets extending above a roof are not true fire walls. Often featuring penetrating openings that are not equipped with fire-rated doors or windows, party walls collapse if the interconnected wall or roof fails during a fire.
- *Freestanding Wall.* A wall exposed to the elements on both sides and the top, such as a parapet wall, a property-enclosing wall, an area wall, and a newly constructed exterior wall left standing without roof beams or floors.

Of the three types of walls—freestanding, nonbearing, and bearing—the freestanding wall is the most unstable and likely to collapse at a fire because it has fewer supporting connections to the structure.

- *Parapet Wall.* The continuation of a party wall, an exterior wall, or a firewall above the roof line. Parapet walls are considered freestanding walls and are less stable during fire conditions than nonbearing or bearing walls. The parapet of an ornamental stone front wall in a one-story commercial building, with large display windows beneath it, is a collapse-prone structure during a fire. One-story shopping center structures with large show windows often have parapet walls resting on steel lintels. The steel lintel beam spans the large show windows and supports the wall above. A parapet wall supported by such a lintel can become unstable and collapse if the steel shifts or warps as a result of expansion caused by the heat of a fire. The impact of a master stream can also collapse a freestanding wall.

- *Party Wall.* A bearing wall that supports floors and roofs of two buildings (fig 2–20). The collapse or demolition of one of the buildings served by a party wall may affect the stability of the adjoining structure. Although it does act as a fire barrier, a party wall is not designed to be a fire wall; fire can spread through a party wall that has wood beams embedded in brick cavities.

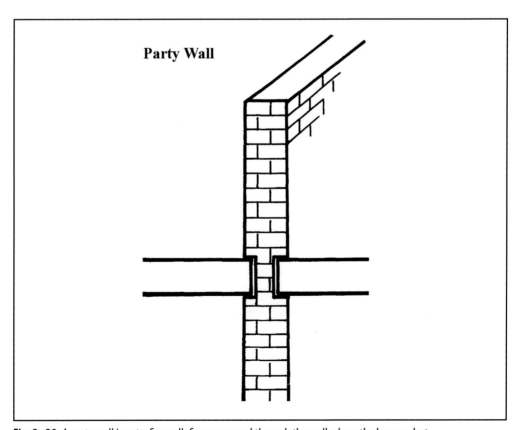

Fig. 2–20. A party wall is not a fire wall; fire can spread through the wall where the beams abut.

- *Spandrel Wall.* That portion of an exterior wall between the top of one window opening and the bottom of another (fig. 2–21). If a brick arch or a wood, concrete, or steel lintel spanning the top of a window, supporting a spandrel wall, is weakened by fire, blasted away by a high-pressure master stream, or removed during overhauling, the spandrel wall can collapse.

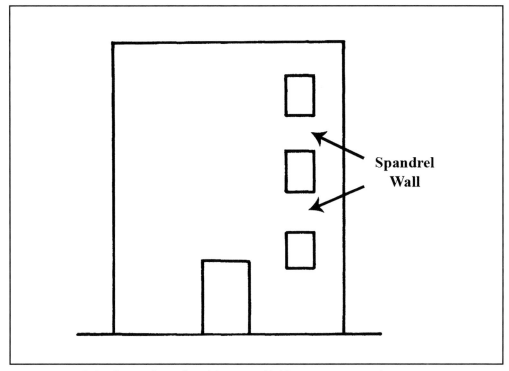

Fig. 2–21. Spandrel walls can collapse if struck by an aerial ladder master stream at close range.

- *Veneer Wall.* A finished or facing brick or stone wall used on the outside of a building (fig. 2–22). A veneer wall is fastened to a backing wall by sheet metal ties or cement. The sheet metal ties or cement bindings are sometimes omitted, defective, or improperly installed, creating an unstable veneer wall. A veneer wall will collapse in a curtain-fall fashion during a fire.

WALL HIERARCHY. A scale of wall stability during destruction by fire, based upon the stability of a wall created by the number of interconnections the wall has with the structure. The fewer the interconnections, the less stable the wall; the greater the number of interconnections, the more stable the wall.

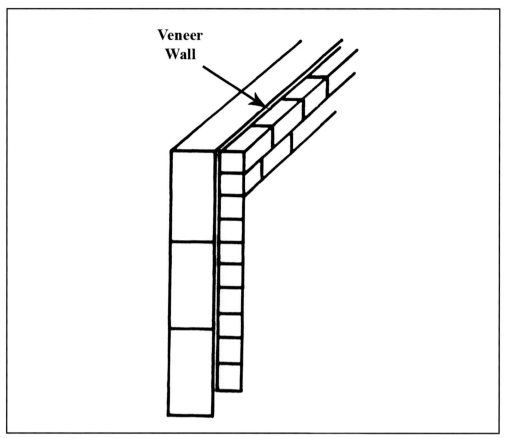

Fig. 2–22. A veneer wall can collapse in a curtain fall configuration if the mortar of sheet metal connecting strips fail.

ZONE OF DANGER. Two types of hazardous areas, defined as follows (fig 2-23):

- *Horizontal Collapse Zone.* The horizontal measurement of the wall. When establishing a collapse zone, firefighters should estimate this measurement in addition to the outward area that the wall may cover if it falls. A miscalculation of the potential horizontal length of a wall collapse could be just as deadly as a miscalculation of the outward area a falling wall will cover with masonry.

- *Vertical Collapse Zone.* The expected ground area that a falling wall will cover when it collapses. It is generally that distance away from the wall equal to the height of the wall. In some instances, heavy stones will fall farther than this distance. A fire department collapse zone of danger may require firefighters to be away from a dangerous wall a distance equal to the height of the wall; or a distance equal to one and one half, or two times the height of the wall.

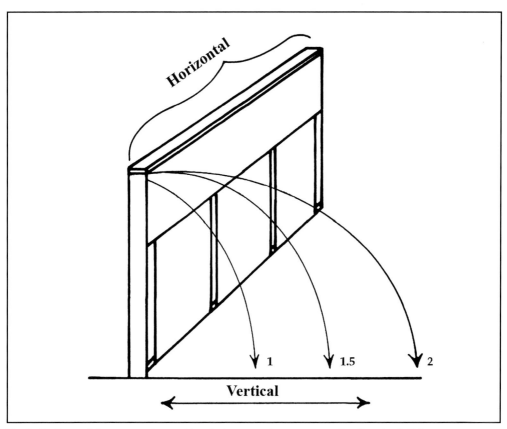

Fig. 2–23. Zones of danger must be considered when operating near the walls of burning buildings.

3

BUILDING CONSTRUCTION: FIREFIGHTING PROBLEMS AND STRUCTURAL HAZARDS

A single engine company pulls up in front of a 20×40-foot two-story, brick-and-joist row house. A man standing in the street is pointing at the smoke that is drifting from two cracked windows of a first-floor apartment. Climbing out of the apparatus cab, the captain orders his crew to stretch a 1¼-inch hoseline and then runs ahead to the front entrance of the building. Inside the public hall, he finds the blistered door to the fire apartment open and thick volumes of superheated black smoke gushing out and sweeping up the interior stairway.

Donning his mask facepiece and crawling into the smoke-filled apartment, he feels for fallen victims and tries to locate the fire. As he moves toward the front of the apartment and the heat becomes less severe, the officer quickly reverses direction and crawls back into a rear room. He glances toward his right and sees the bottom of an upholstered sofa involved in flames. It is barely visible beneath the banked-down layer of smoke. He moves swiftly out of the apartment on his hands and knees, closing the door to keep fresh air from feeding the fire with oxygen. Pulling off his face mask and looking around, the captain notices that the nozzleman already has the hoseline charged, bled of air, and ready to go.

The officer turns to the other three firefighters crouched down in the hall and directs his orders to each of them in turn: "You! Get a six-foot pike pole off the rig and vent those two front windows! . . . You! Get a Halligan and axe, go up through the adjoining building, and open the scuttle cover on the roof of this building!... You! Raise the extension ladder to the second-floor front windows and search the apartment above us!" Looking now at the nozzleman who is crouching on one knee and holding the line firmly in both hands, the captain tells him, "It's to your right as you go in; it's a sofa—the floors are okay."

The nozzleman nods his head and the officer replaces his face mask. Crouched down, the two listen to the crackling flames behind the blistered door and the hissing of the mask regulators forcing fresh air into their gasping lungs. The sound of crashing glass spurs them into action. The captain reaches in front of the firefighter, pushes open the apartment door, and pats the nozzleman's shoulder. The latter opens the nozzle, directs several sweeps of the hose stream at the ceiling, and, backed by the officer, springs into the smoke-filled apartment to attack the blaze. After the fire has been knocked down and the entire building searched and vented, all firefighters return to the smoldering, burned-out apartment to begin overhauling.

The chief outside radios, "Command Chief to Engine 8. How does it look in there? I am going to report this fire under control. Any problems?"

"Engine 8 to Chief. It looks good, Chief. Go with the under control."

"Ten four," responds the chief.

Just then a firefighter calls from another room, "Hey, Captain, we have fire in this wall!"

The officer quickly steps into the other room and sees the plaster and lath partition wall broken open and flames traveling up a bay between two wooden studs. "Quick, get that hoseline in here!" he calls. "Hit that fire, then get upstairs, on the double!" He turns toward another firefighter and orders, "Grab that pike pole and follow me." The two men run upstairs to the room directly above the spreading flames. The firefighter starts to open up the partition wall but is stopped by the officer.

"Never mind the wall," says the captain. "Pull that ceiling down." Driving the hook end into the lath and plaster, the firefighter pulls down a large section of ceiling, revealing flames in the cockloft. Now the nozzleman arrives in the room, carrying the hoseline. "Should I hit the fire, Captain?" he asks.

"No. Wait till the ceiling is opened fully and the entire flaming area is exposed." The firefighters open the ceiling and wall and are just completing fire extinguishment when the Handie-Talkie crackles: "Command Chief to Engine 8. Say, Captain, are you sure that fire is under control? I can see smoke coming out of the top-floor windows."

Controlling his heavy breathing and speaking in a calm, even tone, the captain replies: "It's okay, Chief. We found some minor fire in a concealed space, but we just hit it with the hoseline. Everything is under control."

Five Standard Types of Building Construction

There are many different types of building construction throughout the United States, each one varying in its fire resistance. Classifications of construction types according to building codes are constantly changing. There may be as few as two or as many as 20 fire building types listed in a building code. Firefighters across the country, however, use five basic construction types when considering collapse or the fire resistance of buildings. These are: fire-resistive construction, noncombustible/limited combustible construction, ordinary brick-and-joist construction, heavy timber construction, and wood-frame construction. Each type of building construction has associated with it a specific firefighting problem that increases the complexity of firefighting by spreading flames or smoke throughout the structure or to adjoining structures. To extinguish a fire effectively, and to avoid being caught and trapped by fire spread, an officer must constantly be aware of the specific fire spread weakness of each structure and know how to overcome this firefighting deficiency. The following are the firefighting problems caused by the five basic types of building construction.

Fire-resistive construction

The major fire problem in a fire-resistive building is the central air-conditioning system. In a low-rise building smoke spreads by heat, convection currents, fire pressure, and wind. In a high-rise building you also have smoke spread by the heating, ventilation, and air conditioning system (HVAC) Fire-resistive buildings used for commercial purposes, such as offices and hotels, have one large, central-heating, ventilation, and air-conditioning system which may serve the entire structure with cool air in the summer and heat in the winter. In a structure with a central air-conditioning system, every fire barrier in the building is penetrated. Concrete floors, fire-resistive walls, partitions, and ceilings are pierced to provide duct openings for air supply and return. The entire fire-resistive building may be interconnected by a network of holes, concealed spaces, and voids through which flame and smoke can spread. Though the fire-resistive building was originally intended to isolate and restrict smoke and fire to one floor by the solid enclosure of walls, floors, and partitions, the central air-conditioning system has eliminated this concept of compartmentation, called passive fire resistance. A fire-resistive building with a central air-conditioning system is a honeycomb of openings through which smoke and flame spread freely.

In the past, fire-resistive buildings used as apartment houses did not have a central air-conditioning system, and individual units were installed to serve each apartment. These structures effectively resist the spread of fire and smoke because of the compartmentation provided by fire-resistive walls and floors (fig. 3–1). This is changing. A central air system may connect public hallway on several floors. There may be a central air system connecting several floor of public hallways with ducts and vents. During a fire, the hallway air system may spread heat and smoke throughout several floors in the fire-resistive building. The smoke spread out of a burning apartment can be sucked into the air system and spread to floors above and below. When an air supply duct penetrates a fire-resistive wall, an automatic closing shutter is designed to close when the air inside the duct reaches a specific temperature. This heat-activated fire damper may stop flame spread but does not stem the spread of smoke throughout the system's air ducts, which is dangerous because three-quarters of all fire deaths in the United States are caused by smoke, not flame.

Fig. 3–1. Type 1 fire-resistive construction may allow fire spread by auto exposure from window to window.

Today, some air-conditioning systems are equipped with smoke detectors inside the ducts. When the detector senses smoke from a fire, the entire air-conditioning system shuts down. The fire dampers or shutters do not close, however, and deadly smoke continues to enter the air ducts and drift throughout the entire building by way of the central air-conditioning system. Normal convection currents and the stack effect of a high-rise building move the smoke from floor to floor. A fire in a Nevada hotel resulted in 85 deaths because the central HVAC system was not equipped with smoke detection devices arranged to shut down the system during such an emergency. In addition, the fire dampers did not close properly, and deadly smoke was distributed throughout the entire structure by the central air-conditioning system.

Fire-resistive buildings with central air-conditioning systems should be required by law to have automatic sprinkler protection on every floor, and an emergency smoke control system built into the central air-conditioning system. This change would permit the fire chief to remove any residual smoke from the building after the automatic sprinklers had extinguished the fire. Many building do not have this smoke detector shut off, so during any fire in a modern building,the incident commander should order the building management to shut off the central air system serving the living spaces and the public hallways. To fully protect a fire-resistive high-rise building, there should be an automatic sprinkler system and fire walls and fire partitions to limit the size of a fire in case the sprinklers fail. Floor areas should be limited in size. Experience has shown that firefighters operating hoselines can not extinguish fires in floor areas over 5,000 square feet. The flame and heat become too great.

Noncombustible/Limited combustible construction

The fire problem associated with a noncombustible/limited combustible building is the flat, combustible asphalt roof deck covering that can ignite during a fire (fig. 3–2). In the construction of such a roof, combustible hot asphalt is mopped completely over the top surface for waterproofing; a combustible felt paper is placed over the entire roof; another layer of hot asphalt is mopped over the entire area of felt paper; a combustible rigid foam insulation is laid on top of the asphalt; another layer of hot asphalt is mopped on top of the roof; finally, the roof surface of gravel is spread on top. When a so-called "noncombustible" building was completely destroyed by fire in this type of roof, the post-fire investigation revealed that the main fault was the 2,000 tons of asphalt used to insulate and waterproof the roof deck. Since this fire, the construction type formerly called "noncombustible" has been renamed "noncombustible/limited combustible."

The National Fire Protection Association (NFPA) recommends that the roof of a noncombustible/limited combustible structure be built up as follows: The total weight of asphalt used between the steel deck and the insulation should not exceed 12 to 15 pounds per 100 square feet of roof area. The felt, paper and rigid insulation should be noncombustible. Any hot asphalt mopping or application should not cover the roof deck completely but should cover only the overlapping edges between sections of steel deck, felt paper, or rigid insulation (strip mopping). When a fire occurs in a noncombustible/limited combustible building, as soon as possible after the fire is darkened down, the officer in command should have a firefighter check the roof covering above the fluted metal deck for fire extension. During a fire, heat may conduct through the metal roof deck and ignite the combustible roof covering above. After a fire is extinguished in a building of noncombustible construction, the officer in command should order a fire company to the roof to check the combustible asphalt roof covering. Heat may be conducted through the steel fluted roof deck and ignite that roof covering.

Fig. 3–2. Type 2 construction may allow fire spread in the combustible roof deck.

Ordinary construction (brick-and-wood-joist)

The major fire problem in an ordinary constructed building is the fire and smoke spread throughout concealed spaces. Consisting of brick and wood joist, this construction type has many concealed spaces, voids, and poke-through openings hidden behind plaster walls, ceilings, and floors (fig. 3-3). Unlike concealed spaces in a fire-resistive or noncombustible building, those in ordinary constructed buildings contain large amounts of combustible material in the form of wood lath, wood furring strip, cross bridging, wood joists, and wood 2×4 inch wall studding. Fire sometimes originates inside one of these concealed spaces; more frequently, it starts in a room and burns through a plaster ceiling or wall into a concealed space. Once inside a concealed space, flames and smoke spread throughout the structure, fueled by the interior combustible framework of the concealed spaces.

The largest and most serious concealed space is the common roof space or *cockloft*—the space above the ceiling and below the roof. Thousands of buildings in America are destroyed each year by flames spread throughout this area. Since the turn of the century, the construction industry has built and rebuilt large-area brick-and-joist structures that have a common cockloft. Allowing the spread of hidden fire over the top ceilings of stores and apartments within ordinary constructed buildings, this construction defect often results in the destruction of large shopping centers and apartment houses. Communities that have building or fire prevention codes

Fig. 3–3. Type 3 ordinary construction has many concealed spaces that allow fire spread.

address the problem of concealed spaces by requiring that *firestops* or fire barriers subdivide common cocklofts into smaller areas (not more than 3,000 square feet) and requiring that builders fill in poke-through holes found in concealed spaces. Unfortunately, in many districts there are no such code requirements, and concealed spaces are unrestricted size in the construction of brick-and-joist buildings. The concealed space, however, cannot be eliminated from any construction using wood joists and 2×4 inch studding. In fact, with the increase in renovations of older ordinary constructed buildings, concealed spaces have become larger and more numerous. For these reasons, the problem of concealed spaces in brick-and-joist construction will always present a major hazard to firefighters.

As soon as possible after a fire is extinguished, the firefighter must be directed to open the ceiling above the fire and nearby wall to examine for fire spread to concealed spaces. If fire is discovered in a concealed space, the cockloft space above the top floor must be checked for fire before leaving the scene.

Heavy timber construction

This type of construction was originally designed for textile mills during the last century. These four-to-seven-story brick-walled buildings, featuring large wooden timbers to support heavy machinery, are very stable structures. For a building to qualify as heavy timber construction, a wood column cannot be less than 8 inches thick in any dimension and a wood girder cannot be less than 6 inches thick (fig. 3–4). Instead of fire-resistive ratings for structural members, each interior wood structure must meet size specifications.

Fig. 3–4. Heavy timber construction with large content and structure fuel load

Today, heavy timber construction can be found in modern churches or in storage facilities in the old commercial areas of a city. The major fire problem of this brick-enclosed timber structure is the large wooden interior timber framework. When large timber columns, girders, and wood ceilings are not covered with plaster and become involved in a fire, flames shooting out the windows quickly spread fire to nearby buildings by radiant heat waves. Some heat waves travel distances of 80 feet and ignite combustible portions of nearby buildings. The advantages of not enclosing ceilings or timber in plaster are that there will be no concealed spaces, no obstructions to the firefighter's hose stream, and no hidden flames behind ceilings or walls. If the initial firefighting efforts are unsuccessful, however, and firefighters are forced off the fire floor to begin an exterior attack, the exposed structural wood will become a great disadvantage to operating fire forces. The heavy timber building will experience a large fire, and radiated heat from the windows will ignite the adjoining structures (fig. 3–5).

A fire in a building of heavy timber construction is sometimes called "slow-burning." This term is meaningless, for there is no scientific or fire protection explanation of this expression. It is often stated that in timber construction the surface-to-mass ratio is small. This means the ratio of exposed surfaces to the total volume of wood is small. This statement may be interpreted to mean that a small fire originating in the furnishings or content of a building will not ignite a large piece of timber as quickly as it would a smaller piece of wood. Even large timber structures will be ignited by a content fire, however, if the blaze has burned for a long period of time before firefighters arrive, or quickly if an arsonist has used an accelerate, and if, over the years, the thick floors have become saturated with some combustible liquid used in an industrial process. There will be nothing "slow-burning" about this type of fire in a heavy timber building. Automatic sprinklers have been installed in most of the heavy timber structures that have survived to the 20th century.

Fig. 3–5. Large fire in a Type IV heavy timber construction that creates a radiated heat fire spread problem for surrounding structures.

Wood-frame construction

Wood-frame construction is the only one of the five types that has combustible exterior walls. These wooden outside walls present a major firefighting problem. During an interior firefighting attack at any of the five types of construction, the officer must consider the possibility of fire extension to the six sides of the fire area—above the fire, below the fire and on the four sides of the fire. Wood-frame buildings, however, have a "seventh side" that must be considered: the combustible exterior walls. In some instances, firefighters must position a hoseline to prevent exterior fire spread, for flames lapping out of a window often ignite the combustible exterior of a wood-frame house. A closely grouped development of wood houses will sometimes present the responding fire companies with a fast-spreading fire on the outside walls, in addition to the life-threatening interior fire.

After the interior and exterior fires have been extinguished, a slow-smoldering fire inside the attic and walls of the wood structure will confront the firefighters. States one veteran fire officer, "Wood-frame building fires are a hurry-up-and-wait operation." The combustible exterior of a wood-frame building may be fire-retarded by a noncombustible siding, and fire spread to nearby wood structures may be reduced by providing distances of 10 to 20 feet between frame buildings. Although this allows access for firefighters operating hoselines between burning buildings, it will not protect against fire spread by radiant heat.

After newly recruited firefighters learn about the dangers of fire, the next most important subject they must study for survival is building construction. Collapse is a major killer of firefighters. Knowing how buildings are put up will help firefighters understand how and why buildings fall down. Firefighters equipped with self-contained breathing apparatus (SCBA) and fire extinguishment equipment are protected against smoke and flame, but they cannot be protected against a sudden building collapse unless they have a knowledge of construction and an understanding of its strengths and weaknesses. The NFPA Standard 220 method of classifying construction types has spelled out the basic fire-resistive properties of a structure . This method of classifying a building's fire resistance can assist the builder or the fire inspector in providing performance standards of fire protection and in evaluating the relative fire protection of each different construction type.

What practical meaning does a building's fire resistance have for a working firefighter? Contrary to what some people believe, firefighters cannot use a fire-resistive rating to estimate a burning building's structural stability. There is no test for collapse. A fire-resistive rating is a test for fire resistance. For instance, if a floor has a fire-resistive rating of one hour, this rating does not mean that a firefighter can feel safe from a floor collapse for one hour while battling a cellar fire below; it means only that a small test sample of that structure has passed a controlled laboratory test for one hour. A floor receives a one-hour fire-resistance rating when a small, reproduced section of that floor, approximately 18×14 feet, is placed over the floor furnace of a testing laboratory, loaded to its intended design load, and subjected to a controlled standard test fire from below. If it withstands the test fire for one hour without the top surface of the floor exceeding an average rise in temperature of 250 to 325 degrees Fahrenheit at any one point, it receives a one-hour rating.

Despite such fire-resistive approval, the floor may collapse beneath a firefighter as soon as he or she steps on it during the first few minutes of a fire. There are many possible reasons why this collapse may happen:

- The actual fire may be more intense than the test fire.
- A small-scale sample of a floor cannot predict the way a full-scale floor will be affected by a fire.
- The bearing wall, column, or girder supporting the floor may collapse, causing the floor to fail.
- The fire may burn undetected for longer than the one-hour test period before the firefighters arrive to fight the blaze.
- The workmanship and materials of the actual building floor may be inferior to the workmanship and material used in the test floor.

Therefore, to a firefighter working on a floor above the fire, an hourly fire-resistive rating should mean absolutely nothing.

Collapse Hazards of Each Standard Construction Type

Any structural part of a building may collapse, and it is impossible for a chief, company officer, or firefighter to predict what part will fail during a fire. Firefighting experience over the years, however, has shown that the same parts of a certain construction type frequently collapse and injure firefighters. For example, wood buildings often experience wall and floor collapse simultaneously; in ordinary construction types (brick-and-joist) often burning floors collapse first, with the wall collapsing some time later. Different structures have different design and maintenance weaknesses, which lead to collapse during fires. To fight the fire more effectively, a fire officer should know the structural weakness peculiar to each one of the five construction types. At the risk of oversimplifying the complex danger of building collapse, but with the intent of pinpointing recurring collapse danger, the following sections describe the specific structural hazards associated with each one of the five basic construction types.

Fire-resistive construction

There are two basic types of fire-resistive construction: reinforced concrete buildings and structural steel buildings. Both are designed to resist a fire which burns out an entire floor without spreading flames to other floors or collapsing the structure. Of all five construction types, fire-resistive buildings are the most stable and have the best collapse record, from the viewpoint of death and injury to firefighters. A fire-resistive building, however, does suffer structural failure during serious fires, and its collapse danger lies in concrete. In reinforced concrete buildings, heated concrete ceilings collapse on top of firefighters; in steel skeleton buildings, heated concrete floors crack and sag. Concrete ceiling collapse in a reinforced concrete fire resistive building is caused by spalling, the rapid expansion of heated moisture inside the concrete. Small amounts of moisture, normally trapped inside concrete, expand when heated by fire and create an internal pressure within the concrete. This pressure can cause heavy sections of concrete to crack away from a ceiling and collapse down into a fire or on top of a firefighter. This type of collapse occurs in a building without a suspended ceiling, where the concrete ceiling above the fire is directly exposed to flames below (fig. 3–6).

Fig. 3–6. Type I reinforced concrete structure suffers concrete spalling during a fire.

In steel skeleton construction, the under side of each floor is not concrete; each floor consists of a light-gauge steel sheet, called a fluted metal deck, which supports several inches of concrete floor above it. Heat from a fire reaching the under side of the fluted metal deck is conducted through the concrete floor directly above. The moisture in the concrete above the steel is heated, and the internal pressure develops in the concrete above; however, the bottom of the concrete cannot collapse downward because of the supporting steel, so the expanding moisture explodes the concrete floor upward. The floor cracks and suddenly erupts upward several inches. If the fire is severe, the steel beams supporting the fluted metal deck start to expand, and if steel beams are restricted, they start to sag and bow downward, causing the concrete above to also sag and bow downward. When heated, a concrete floor can crack and allow heat, smoke, and flame to spread to the floor above. Fire experience has shown at a serious fire in fire resistive building the composite, concrete, and fluted metal deck floors will be damaged because the steel expansion is greater than the concrete when heated by fire. The composite concrete and fluted metal deck is a collapse danger (fig 3–7).

Fig. 3–7. Type I steel skeleton construction. The heat from a fire reaching the underside of a corrugated steel deck is transmitted upward to the concrete floor above. The concrete floor may buckle upward between 6 and 12 inches. The floor may also bow and sag downward as well.

At a fire in New York City involving a high-rise, steel skeleton, fire-resistive building, 20,000 square feet of spalled concrete flooring had to be removed and replaced and over 100 steel beams were replaced. During the fire, none of the cracked and broken concrete floor collapsed into the floor below where the fire was raging. However, because of this possibility, firefighters could not be ordered into the floor above the fire to operate. The floor was cracked and heaved up at crazy angles and large sections of the damaged floor sagged downward.

Noncombustible/limited combustible construction

Noncombustible construction is becoming the nation's most widely used construction type, replacing brick-and-wood-joist as the most economical method of building such structures as one-story, large-area, commercial occupancies. There are three basic types of noncombustible buildings:

1. The "steel building" metal-frame structure covered by metal exterior walls
2. The metal-frame structure enclosed by concrete block, nonbearing exterior walls
3. The concrete block bearing walls supporting a metal roof structure

On all three types, the steel roof support system may be one of the following: a system of solid steel girders and beams, lightweight open-web bar joist, or a combination of both. The collapse danger to a firefighter from a noncombustible building is roof cave-in, the collapsing material, and the unprotected steel open-web bar joist. The main advantage of the lightweight steel roof support is its noncombustibility—the bar joist does not add fuel to the fire. The main disadvantage of the open-web bar joist is its susceptibility to damage by a fire in the combustible contents inside the building. Tests have shown that unprotected lightweight open-web bar joist can fail when exposed to fire for 5 to 10 minutes. This possibility makes it extremely dangerous for a firefighter to operate on a roof supported by steel open-web bar joist which is being heated by flames. The open-web bar joist is the main structural hazard of noncombustible construction. Each year heavy snow falls collapse roof structures of noncombustible/limited combustible buildings. These metal roofs are not designed to support firefighters during fire. The floors in the World Trade Towers were open-web bar joist construction. The New York City Building code prohibits steel open bar joist construction in high-rise commercial buildings since 9/11.

Ordinary construction

The ordinary constructed building called *brick-and-joist* has exterior bearing walls of masonry with wood floors and roof. This construction method was used to build most of the public, commercial, and multiple dwellings throughout the country. The structural hazard of an ordinary constructed building is the parapet wall, the portion of the masonry wall that extends above the roof level. Located at the front, side, or rear of a structure, the parapet walls usually collapse during firefighting. They are weakened by exposure to the elements—rain washes away the mortar between the bricks, freezing causes cracks in the wall. The wall begins to lean inward or outward. In some instances, the mortar between the bricks completely loses its adhesive qualities and turns to sand, and the bricks are left resting on top of each other, unconnected by mortar. The slightest lateral force will topple the bricks over into the street or back on to the roof. Heavy-caliber hose streams, sweeping the top of a burning roof, can wash tons of loose bricks from a parapet wall down on top of firefighters in the street below.

At one-story buildings that have large display windows, decorative or ornamental stone parapets at the front wall are balanced on top of a steel I-beam, acting as a lintel. If heat from flames coming out of a vented display window distorts the supporting steel I-beam, the ornamental stone wall can collapse on to the sidewalk in front of the fire building. A Vermont firefighter was killed by a parapet wall when opening a set of double truck entrance doors to provide an opening for exterior hose stream attack. As he went up to the doors, the parapet masonry wall suddenly fell, crushing the firefighter. (See NIOSH report #98F-20.) At another fire, a 30-pound coping stone falling from a parapet wall struck and killed a firefighter who had stopped momentarily in front of the fire building to don his mask. At another incident, six firefighters were killed when 100 feet of parapet wall collapsed on top of them. The collapse danger of the parapet wall is one of the reasons why the area directly in front of a fire building is so dangerous, and why firefighters are ordered to move away from the front of the building altogether.

Heavy timber construction

Falling masonry walls that crash to the ground and spray bouncing chunks of bricks and mortar along the street or pavement are the structural hazards of heavy timber buildings. This type of construction does not collapse during the early stages of a fire when interior firefighting is taking place. Massive masonry walls, large timber girders, and columns characteristic of this construction are quite stable during the growth period of a fire. If, however, the fire is not extinguished during the early stages and flames grow beyond the firefighters' extinguishing capability, an explosive fire will occur, requiring firefighters to withdraw to safety. Radiated heat will spread fire across streets and alleys to nearby buildings, and firefighters, under the protection of hose streams, will be forced to reposition apparatus to avoid blistering heat waves. Withdrawal to protect exposures is the strategy used at a fire of heavy timber construction when the initial attack fails. Heavy timber construction becomes a fast-burning fire once the timbers become engulfed in flame. At this point, control of the structure fire is impossible. Exposed buildings must be wetted down as the heavy timber building burns. After several hours, its floors will collapse, and the freestanding walls will fall into the street and on to the roofs of lower buildings nearby. The roofs of smaller structure will also collapse under the impact of the falling walls of heavy timber buildings. Very few firefighters are killed by the falling walls of heavy timber buildings, because the radiated heat from the fire forces firefighters away from the front of the building. They would be seriously burned by radiated heat if they were close enough to be struck by a falling wall.

Wood-frame construction

The structural hazard of a wood-frame building is the combustible bearing wall constructed of 2×4-inch wood studs. A wood-frame building is a bearing wall structure. The two side walls are usually bearing (that is, supporting a load other than their own weight); the front and rear walls are usually nonbearing. The structural supports of the side bearing walls are only 2×4 inches in size, yet they support wood floor joists 2×10 inches in size and roof joists also 2×10 inches. Firefighters should know that wood-frame buildings use smaller structural members to support larger structural members, and the weak link in this design is the smaller structural supports—the 2×4-inch bearing walls. Flames coming out of several window openings of a side bearing wall should be treated with more caution than flames coming out of windows of the front or rear nonbearing wall. If flames are coming out of the windows, they are also destroying the wall in which the windows are placed. Fire burning through or against a side wall is more likely to collapse the building than fire burning through several floors or the roof. The smaller, 2×4-inch bearing wall supports will burn away faster than the 2×10-inch floor or roof joists. Failure of a bearing wall will trigger simultaneous failure of the floors and roof.

Lessons Learned

In the fire service, as in other fields, this is the "age of information." Progressive chiefs and fire officers are realizing that information is the key to successful firefighting. How a structure contributes to fire spread and the particular structural hazard of a building are some of the most important items of information that a fire officer can know to combat a fire efficiently and safely.

4

MASONRY WALL COLLAPSE

At two o'clock on a bitter cold morning, firefighters are working at the scene of a six-story, brick-and-joist factory fire. No smoke or flame is visible in the 75×100-foot fire building, but the hurried movements of firefighters indicate the severity of the situation. The aerial ladder is raised to the roof of the building, which stands alone, surrounded by empty lots. The first hoseline stretched inside is snaking around the sidewalk with water rushing through it. As three firefighters drag folds of the second hoseline, an officer appears at the entrance of the burning structure and shouts at them to hurry. Above the dim street light, sparks and a dark cloud of smoke can barely be seen rising over the rooftop.

The chief tells his aide to check the rear of the building to see how much fire there is. Then he contacts the first-arriving engine by Handy-Talkie, "Battalion 1 to Engine Company 4, what do you have in there?"

"Engine Company 4 to Battalion, we have at least two floors of fire. 1 think you're going to need a second alarm."

"Battalion I, 10-4. Battalion 1 to Ladder 5, how does it look from the roof?"

"Ladder 5 to Battalion, we have fire on several floors traveling up through a shaftway. You are going to have fire on all floors if this shaft is not enclosed."

"Battalion 1, 10-4." Just then the battalion driver comes rushing back.

"Chief," he says, "We have fire in the four lower floors, and the two top floors are about to light up. There is no problem yet with the surrounding buildings. A 35-foot rear yard separates the fire building from Exposure 3, a four-story apartment house, but we'll have to get someone in there just in case. We have open areas on each side of the fire building."

"Transmit a second!" orders the chief. "Tell the ladder assigned on the second alarm to set up the aerial stream in the lot on the Exposure 2 side of the fire building. Have an engine company set up a portable deluge nozzle in the rear yard behind Exposure 4—the row houses—and have the second-alarm chief check out Exposure 3. I am going into the building. Stay by the command post and relay this information to the deputy chief when he arrives."

As the chief enters the dark hallway, the second hoseline, charged, coils upward and pushes him into the wall. He moves the line aside and continues on. Just ahead he sees crouching firefighters from the first engine company silhouetted against the orange flames in the first-floor doorway. Following the second hoseline, the chief climbs the stairs two at a time, feeling a strong updraft of air on the back of his neck. At the doorway on the second floor, he meets firefighters operating the second hoseline. "Can you move this hoseline into the floor area?" asks the chief.

"We pushed several feet inside the doorway and were forced out by the heat," responds the company officer. The chief proceeds quickly upstairs to the third floor. Here, a forcible entry team has just opened a heavy, metal-clad door to the factory floor area that is already fully involved. The room is lit with an orange glow, and the entire open joist ceiling has fire between each joist bay. Heavy machinery, metal barrels, workbenches, and storage racks of steel rods can be clearly seen. In the center of the floor, fire roars upward in an open elevator hoistway shaft with the safety gates open. Pieces of elevator car machinery are crashing down the shaftway. The chief, turning to the firefighters on the stair landing, orders, "Close the door to this floor. Forget the floor above and come back down to the floor below. This is going to be an outside job." As the chief leaves the fire building, he meets the deputy chief outside.

"What do you have?" asks the deputy chief.

"Four floors of fire, fully involved," replies the other. "Lines are operating on the first and second floors, but they are going nowhere. Fire is spreading up an open elevator shaft, and there is heavy machinery on the third floor. I think we should get the firefighters out of there! The exposures are no problem at this time. Exposure 3 in the rear is a four-story apartment house, and I sent the second-alarm battalion chief there."

"Okay," nods the deputy chief. "Get everyone off the roof and out of the building. I'll start setting up the outside streams. Let me know when everyone is out—I won't start the exterior streams till I get the word from you." The deputy chief considers the situation. Unlike the large city nearby, he thinks, this city's fire department does not have the "bottomless well" of firefighters and equipment on which to draw. Of this department's nine engine and four ladder companies, the equivalent of a third-alarm assignment, he already has the first alarm inside and the second setting up master streams outside. He turns to his aide and says, "Tell the communication center to start Phase 1 of the mutual-aid plan. Start the volunteer companies moving into the city."

Operating in the back yard of Exposure 4 with his two men, the captain of Engine 8 removes the 1¼- and 1½-inch stacked nozzle tips from the portable deluge nozzle about to be supplied by the 3½ inch diameter hose. The master stream of water rises up from the 1¼-inch nozzle tip remaining and is quickly directed by firefighters into a window of the fire building, now burning from the first floor through to the roof. All firefighters are positioned outside.

Suddenly the interior floors of the building begin to collapse. The chief operating inside Exposure 3, the four-story apartment house, calls on the radio, "May Day! May Day! The side wall facing Exposure 4 is separating from the rear wall! It is about to fall! The company operating in the back yard of Exposure 4—get out of there!"

Quickly realizing his company is in danger, the captain tells his two men, "Secure this stream and let's get out of here." They turn to run from the wall and see that they are encircled by a 6-foot chain-link fence. The first firefighter hits the fence near where it is attached to the wall of the row house and scrambles over the top. The second firefighter follows him, but his foot slips and he falls backward. He tries again and this time the captain pushes him over the top of the fence.

"Mayday! Mayday!" the radio keeps blaring. "Get out of the yard!"

The captain grabs a metal strap anchor securing the drainpipe on the row house wall and starts to pull himself up over the fence. The firefighter on the other side of the fence cries out, "Captain! Watch out! Here comes the wall!" The officer turns toward the row house and crashes through a rear basement window, moments before the massive brick wall comes thundering down behind him with an earthshaking thud.

Types of Masonry Wall Collapse

There are three ways in which a masonry exterior building wall may collapse. The wall may fall straight out in a monolithic piece at a 90-degree angle, in a manner similar to a falling tree; the wall may crumble straight down in a so-called *curtain fall*; or the wall may collapse in an inward/outward fashion, with the top falling inward and the bottom outward.

90-degree-angle collapse

A 90-degree angle collapse is the most common type of masonry wall failure that occurs at fires (fig. 4–1). The wall falls straight out and the top of the collapsing wall strikes the ground, a distance equal to the height of the falling section measured from the base of the wall (fig. 4–2). A 50-foot section of wall collapsing in a 90-degree angle fall will cover at least 50 feet of ground with brick. Bricks and steel lintels may bounce or roll out even farther. Here's how a multistory brick exterior wall can collapse at a fire in this manner. A fire spreading uncontrolled throughout a brick-and-joist structure causes the interior collapse of all floors. The pile of compressed rubble created by the fallen interior exerts an outward or lateral force against the inside of one of the still-standing brick enclosing walls. As the wall experiences this lateral force, a vertical crack or separation appears at a corner, starting at the top and progressing downward. The wall begins to lean outward at the top, separating from the other enclosing walls, and falls straight out at a 90-degree angle.

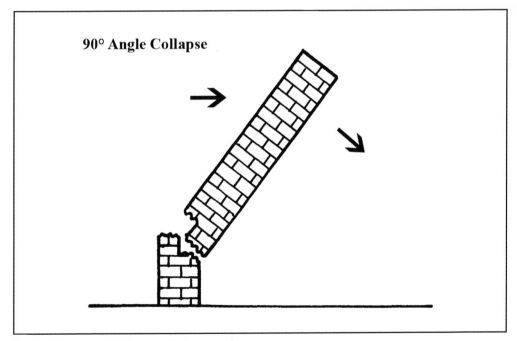

Fig. 4–1. Masonry wall falling at a 90-degree angle

A freestanding wall of a burning structure can also become unstable and fall outward during extremely cold weather. If the inside surface of the masonry wall is heated by the fire while the outside surface remains cold, unequal expansion of the masonry occurs; the heated inside of the wall expands as the outside contracts, causing the wall to lean outward and possibly to fall. Large accumulations of ice forming on the outside of a wall can cause it to become unstable and fall, regardless of the temperature of the inside surface. The force of a large-caliber stream of water can also be a destabilizing factor capable of causing a masonry wall to collapse. Directed from one side of a fire building against the inside of a freestanding wall, a high-pressure aerial stream can be powerful enough to cause the wall to collapse outward on to firefighters operating on the outer side of the building. Masonry walls often separate from the other enclosing walls at corners where they intersect. If there is no brick bonding of the intersecting walls by either overlapping brick bonding or metal reinforcing rods, the wall may split apart at this point. Vertical cracks allowing walls to separate and fall at a 90-degree angle may also be the result either of structural movement caused by uneven settling of the foundation prior to the fire or of a combination of window openings and cracked spandrel walls over each window.

Fig. 4–2. A masonry wall falling at a 90-degree angle

Curtain-fall collapse

In a curtain-fall collapse, the exterior masonry wall drops like a falling curtain cut loose at the top (fig. 4–3). The wall crumbles and falls straight down, with bricks and mortar forming a pile on the ground near the base of the wall. The collapse of the brick veneer, brick cavity, or masonry-backed stone wall often occurs in a curtain-fall manner. If the metal ties holding a brick veneer wall to plywood backing are destroyed by fire, or if mortar bonding between an exterior finished stone wall and a masonry-backed wall is washed away by hose streams, large sections of brick or stone veneer may fall off the building's exterior. Firefighters entering, leaving, or operating near the doorways beneath the curtain-fall collapse may be killed or seriously injured by falling brick. Another situation of potential curtain-fall collapse occurs when fire has collapsed the interior of a multi-story brick-and-joist structure and the remaining freestanding walls have many window openings with brick arches serving as lintels. If one of the masonry walls starts to fall and the brick arches spanning the tops of the wall openings crumble and fall apart, the wall will fall downward rather than straight out (fig. 4–4).

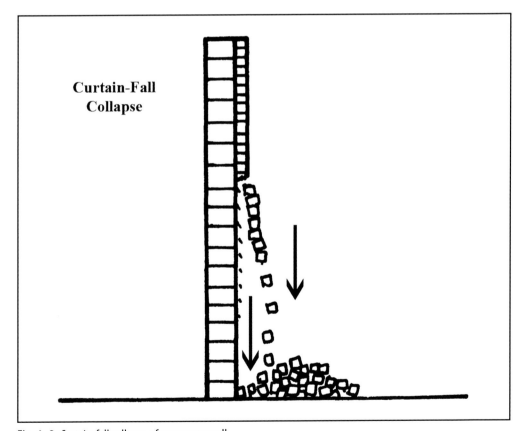

Fig. 4–3. Curtain-fall collapse of a masonry wall

Fig. 4–4. An example of a curtain-fall collapse of a masonry wall

Inward/outward collapse

When a masonry wall becomes unstable and begins to lean inward, it does not always mean that the wall will fall inward (fig. 4-5). Firefighters operating ground streams must still maintain a safe distance between themselves and the unstable wall, because when a section of the broken wall falls inward, the lower portion of the wall may kick outward, or the upper portion may initially fall inward but then slide down and outward into the street, bottom first (fig. 4-6). Known as an inward/outward collapse, this type can be caused by a force directed against the inside surface of the collapsing wall.

An explosion or the outward impact caused by the collapse of a roof or several floors in a " progressive pancake" fashion—the upper floors falling on top of the lower ones can cause an inward/outward collapse of several enclosing walls simultaneously. An example of an inward/outward collapse is a masonry wall failure caused by the collapse of a bowstring timber truss roof. When the design of the timber truss includes hip rafters sloping down from the front and rear bowstrings, one end of the hip rafter is tied into the outermost truss section and the other end into the masonry enclosing wall. If the bowstring trusses are weakened by fire and the roof fails, the load of the falling roof is transferred from the truss hip rafter supports to the front and rear masonry walls. In some instances, the load transmitted to the masonry wall through the hip rafters cracks the front wall and collapses it in an inward/outward manner. The top section falls inward and the bottom section outward, into the street. The extent of the area in front of the fallen structure that gets covered with tons of brick and steel lintels, depends upon the amount of internal force transmitted against the inside of the enclosing wall by the falling roof.

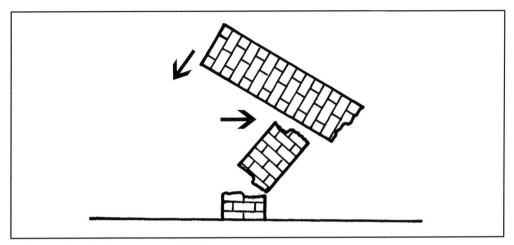

Fig. 4–5. Inward/outward masonry wall collapse

Fig. 4–6. An example of an inward outward wall collapse

A secondary collapse of the front masonry wall following a bowstring truss roof collapse is extremely dangerous. Firefighters anticipating a roof collapse may be caught off guard and be struck or buried by the secondary front or rear wall collapse. When planning for a bowstring timber truss roof collapse, firefighters should also consider a front or rear wall collapse. In such cases one-story enclosing walls have been driven out into the street for a distance greater than the height of the one-story wall. A fire officer can never predict the way in which a wall will collapse, so when a safe distance is established between the unstable wall and the firefighters in the command, the officer must expect the worst: a 90-degree-angle wall collapse with chunks of bricks and steel lintels thrown out farther than the falling wall. As a general rule, the *collapse zone*—the area adjacent to an unstable wall that firefighters should not be allowed to enter—is a distance equal to the height of the wall. This may be increased to one and one half or twice the height of the wall if a truss roof or explosion triggers the wall collapse or if ordered by the incident commander. At fires where the officer in command suspects an explosion or other factor to cause a wall to collapse out into the street for a distance greater than the height of the wall, a chief or fire officer should keep everyone away from the front wall of the fire building altogether, positioning heavy-caliber streams in a flanking position, that is, on either side of the front wall, beyond the outer perimeter of the building's width. If it is absolutely necessary to operate a master stream inside a collapse zone, and the wall appears unsafe, the portable deluge nozzle or aerial stream should be secured to direct the stream effectively and safely, and then left unattended.

Wall Attachment

A *facade* is fancy name for the front wall of a building. There are decorative attachments to a wall that increase the collapse danger. They are: a parapet, marquee, canopy, and cornice. These wall attachments are architectural features that can collapse themselves or collapse and pull down the entire supporting wall.

Incident commanders must size-up the front wall of a burning building and identify these dangerous attachments. First, let's define some construction terms.

- *Facade* is a general term that indicates the front of a building and could include a parapet wall, a marquee, a canopy, and/or a cornice:
- A *parapet* is a continuation of an exterior wall above the roof level. The portion of a front wall rising above a roof is the parapet.
- A *marquee* is a large metal structure attached over the entrance to a theater or store. A marquee extends from the front wall out to the street and is used to display signs about the show or the store.
- A *canopy* is a cloth, wood, or metal covering over a building entrance designed to protect people from the weather.
- A *cornice* is a horizontal ornamental construction along the entire front wall of a building, usually near the top.

Fire officers responding to a fire in a building that has any one of these structural attachments as part of its facade must consider it a collapse danger. During the firefighting operation, a parapet, marquee, canopy, or cornice must be monitored for failure. Firefighters should be kept away from these wall attachments. A parapet wall supported by a heated, twisting steel beam could collapse suddenly like a giant stone wave. A marquee filling up with water from hose streams could collapse and pull the parapet wall on each side down with it. A canopy with firefighters or equipment on its top could collapse and pull a supporting brick wall down with it, crushing firefighters below. A cornice weakened by fire could suddenly collapse at one end, and swing down across the front of the building knocking firefighters off ladders or fire escapes. A canopy collapse caused one of the greatest tragedies in the history of the fire service during a fire in a burning meat packing plant in Chicago in 1910. During the fire, a brick wall and a canopy attached to it collapsed. Twenty-one Chicago firefighters were crushed to death beneath the crumpled metal canopy and bricks. A similar collapse occurred in New York City. Six firefighters were killed when the weight of a metal canopy pulled a brick wall down on top of it. The falling bricks landing on top of the corrugated metal canopy caused it to collapse on firefighters operating hoselines below the shed.

A marquee collapse caused another tragedy in New York City. A heavy marquee attached to a parapet wall collapsed during a fire in a furniture store. The falling marquee pulled the brick wall on each side down too. Again six firefighters died.

A parapet wall collapsed during a fire in Altoona, Pennsylvania. The front wall of a burning auto electric store collapsed on a firefighter as he exited the building after searching for a victim.

A cornice collapse in Los Angeles County killed a firefighter A 160-foot long cornice collapsed suddenly during a fire in a row of stores. The cornice collapse started at one end of the storefront and killed the firefighter operating a hoseline at the other end.

Parapet walls

Parapet walls are unstable because they are freestanding. A freestanding wall is considered by engineers to be the least stable of three basic types of walls (freestanding, nonbearing, and bearing) because it has fewer connections to a structure. The more connections a wall has to the structure it is part of, the more stable the wall.

A parapet wall built over a one- or two-story commercial building with large display windows beneath it is a collapse-prone structure because the parapet walls are often supported by steel beams. A steel beam spans the large windows and supports the parapet wall above. A small shock during a fire can topple a parapet down on to a sidewalk. An explosion, the impact of a master stream, or an aerial ladder can also cause a parapet wall to collapse. A parapet wall is a vertical cantilever beam. A cantilever beam is supported at only one end.

Marquee

Ironically, a parapet wall often supports a marquee, canopy, or cornice. This is an unstable structure supporting an unstable structure. A marquee beam goes through a parapet wall and is connected to a roof or floor beams behind the facade. In addition, the marquee may also be connected to the parapet by steel cables.

A marquee is cantilever beam supported at only one end that is connected to the parapet wall. A cantilever beam is considered by engineers to be the least stable of the three basic beam designs (a cantilever beam, a simple support beam, and a continuous support beam). A cantilever beam is supported at one end. A simple support beam is supported at both ends. A continuous support beam is supported at both ends and in the center.

If a marquee collapses during a fire, it can pull the facade wall down with it. For example, if the beams behind the facade that support the marquee are destroyed by fire, the marquee can suddenly collapse downward, pulling the wall it passes through, outward. Another cause of marquee collapse is overloading due to water build up inside the marquee. Because it is hollow. a marquee is like a swimming pool hanging off the front of a building. If the drains from the hollow portion of the marquee are clogged during a fire, it will collapse due to the excess weight. And, if the marquee falls, it can pull the front wall down with it.

Canopy

A canopy is more of a collapse danger than a marquee. A canopy is a cantilever beam, like a marquee; but it is lightweight and constructed of small pieces of wood or metal held together by bolts, cables and small pieces of framework. Like a truss, a canopy is only as strong as its weakest connection. A marquee is one large continuous beam supported only at one end; a canopy is structural composition of no-continuous beams supported at one end. If a bolt, cable, or framework connector of this non-continuous beam fails, it can trigger a complete canopy collapse.

A structure exposed to fire usually fails at one of its connections. A canopy, which has many points of connection, has a greater chance of failure than a marquee, which has few connections. Another reason a canopy can collapse is because it is combustible. Some canopies have a wooden framework of rafters supporting the shed roof. The most dangerous type of canopy is a metal or wood shed suspended over a truck loading area. It is designed to protect workers and products from the weather. The corrugated tin or wood canopies found on buildings in the oldest sections of town are covered with tarpaper and thus extremely combustible.

Old skylights, originally designed to provide light on the platform below, are sometimes tarred over. They look like scuttle covers on top of the canopy, but collapse when stepped on. Adding to the collapse danger of a canopy is a system of tracks and rails suspended from the underside. Heavy products unloaded from trucks or railroad cars are attached to the rail system and pushed inside the building. The weight of the rail system heightens the canopy collapse danger.

Firefighters must realize a canopy is not a porch. It may look like a porch, but it is not. It does not have the same load-bearing capacity. A canopy is supported on one end only. A porch with columns is supported at both ends. Firefighters should not stretch hose streams on top of a canopy roof or place ladders on top of a canopy roof.

The lesson learned from the tragic Chicago and New York City canopy collapses is that the area beneath a canopy should be considered inside the collapse danger zone. When there is a danger of collapse due to a large body of fire, withdraw firefighters not only from the burning building but from beneath the canopy as well.

Cornice

A cornice is a decorative horizontal overhang, projecting outward along the top wall of a building. If a parapet wall is defined as a portion of an exterior wall extending above the roof, a cornice can be defined as a portion of the roof extending out beyond an exterior wall. A cornice is also a cantilever structure, like a parapet, marquee, and canopy, but there is one important difference. A cornice burns and spreads fire. A cornice may be constructed of wood or combustible plastic and it may have a wood framework inside, and/or wood shingles outside. A fire officer should also be concerned about horizontal fire spread from one end of a building to the other via a cornice.

Flames blowing out of a window below a cornice will spread to the cornice. Flames will then spread along the underside of the cornice or inside its framework, possibly to an adjoining building. After flames destroy a cornice, it can collapse off the facade. Also, when a cornice has been weakened by fire, it can collapse due to a sudden impact like being struck by a powerful master stream. If a cornice collapses suddenly it will crash down like a wave of death and destruction. Firefighters operating below can be crushed under a falling cornice.

Lessons Learned

Firefighters operating outside a burning building are exposed to the dangers of wall collapse. After a roof and/or floor collapse occurs because of fire, the danger of exterior wall collapse is greatly increased. The collapse danger zone is defined as the distance outward from the foot of the wall equal to one and one half or two times the height of the wall. The officer makes the decision about the collapse zone distance. When the collapse danger zone is greater than the reach of a hose stream, the officer should consider operating the hose stream from a flanking position; that is, on either side of the front wall beyond the outer perimeter of the building's width or setting up an umanned monitor nozzle (fig. 4–7).

A fire officer cannot predict if a masonry wall will collapse in a 90-degree-angle, a curtain-fall, or an inward/outward type of configuration. The officer should, however, expect the worst: a 90-degree-angle wall collapse with large chunks of brick and mortar thrown out farther than the falling wall.

When a wall has a parapet, marquee, canopy, or cornice it becomes more deadly. The area beneath a marquee, canopy, or cornice is considered to be inside the burning building. When the order is given to withdraw from the building, the firefighters must be withdrawn from beneath the marquee, canopy, and cornice as well.

Collapse of Burning Buildings | Second Edition

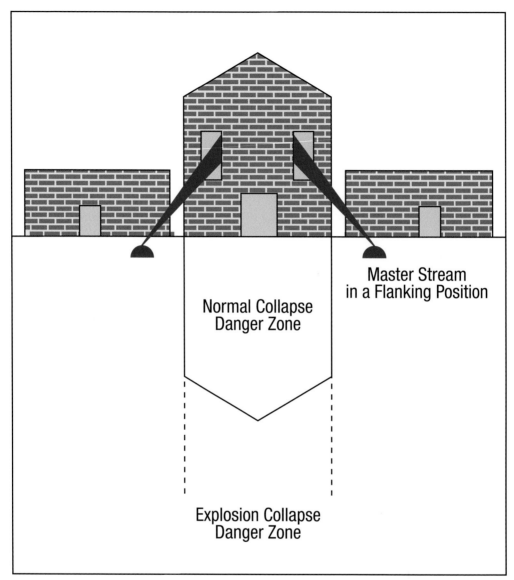

Fig. 4–7. Flanking positions protect firefighters against wall collapse.

Note: For more information about wall collapse, search the Web for *Firefighter Fatality Investigation NIOSH F2001-09* and *2004-37*.

5

COLLAPSE DANGERS OF PARAPET WALLS

A row of five retail stores is heavily engulfed by a late-night fire. As flames shoot through the wood-and-tar roof of the one-story, brick-and-joist structure, fire blows out of the large broken display windows and black smoke pours through the open doors of each end store (fig. 5-1). Their interior operations barred by maze-like partitions and concealed spaces, firefighters have retreated to the sidewalk to set up a defensive attack.

Fig. 5–1. A parapet wall supported by a steel lintel with flame flowing out the opening below the steel lintel is distorting the steel support.

As several members try unsuccessfully to drive back the flames with hoselines, another member uses a 10-foot hook to remove decorations and partitions from display windows, letting streams penetrate deeper into the burning stores. Extending 6 feet above the roof of the structure is a large brick parapet wall, 125 feet long and 18 inches thick. Below the wall, each of the five stores has one large display window and an entrance door. Separating the stores are brick columns, extending from the sidewalk to the underside of the parapet wall. The parapet wall is supported by a large steel beam which crosses above the columns and acts as a lintel, spanning the display windows over each store. Flames blowing out of the broken display windows are heating the steel lintel.

When fire conditions have stabilized and the outside fire attack has been established, a firefighter operating one of the hoselines points up at the structure, turns to his officer, standing nearby, and asks, "Captain, does that section of parapet above the paint store look like it is leaning outward, or am I imagining it?"

The officer looks up at the wall. "It seems okay to me," he says, "but I'll keep an eye on it."

"Say, Captain," says another firefighter. "I think he's right—that wall does look as if it is leaning outward. See, at the top edge of the parapet—directly over the paint store—it's wavy, and yet the rest of the wall, over the other stores, is a straight line."

"Yes, you're right," replies the officer. "Back the line off the sidewalk between the parked cars. I'll tell the chief." He calls on his radio, "Engine 8 to Battalion 2."

"Battalion 2 to Engine 8, go ahead with your message."

"Section of brick wall is leaning out," he says.

"I'll have my aide back the other lines off the sidewalk. Let me know if it gets any worse."

Walking back to the command post, the chief again speaks into his radio: "Battalion 2 to Engine 8 pump operator, shine your apparatus spotlight on the parapet wall over the center paint store." He then tells his driver, "Go over there and make sure no one walks under that wall near the paint store."

Following orders, the driver stands one store down from the paint shop, where he can observe the entire sidewalk. Five master streams shoot out at the structure from between parked cars. With its stream operating on the roof at full blast, a snorkel slowly moves above the overhead utility wires. With all flames now driven back, white steam and smoke lap out of each store, but a large black cloud of smoke drifts up into the night sky from the roof behind the parapet wall.

A firefighter uses a pike pole to pull down a smoldering advertising sign above one of the stores. Seeing him, the battalion aide shouts and waves the firefighter back into the street. Suddenly the parapet wall collapses over the center store, pulling down with it the wall on either side. In one thunderous wave, the massive wall breaks away from its steel support, collapsing outward toward each end of the row of stores. The firefighter positioned on the sidewalk starts to run into the street but slips on broken glass and falls down. Looking up at the tons of falling masonry rushing toward him, he scrambles up into a crouching position and leaps on to the hood of a car parked nearby, seconds before massive chunks of bricks and mortar crash down on to the sidewalk beneath him.

Parapet Wall Collapse

What exactly is a parapet wall? Building codes define a parapet wall as the continuation of an exterior wall, a fire wall, or a party wall above the roof level (fig. 5–2). Firefighters know it as the waist-high wall that encircles a roof. A masonry parapet wall is one of the most dangerous walls that a firefighter can encounter at a fire. Every several years in the United States, firefighters are killed or injured by parapet wall collapse (NIOSH FIRE FATALITY #98F-20). The collapse danger of a parapet wall comes from the decorative front parapet built over a row of stores. Because of the hazards of falling objects, the most dangerous area outside a burning structure is the sidewalk directly in front of the building. If the building has a parapet wall, however, the sidewalk is even more deadly because of the possibility of parapet wall collapse.

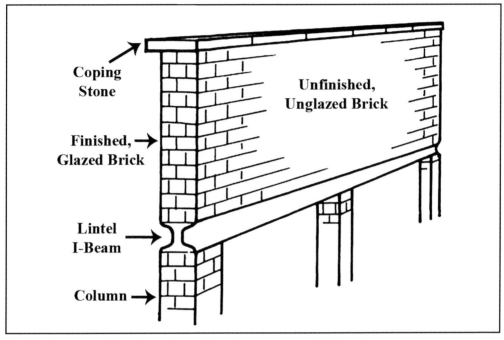

Fig. 5–2. A parapet wall above large show windows is a dangerous wall during a fire.

Generally, high, ornate parapet walls collapse more frequently than lower, less decorative ones. Parapets are found on rooftops and setbacks of multistory buildings. The most dangerous parapet, however, is the one constructed as the front wall of a one-story structure above several large display windows, the sort found in supermarkets, shopping malls, and rows of stores. This type of parapet is supported on top of a steel I-beam which acts as a lintel, spanning the big windows below. The steel lintel is supported only by narrow masonry or cast-iron columns rising between the large windows, and movement of the structure during a serious fire can topple this type of parapet wall. Further examination reveals that there are five separate additional design defects in the construction of a parapet wall.

Freestanding wall

The parapet described in the scenario above is a freestanding wall (fig. 5–3). It does not support any other part of the structure, it is balanced on top of a steel I-beam, and, as the term implies, it just stands in place. Because a wall receives its stability from other portions of the structure to which it is connected, a freestanding parapet of an exterior wall is less stable than either a nonbearing wall or a bearing wall. The more connections a wall has to other portions of the structure, the more stable it becomes, but the more damage it will cause to other portions of the structure when it collapses. A front, freestanding parapet wall over a store can be considered a cantiliver beam. It is a vertical cantilever wall supported from one end, the bottom.

Fig. 5–3. A freestanding parapet wall. A master stream striking the back of this wall could cause it to collapse outward onto the sidewalk.

The steel lintel

In a one-story building with several large display windows and a 50-to-100-foot parapet wall above the windows, the wall rests on top of a steel I-beam. This steel lintel supports several tons of brick and mortar above. If the large display windows must be broken to vent a fire, the heat from the flames blowing out of the window will be hottest near the window opening directly below the steel lintel. Here the combustible gases coming from inside the store are mixed with oxygen in the air. If the process continues for a long time, the heat of the flames can cause the steel lintel to expand, twist, or buckle. These movements could, in turn, cause the parapet wall above to become unstable and suddenly collapse.

Steel reinforcement rods embedded in a parapet wall

A parapet wall may be constructed of solid brick without any steel reinforcements, or it may be tied together by steel reinforcement rods embedded within the masonry or fastened to the back of the wall, visible only from the roof. The presence of steel reinforcement determines how the wall will collapse. If a parapet wall is solid brick without steel reinforcement, it may fall in one small section, with the rest remaining on top of the steel lintel support. But if the parapet wall is tied together with reinforcement rods, the entire wall may collapse, even if only one small section of it is unstable (fig. 5–4). Ornamental cast stone and decorative terra-cotta parapet walls are often attached to steel reinforcement rods and angle irons, making them the most susceptible to total collapse during firefighting operations.

When planning for a parapet collapse at a serious fire where a wall appears unstable, firefighters must estimate the horizontal as well as the vertical collapse danger zone. What may appear to be a small, localized collapse danger of a parapet wall could result in the total collapse of a 100-foot parapet tied together by steel reinforcement rods.

Fig. 5–4. This parapet wall supported by steel lintel collapsed suddenly during a fire.

Coping stones and cornices of a parapet wall

A coping stone, or capstone, is placed along the top of a brick parapet wall to keep rainwater from seeping down into the parapet wall. Depending upon its composition—terra-cotta, cement, or slate—it can weigh between 5 and 50 pounds. When a parapet wall is not properly maintained over the years, mortar used to cement the coping stone to the wall can lose its adhesive qualities. When this happens, the stone will be held in place by gravity alone and will merely rest on top of the wall. During firefighting operations, coping stones can easily be knocked loose by sweeping master streams or by retracting aerial platforms, snorkels, or aerial ladders. Firefighters have been killed or seriously injured after being struck by falling cement capstones.

Unfinished brick of a parapet wall

Exposed to wind, rain, and snow on both sides, parapet walls deteriorate more rapidly than masonry walls that are exposed on one surface only. Though the front of a parapet is made of finished brick glazed with a water-resistant coating, the back is unfinished, unglazed, and often constructed of poor materials with poor workmanship. It is less resistant to moisture and erosion. If water absorbed by the eroded back surface freezes during a fire, the expansion of the freezing water can cause the entire parapet wall to lean outward and become unstable. Over the years, the New York City Fire Department has had several deadly experiences with collapsing parapet walls. The worst occurred on April 4, 1956, when six firefighters were killed by a collapsing parapet wall while they were operating hand lines and cellar pipes in the first floor and cellar of a one-story furniture showroom. On October 4, 1965, a parapet wall section collapsed on a company that was directing a hose stream into the doorway of a one-story supermarket; one firefighter was paralyzed. At a second-alarm fire on June 23, 1977, a captain, who had paused to don his face piece before entering the fire building, was struck on the head and killed instantly by a falling cement coping stone.

Movie Marquees and Canopies

Movie marquees, canopies, metal overhanging sheds, and large illuminated advertising signs are sometimes attached to parapet walls. These heavy, cantilevered attachments have fastenings or anchor bolts that extend through the parapet and are connected to roof beams directly behind it. A collapse can occur if a fire burns through the roof and destroys the wooden roof beams directly behind the parapet or weakens the anchors or fastenings connected to the wooden beams. Then the support of the heavy, overhanging marquee or canopy shifts from the roof beams to the parapet, and if the marquee, canopy, or sign collapses, it can pull the entire parapet wall on both sides down with it.

In the 1956 collapse that killed six firefighters, the parapet wall supported a marquee, whose weight was determined to be one of the causes of the collapse. The firefighters who perished were not under the marquee but were operating well to one side of it, but when the marquee collapsed, the entire parapet wall was pulled down with it (fig 5–5).

Steel Expansion

Some large, one-story, brick-and-joist buildings have steel girders that support wooden roof joists and span the distance from the back wall to the front parapet wall. When exposed to heat during a serious fire, these girders can expand and push out the parapet wall. The freestanding parapet can then become unstable and collapse on to the sidewalk.

In New York City on November 26, 1977, a fire burned out of control in a row of stores comprising a 125-by-100-foot brick-and-joist structure. Four steel girders of the building ran from front to rear. Thirty minutes into the operation, the front parapet wall was pushed out slightly by an expanding steel girder. About 35 minutes after the bulge was discovered, the entire 125-foot stone parapet wall crashed to the sidewalk in one wave.

Fig. 5–5. This marque and parapet wall collapsed and killed six FDNY firefighters.

Timber Truss Roof Collapse

The collapse of a timber truss roof that has sloping hip rafters connected to a parapet can cause the collapse of the parapet wall. Heavy timber bowstring trusses often support roofs of one-story commercial structures, which have front and rear masonry parapet walls (fig. 5-5) The high parapet conceals the round rooftop. If a serious fire occurs in the roof space of a timber truss roof which is not subdivided by fire curtains at each truss, flames will spread through the open-web members of all the trusses and involve the entire underside of the truss roof. The large wooden trusses will be weakened by the flame and will begin to lose strength simultaneously. As all the trusses begin to fall, the weight of the roof will shift from the trusses to the front and rear parapet walls by way of the hip roof rafters or *purlins*, which extend from the end bowstring trusses to the front and rear parapet walls. This shifting of the roof load from the trusses can cause a violent front parapet wall collapse seconds after the entire roof caves in. The front parapet wall receiving the sloping hip rafters breaks at the middle and is driven out into the street as a result of the roof collapse. The entire front wall will collapse in an inward/outward type of collapse: the upper part of the parapet wall falls inward, and the lower part of the wall falls outward (fig. 5-6).

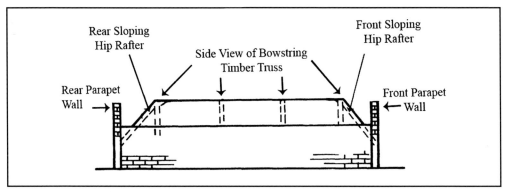

Fig. 5–6. A side view of a building with a bowstring truss roof and a parapet wall

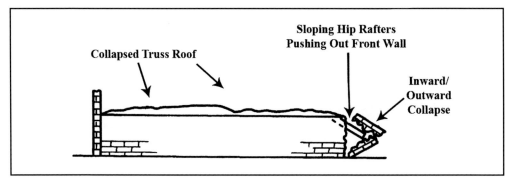

Fig. 5–7. A side view of a parapet wall collapse caused by the truss roof collapse

When a serious fire takes control of a large underside of a bowstring truss roof that has sloping hip rafters extending from end trusses to the front and rear parapet walls, the officer in command should anticipate a roof collapse followed instantly by an explosive front and rear parapet wall collapse. A 1980 fire involving a one-story 75-by-100-foot bowstring timber truss roof building in New York City collapsed the roof and drove the front brick wall and parapet out into the street. The collapsing wall broke a large wooden utility pole in front of the structure and caused it to collapse. No firefighters were injured, as all crew members and apparatus had been moved away from the front of the wall to a flanking position. The collapsing wall would have killed anyone on the sidewalk or in the street.

Explosions and Backdrafts

Explosions and backdrafts can cause the collapse of parapet walls. An explosion in the concealed roof space of a structure housing a row of stores can blow out a parapet wall; in addition, an explosion inside the store can create a violent tremor throughout the structure, lifting roofs above the supporting walls, exploding windows outward, and toppling parapet walls off their steel lintel support.

Snorkels, Aerial Platforms, and Aerial Ladders

When firefighters want to lower an aerial device that is several inches above a parapet wall or resting on it, they must make sure the apparatus is clear of the parapet wall before the device is lowered. If it is not, the apparatus may knock down a section of bricks and coping stone, and firefighters could be struck by the falling pieces. If the top of an aerial device is obscured by smoke or darkness and it is raised near the top of a parapet, the firefighter operating the controls should assume that the tip of the apparatus is resting on the wall. When the boom or aerial must be repositioned, the firefighter should first activate the control to raise the device, lifting the tip of the aerial several feet above the parapet, and then activate the retracting control. Firefighters at street level should not work directly beneath aerial apparatus that is being repositioned.

Lessons Learned

A parapet wall does not collapse during the early stages of a fire when only the content is burning and the structure is not yet involved. If interior lines are unsuccessful and the building becomes involved, however, the officer in command should start to inspect the parapet wall for signs of failure, especially if the structure is a row of stores or a supermarket with large display windows. When a fire has been extinguished, firefighters often become less concerned about observing and reporting structural collapse hazards—but we should not drop our guard. Parapet walls often collapse during the overhauling stage of a fire, after the structure has been weakened and partially destroyed by fire and large caliber master streams. The parapet wall collapse that killed six FDNY firefighters occurred after the fire had been brought under control.

There is a myth in the fire service that a firefighter can avoid being crushed in a masonry wall collapse by jumping into a doorway or an open window. True, there have been collapses where walls have fallen outward and firefighters standing in front of a door or window opening have miraculously survived. But what actually happened was sheer luck—they were standing in the spot where there was an opening, and the falling wall passed around them. They did not jump into the doorway or move to escape. A professional fire department does not recommend miraculous escapes as a substitute for safe fireground procedures. When a wall is in danger of collapse, firefighters should withdraw beyond the established collapse zone. Moreover, if a parapet wall collapse is caused by a failing timber truss roof and a firefighter jumps into the doorway to escape the falling wall, he will be immediately crushed to death by the roof collapse.

The sixteenth edition of the *Fire Protection Handbook*, published by the National Fire Protection Association, states that "ordinary masonry, assuming no explosion or internal pressure, will fall within a distance from the wall equal to one-third its height, but bricks may bounce or fall further." Fireground experience has shown this statement to be incorrect—it is a serious underestimation of the distance a wall may collapse. Firefighters operating a hose stream at a distance of one-third the height of a wall will be buried beneath the falling masonry wall. The quoted statement was eliminated in updated edition of the handbook. Today the fire service knows an ordinary masonry wall will fall and cover, with brick or masonry, a distance from the wall equal to one, one and one half, or twice its full height, and bricks may bounce or roll farther. If the masonry wall falls because of explosion or internal pressure, the wall will collapse outward for a distance equal to twice its height with force, and bricks may bounce or roll out farther. Because the length of a parapet wall is usually greater than its height, the officer in command must consider the amount of horizontal wall that may collapse during a fire. All firefighters—not just chiefs and other officers—are responsible for reporting signs of structural failure. The firefighter should report any condition that might indicate structural failure to his officer, who in turn must first act to safeguard firefighters and then notify the officer in command. Remember, the officer in command cannot predict burning building collapse, and a firefighter can change the overall fire strategy by advising his commander of a structural danger.

Note: For information about parapet wall collapse, search the Web for: *Firefighter Fatality Investigation NIOSH 98F020*.

6

WOOD FLOOR COLLAPSE

A responding engine company turns into a street lined with vacant, brick-and-joist row houses. The last house on the block, a three-story, 20-by-60-foot structure, is fully involved in fire. Smoke pushes out around the sides of the tinned-up windows on the first floor. Through an opening on the second floor, flames can be seen spreading along the underside of the ceiling. On the third floor, smoke is being emitted from the windows.

The pumper stops near a hydrant several hundred feet up the street from the fire building. A firefighter dismounts from the side jump seat, runs to the back step, and grabs a fold of large-diameter hose connected to a hydrant fitting with a hydrant wrench attached. He pulls off several feet of hose, encircles the hydrant with it and shouts to the driver to go. As the engine moves up the street, the large hose tightens around the hydrant and the supply-line hose jumps out of the hose bed, length after length. When the pumper stops in front of the fire building, the cab door flies open and the officer gives the order to stretch. Two mask-equipped firefighters exit the side jump seats and start advancing a 1¾-inch preconnect toward the building. The motor pump operator, running to the rear of the engine, uncouples the supply line lying in the street and connects the male end to a gated inlet in the pump, which was prefitted with a reducer. He opens the inlet gate and gives the signal to the hydrant man to start the water. As the water flows, the pump operator awaits the order from his officer to start water into the attack hoseline.

The ladder company chauffeur swings the tiller rig into the street from the opposite direction. He brings the truck nose to nose with the pumper, giving the aerial maximum coverage of all side and front windows of the fire building. The officer and a two-man forcible entry and search team race to the entrance door, where the engine company is preparing for the fire attack. One firefighter strikes the one-piece, tinned up entrance doorway in the center with a full swing of the back of an axe. Then, using the adz end of a Halligan tool as a hook, he pulls the top-left corner of the tin away from the doorway frame to which it is nailed. He pries the tin away on the left side, starting at the top and working downward. When the left side is free, he pulls out the tin and pushes it to the right side, exposing half of the door entrance.

With smoke enveloping him, the truckman steps into the doorway and swings the Halligan tool like a bat, banging the pried-open tin all the way back from the door. He backs out just before the flames explode out of the now opened doorway. The other forcible entry team firefighter, equipped with a 6-foot hook, runs to the rear of the fire building to vent for the advancing hose team. Starting at the window on the leeward side, he swings the 6-foot hook over his head into the center of a tinned-up window, piercing the tin at the center near the top half. He pushes the metal end of the hook through the small hole in the tin, and turns the hook sideways to catch the tin; he then begins a rapid in-and-out motion. Each outward pull brings the tin away from the window frame to which it is nailed, until the tin falls from the window and flames shoot out. The firefighter begins the same process on the next window.

Meanwhile, the chauffeur of the ladder company has raised the aerial to the roof of the three-story fire building, and the firefighter assigned to ventilate the roof is climbing the ladder. Arriving on the roof, he encounters a scuttle cover and a skylight. He is starting to break the panes of glass in the skylight when the entire metal frame and glass collapse into the interior stairway below. Smoke and flame rush up out of the roof opening. The soldered sections of the tin skylight frame had been melted by the fire.

"Good thing the engine company has not made it to the top floor yet," he thinks to himself. Moving over to the scuttle cover on the roof, he uses the Halligan tool to pry up one end of the tarred, flat square cover placed over an opening which provided access to the roof from the top floor. The cover breaks loose from the latch below. As the firefighter pulls off the cover using the tool's adz end, flames rush out of the opening. In the street, engine company members are crouched with a charged line, driving flames back into the front door. At that moment the chief arrives on the scene. Suddenly, thick black smoke billowing out of the second-floor windows explodes into fire. Flames shoot out front and side windows above the engine company on the first floor. A second later, the third-floor windows start to blow out flames and smoke.

The chief checks his clipboard with a computer alarm response sheet attached. "Where are the second-due companies?" he asks his aide. "They should have been here by now!" Before the aide can reply, the chief orders, "Go radio the dispatcher and check on the second-due engine and ladder. Also transmit a signal for a working fire."

Just then the Handy-Talkie crackles, "Ladder 2 roofman to Chief."

"Battalion 1, go ahead."

"Fire is burning front to rear on all floors. The roof has been vented. I think you should pull the engine company out. There's just too much fire in this old building."

"Battalion 1, 10-4. Get off the roof."

"Battalion 1 to Engine 8. Back the line out of the first floor."

"Engine 8 to Battalion 1. Chief, we have this fire almost out—we have only one more room to go."

"Battalion 1 to Engine 8. I don't care; back that line out." Engine 8 does not respond.

The battalion aide runs up to the chief. "Chief, the second engine broke down responding, and the ladder company is operating at another fire. The replacement units will be delayed."

Inside the burning building, the engine officer directs two firefighters to go out and tell the chief the fire is almost out and only one room of fire remains. Just then the chief transmits, "Battalion I to Engine 8. Did you get that last message? I said back that line out! The two floors above you are fully involved."

Engine 8 still does not answer. Flames are now reaching into the afternoon sky over the rooftop. The aerial ladder is being retracted from the roof. Two engine company firefighters and ladder company members exit from the smoky first-floor doorway. Just then there is a loud cracking noise and the third-floor timbers are visible as they collapse past the second-floor windows and crash violently on to the second floor. Then the second floor gives way and collapses down into the first floor with a loud, sickening rumble. When the smoke clears, the street-level entrance doorway is completely filled with wood joist ends, sections of lath, plaster, and brick.

"Dammit! I just lost a company!" shouts the chief as he throws the clipboard on the ground and runs toward the entrance where the still charged hoseline, on the ground, leads into the doorway packed tight with collapse rubble. The silence is suddenly broken. "Urgent! Urgent!" comes a voice on the radio. "Engine 8 to Battalion 1. Chief, did my two firefighters make it back out to you? The nozzleman and I were able to dive out the rear window before the collapse!"

"Battalion 1 to Engine 8. Yes, they made it out; they are okay," replies the chief with a sigh.

Floor Collapse

Floor failure is the leading cause of firefighter death by collapse. Firefighters use floors as platforms from which to launch interior searches and hoseline attacks. They depend on the floors inside a burning building to support them and their action. When floors do not support firefighters and collapse unexpectedly, there is often a loss of life or serious injury. When floors collapse, firefighters can be caught and trapped in the broken smoldering building and asphyxiated, crushed beneath the ton of rubble, or burned to death. Floor decks, floor beams and floor supports, such as, columns and girders, collapse during fires causing catastrophic, progressive, multilevel floor failures.

Case study of a floor collapse

A deadly floor collapse in the 1990s killed four firefighters in Brackenridge, Pennsylvania. The four firefighters were killed when a floor in a two-story 75×75 foot structure, built in the 1930s collapsed; First Lieutenant Rick Frantz, firefighters David Emanuelson, Michael Cielicki Burns, and Firefighter Frank Veri Jr. died when they were caught and trapped by a ball of fire after the floor failed.

The floors were 4-inch-thick concrete, supported by steel columns, steel girders and steel floor beams with masonry walls. This building originally was occupied as an auto dealership selling new cars. However the auto dealership went out of business and the occupancy changed. A furniture refinishing company moved into the cellar of this noncombustible structure built of steel and concrete. At the time of the fire, there was a large amount of furniture storage, a wood refinishing workshop, and a paint-spraying booth near the front of the cellar directly below the first floor entrance, and large amounts of flammable paints, lacquers, varnishes, and thinners in 55-gallon drums.

The fire started in the workshop. Upon arrival two attack hoselines were stretched to the burning cellar from the rear doors (exposure "C") to attack the fire. Heavy smoke and heat prevented the lines from advancing towards the front of the cellar where the fire was burning. A third hoseline was ordered stretched into the first floor, front entrance (exposure "A") to prevent vertical fire spread up an interior stairway located in the center of the building. This stair extended from the cellar to the first floor and smoke and heat were extending up the stair enclosure. As firefighters stretched the line through the first floor front entrance, a large section of the first floor collapsed behind them. The floor collapse cut off their escape. As the floor collapsed, it crushed large drums of flammable liquid in the basement and caused a massive fire-ball explosion, killing the four firefighters on the first floor.

The structural hierarchy

A post-fire analysis revealed the series of events which caused the collapse. The first structure to fail was a massive steel column. The unprotected steel column twisted and sagged, causing a steel girder to warp and move backward, pulling floor beams out of the foundation and collapsing the first floor. This collapse was another example of the *structural hierarchy* effect during a collapse (fig. 6-1). The structural hierarchy effect means, the destructiveness of a collapse depends on the first structure to fail, and where this structure is positioned within the building supporting system. The structural hierarchy principle was first identified as a factor in floor collapse at the Boston Vendome collapse, of June 17, 1972. This floor collapse killed nine Boston firefighters. Firefighters must know that when the first structure to fail is high up on the structural hierarchy, the more widespread and deadly the collapse will be. For example, the "hierarchy" of a floor system has the bearing wall and column highest on the scale. Progressing downward in the structural hierarchy of a floor system, the girder come next, and a floor beam next, followed by the floor deck. So if a column fails, it affects the girders and floors and the deck. A column failure will have more consequences than if a girder fails. If a girder fails, it will create more destruction, than if a floor beam collapses but not as much as if a column fails. And if a floor beam fails during a fire, it can have more impact than if a floor deck fails but not as much as when a column or girder fails.

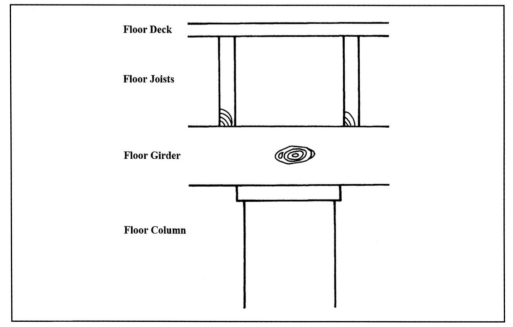

Fig. 6–1. The seriousness of a floor collapse depends upon the first structure to fail. A column failure is more serious than a girder collapse, a girder collapse more serious than a beam, and a beam more serious than a deck.

Floor Construction

A floor can be constructed of masonry, steel, or wood. Floors in many modern and renovated buildings are lightweight construction. Lightweight constructed floors can collapse during the early stages of a fire. A floor of wood truss, wood laminated I-beam, lightweight cold-formed steel C-floor beams, and open-web bar joists are considered lightweight truss floors. Fire resistance of a floor is directly related to its "mass"—bulk and spacing in the floor system. These lightweight floors have less mass and greater spacing between joists. Instead of traditional wood floor beams, which are 16 or 18 inches on center, lightweight floors can be 2 or 4 feet apart. The mass of the lightweight floors is reduced when the inside or middle section of the floor truss is replaced by thin bars or wood web pieces. The I and C floor beams are hollowed out or thinner than the top and bottom beam sections. This reduction of mass and increased spacing reduces fire resistance and allows floors to burn through and fail faster than conventional solid wood floor beams.

A floor of wood I-beam collapsed during a cellar fire in a private dwelling killing Deputy Chief Steven Smith of the Wea Township Community Fire Department in Lafayette, Indiana, on August 25, 2006. Chief Smith was this nation's first firefighter killed by a collapsing wood I-beam construction floor. The lightweight wood truss has killed 20 firefighters since 1984. We now have the first firefighter to die in the collapse of a wood laminated I-beam; and the fire service is holding its breath on the sheet metal C-shaped floor supports.

New buildings have lightweight floor construction. However, older and existing buildings still have solid wood floors. The conventional wood floor used in older brick–and–joist, and wood-frame construction, is a solid wood joist system 2×8-inch, 2×10-inch, or 2×12-inch floor beams spaced 12, 16, or 24 inches on center. The solid wood floor supports an under-floor of rough wood board or plywood, which in turn is covered by a finished floor of wood or tile. All types of floor systems, lightweight and conventional solid wood, can collapse when attacked by fire, but the lightweight constructed floors may collapse faster when exposed to flame and heat.

Types of Floor Collapse

The three ways all floors fail during fire are:

1. Floor deck may collapse, where only the wood deck may burn through and collapse, leaving the supporting joists intact. An example of a floor deck collapse occurred when FDNY Fire Lieutenant John Clancy died December 31 1995. The floor deck of an entrance landing on a two-story ordinary constructed private dwelling collapsed (fig. 6–2). This was a vacant building, and the fire was in the cellar when Lieutenant Clancy stepped inside a smoke–filled, first-floor, side doorway. The floor deck collapsed sending him into the burning cellar. Visibility was zero as the door opening was completely filled with smoke.

Collapse of Burning Buildings | Second Edition

Fig. 6–2. Lieutenant John Clancy died when this floor deck collapsed.

2. Floor beam collapse, where several floor joists fail, causing a localized failure of a section of floor within room is a more deadly collapse (fig. 6-3). An example of a, recent, deadly floor beam collapse occurred when FDNY Lieutenant Howard Carpluk and firefighter Michael Reilly were killed at a Bronx fire on April 27, 2006, when the conventional solid wood floor beams of a one-story ordinary constructed strip store collapsed. The most deadly floor beam collapse occurred on October 17, 1966, when 12 FDNY firefighters were killed in the Wonder Drug store located on 23rd Street in Manhattan.

3. A multilevel floor collapse describes a progressive floor failure. Here a floor collapse triggers the subsequent collapse of floors below and of one or more enclosing walls (fig. 6-4). This is the most deadly type of collapse. The two most deadly multilevel floor collapses in the history of the fire service occurred June 17, 1972, when the Vendome Hotel collapsed and killed 9 Boston firefighters, and on September 11, 2001, when the terrorists attacked the World Trade Center. The two towers collapsed and killed 2,749 people, including 343 New York City firefighters.

Fig. 6–3. An example of a floor beam collapse

A multilevel floor collapse most often happens in burning buildings that have columns and girders supporting floors. A building with a frontage of 25 feet or more usually has columns and girders. A building 25 feet or less can have floor beams supported by bearing walls on each side. Floor beams supported at each end by bearing walls are called "simple" beams—beams supported at both ends. A building, over 25 feet frontage, with a system of columns and girders may have floor beams supported at each end by bearing walls; however, at the center of the floor span, the beams are supported by a girder and columns. These are called "continuous" beams. And in these buildings, if the column or girders fail, there can be a progressive collapse of the floors and walls of the building. The chances a multilevel floor, or, so called, "progressive collapse," occurring are great when the first structure to fail is a column, girder, or bearing wall.

Figures 6-5 through 6-9 illustrate types of floor collapses.

Fig. 6–4. An example of a multilevel floor collapse

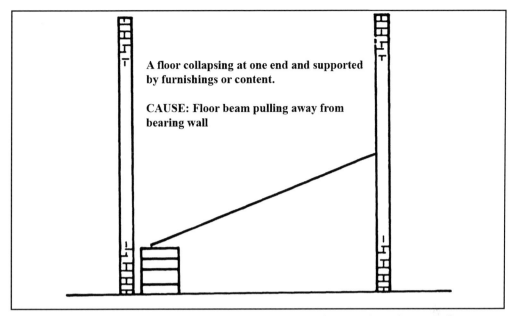

Fig. 6–5. A supported lean-to floor collapse can be caused by a floor pulling away from a bearing wall.

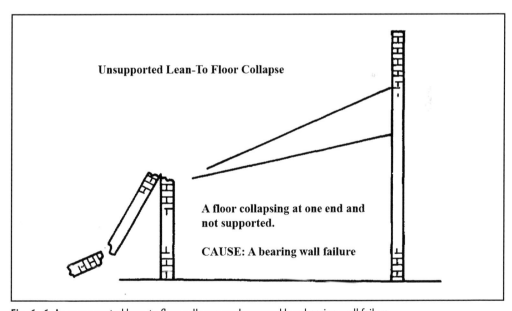

Fig. 6–6. An unsupported lean-to floor collapse can be caused by a bearing wall failure.

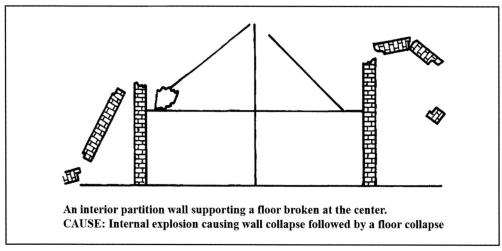

Fig. 6–7. A tent floor collapse can be caused by an explosion, resulting in bearing wall failure followed by floor collapse.

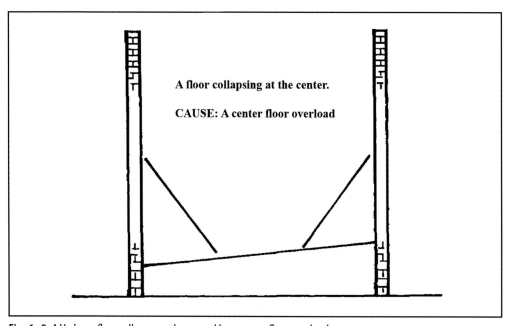

Fig. 6–8. A V-shape floor collapse can be caused by a center floor overload.

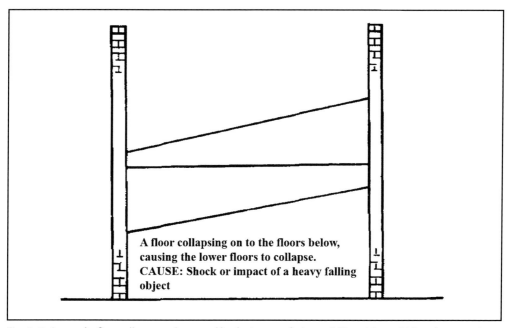

Fig. 6–9. A pancake floor collapse can be caused by the impact of a heavy falling object, which makes several floors collapse.

Managing the Risk of Floor Collapse

There are strategies and tactics that can protect firefighters from the three types of floor collapse listed above. These are:

1. **Floor deck collapse: Sound the floor.** When searching in smoke, use a tool to probe the floor in front as you move forward or when advancing and directing a hoseline, keep one leg outstretched to feel for floor openings or weakening. Sounding the floor can protect a firefighter from floor deck collapse.

2. **Floor beam collapse: Use the reach of the hose stream.** When encountering a weakened section of floor, use the reach of the hose stream to avoid the danger area. Using the reach of a hose stream can protect firefighters from a floor beam collapse.

3. **Multi-Level floor collapse: Withdraw firefighters.** When there is a danger of multilevel (progressive or disproportionate) floor collapse, withdraw firefighters from inside the burning building and set up defensive master streams around the building outside a collapse zone. When floors collapse they can sometimes cause a secondary enclosing wall collapse. So firefighters must be withdrawn outside the collapse zone or positioned to flank the fire, away from all four walls. When there is a multilevel floor collapse, firefighters outside the building can be killed by the secondary wall collapse. The National Institute of Standards and Technology (NIST)

in its final report on the World Trade Center tragedy estimates 160 FDNY firefighters where killed outside the towers. The falling floors and walls killed firefighters in the streets around the towers.

Fire Service Understanding of Fire-Resistive Ratings

Floors may have fire-resistive ratings of one, two, three, or four hours. However firefighters must realize that a fire-resistance rating has nothing to do with collapse. Fire-resistive rating only indicates how long a small test sample in a laboratory has resisted fire spread. The fire service has always understood the fire-resistive hourly rating of a floor does not have much significance on firefighting strategy. How long firefighters are allowed to remain inside a burning building is not determined by the fire-resistance rating of the floors. A fire chief would never keep firefighters inside a burning building for one or two hours because the floor had that rating. A floor with a two-hour fire-resistive rating may spread fire and collapse much sooner than two hours. The floor can collapse upon arrival of the firefighters. Fire-resistance ratings have little significance to firefighters because we do not know how long the fire has burned before our arrival. The fire could have burned for four hours before discovery.

Floor Testing

The fire-resistance tests are conducted on small portions of the floor. A floor can receive a one-hour rating when a small reproduced section approximately, 18×14 feet, is placed over a top surface of a testing oven, loaded to its intended design load, and subject to a controlled standard test fire from below. If it withstands the test fire for one hour without the top surface of the floor exceeding an average rise in temperature of 250 degrees Fahrenheit to 325 degrees Fahrenheit at any one point, it receives a 1-hour rating. If the floor collapses after the test ends it still gets a 1-hour fire resistance rating. There is no test for collapse resistance.

Another reason that fire-resistance ratings have little significance with the fire service is that large sections of a building's floor may not react the same as a small test sample. A 60-foot floor beam will not resist fire as long as a smaller 20-foot test sample. Also, renovations to a floor may negate the fire-resistive rating; and the workmanship of test sample and the actual floor construction at the building site often varies. Changes of the construction design and real-life construction can vary. Also, requirements for installation are not always followed and the real fire may reach higher temperatures than the test fire.

Builder's Understanding of Fire-Resistive Ratings

The significance of the hourly fire-resistance rating of a floor means little to builders. According to the NIST, "to a building architect and engineer, a fire-resistive rating expressed in hours does not mean that the structure will sustain its performance for that length of time in a real fire. The architect or engineer knows the actual fire performance may be greater or less that what is achieved in the test fire. The building professional will only say that a floor rated at two-hour fire resistance will block fire longer that than a floor rating of one hour." It will not block fire twice as long, just longer. And this is true only if the floor has been installed properly and is the same size as the test sample.

Time Limits for Fire Attack and Fire-Resistive Ratings

One way an incident commander can use a fire-resistive rating is to use the hourly designation as a "time limit" for interior operations. Over the past 25 years there has been an increase in the time firefighters spend inside burning buildings during firefighting operations. For example at the 1990 One Meridian fire in Philadelphia, the firefighters were battling the blaze inside the burning high-rise for 11 hours before the chief ordered all firefighters to evacuate the building due to structural damage and a danger of collapse. The floors had a 2-hour fire rating. Was it a good decision in the 1970s to conduct firefighting by New York City firefighters inside the 1 New York Plaza fire for four hours when the floors were rated for two hours? In the 1980s was it safe to have Los Angeles firefighters inside the First Interstate building fire for six hours when the floors had a 2-hour fire rating? I do not think so. The NIST investigation final report states the World Trade Center towers survived the impact of the terrorist planes. The ensuing fire caused the 110-story buildings to progressively collapse. Tower 2 collapsed after 58 minutes of burning. Tower 1 collapsed after an hour and 42 minutes of burning. The floors of all World Trade Center buildings had fire-resistive ratings of 2 hours.

The fire service must rethink strategy that has firefighters inside unoccupied, burning buildings five or ten hours that have fire-resistive ratings of only 1 or 2 hours. Should we depend on the support of floors with fire-resistive ratings that, even the construction industry states, may not resist fire for the approved hourly rating? Unless the incident commander determines otherwise, during an uncontrolled fire in a high-rise building, there should be a "time limit" for interior operations. The time limit for interior operations might be the floor fire-resistive ratings. For example, if the floor has a fire-resistive rating of two hours and the fire is burning, uncontrolled, for this time, withdrawal of firefighters and defensive firefighting might be considered. This could be a guideline. Fire-resistive floor rating could also be a guideline for occupant evacuation. For example, all occupants should be able to leave a burning building within the hourly fire resistance rating of the floors. If the floors have a 2-hour rating, everyone should be able to be evacuated from the building within 2 hours. There should be sufficient exits capacity to allow all occupants to leave a high-rise office building within the maximum fire-resistance rating of the floors.

The Philadelphia fire chief's decision at the Meridian fire was a historic action. When the chief ordered the firefighters to withdraw due to the structural dangers, he ordered outside master streams from surrounding buildings, and continued to supply the partial sprinkler system. This landmark fire strategy decision (the first time a fire chief ordered firefighter out of a high-rise burning building) established a new benchmark for the fire service—the evacuation of all firefighters from a burning high-rise building due to structural danger. This benchmark must be included in the command and control decision making for future fire chiefs. This benchmark decision was per the priorities of firefighting, which are life safety as first priority (this includes firefighters), incident stabilization is the second priority, and property protection is the third priority. After the firefighters withdrew from the burning high-rise, nine sprinkler heads stopped the uncontrolled fire that spread from the twenty-second floor to the thirtieth floor. After years of litigation, the 38-story, burned-out structure was declared structurally unsound and demolished.

A progressive multilevel floor collapse of a burning, fire-resistive, high-rise structural steel building was something many fire chiefs (myself included) could not believe could happen until we saw the World Trade Center towers crumble down in 10 seconds. After 9/11, the fire service must acknowledge new lightweight building construction calls for new firefighting strategy. Instead of the standard "controlled burn" interior attack strategy we use for high-rise fires, the fire service must also consider full evacuation of high-rise buildings, and withdrawing firefighters from unoccupied burning high-rise buildings. When fires rage out of control for hours, despite the firefighters' efforts, structural engineers should be called to evaluate the buildings stability. If firefighters are withdrawn, continue to supply a sprinkler system, and use master streams from adjoining high-rise buildings.

Floor Collapse Causes

There are three reoccurring factors that contribute to floor collapse in a burning building. Vacant buildings, renovated buildings, and buildings overloaded with heavy machinery or dense content, such as baled paper or textiles are risks associated with burning building floor collapse.

When a building becomes vacant or unoccupied, the maintenance stops and the building deteriorates. The floors can be quickly weakened when exposed to the freeze-thaw cycle of weather changes. Experience shows vacant buildings are "fire breeders" and collapse hazards. Fire breeders are buildings where arsonists can start a fire and remain hidden from view. Floors of vacant buildings that are rotted by the elements and experience several fires are collapse dangers.

Renovated buildings sometimes have floor construction that is not as stable as the older floor. When a building is renovated, the floor supports can be changed to lightweight or substandard construction. Floors may be further weakened by removing supporting walls or columns. A partition that was indirectly supporting a floor may be removed. And now the span of the floor beams may be greater and the danger of collapse exists.

The third factor contributing to floor collapse is overloading. Loading floors with merchandise that can absorb water, or heavy machinery, makes a floor vulnerable to failure during a fire.

Floors Collapse and Time

Firefighters must know when the floors are most likely to collapse during a fire. There is a time during a fire's growth when floor collapse danger increases. There are three stages of a fire:

1. The growth stage
2. The fully developed stage (active flaming) after flashover
3. The decay stage after most of the fuel is consumed and/or extinguished

The most dangerous time for floor collapse is during the end of the fire, in the decay stage, after it has been extinguished. The history of multilevel floor collapse, the most deadly type of floor collapse, tells us the salvage and overhauling time of a fire is when collapse danger is highest. A building may collapse at any time during a fire, however, experience shows that this final stage of a fire, when salvage and overhauling is conducted by firefighters, is most dangerous. At this time, the building structure has been destroyed by fire, the impact of powerful hose streams have weakened the structural supports, and the buildings content and structure may have absorbed tons of water from the hose streams. When firefighters perform salvage and overhauling, they add weight and vibrations to the weakened building. They must move heavy objects, pull down ceilings, and cut open walls and floors. After a major fire when master streams have been used, the building should be inspected before salvage and overhauling begins. During the inspection, if the building appears in danger of collapse, overhauling should not be undertaken. Instead the incident commander should use outside master streams to cool down the smoldering fire. This is called "hydraulic overhauling" or "defensive overhauling." In this strategy, firefighters are not sent back in to overhaul. They are sent back to quarters. A "watch line" of one or two firefighters and a supervisor remain on the scene to pour tons of water on the smoldering hot spots from a safe distance.

Lessons Learned

Floor collapse is the nightmare of all chief officers. At most floor failures there are no warning signs, no time to act and withdraw firefighters to safety, and no satisfactory explanation of the incident. A sudden floor collapse without warning sign makes the concept of firefighting strategy seem useless. It is this fear that causes fire chiefs to withdraw firefighters from interior firefighting operations.

Three defensive strategies to safeguard firefighters during floor collapse danger are:

1. For deck collapse, use a tool to probe ahead or keep one leg outstretched and support the weight of your body with the back leg.
2. For a floor beam collapse, use the reach of a hose stream to stay away from the weakened floor or order defensive outside attack.
3. For a multilevel floor collapse danger, firefighters must be withdrawn from the building and away from the walls of the building too. Multilevel floor collapse can cause progressive collapse of the walls.

A collapse zone must be considered for wall collapse after the floors fail.

Note: For more information on floor collapse, search the Web for: *Firefighter Fatality Investigation F2004-05*.

7

SLOPING PEAK ROOF COLLAPSE

Two engines, a ladder, and a battalion chief respond to a fire in a brick-and-joist household furniture storage building. Upon arrival, they see smoke pushing out from around the frame of a closed overhead truck entrance door. The structure is one story in height and measures 35 by 105 feet. Exposure 1 is the street in front of the fire building. Proceeding clockwise, exposure 2 is a truck parking lot, exposure 3 is an open lot which slopes downward in a steep decline, and exposure 4 is a 3-foot alley and a three-story, wood-frame residence. One engine company stretches a hoseline to the front of the fire building, while the ladder forces open the overhead entrance doors. The truck captain orders a portable ladder placed against the exposure 4 side in the alley, to provide access to the roof. Then he climbs to the roof, followed by a firefighter who has a power saw slung over his back, and another who carries a flathead axe and a pike pole.

Stepping from the ladder and over a 2-foot parapet wall, the trio walk toward the rear of the roof along a narrow, 18-inch level space where the sloping asphalt shingle, the roof, and the parapet wall meet. (The level roof surface along the parapet is created by metal flashing.) The walking space is filled with 6 inches of slush, and, as the firefighters trudge through it, freezing rain falls and a strong wind blows across the roof. About a third of the way across the roof, the officer stops and points to the rooftop.

"Make the vent cut here," he says.

The firefighter, carrying the power saw, cautiously starts to climb up the sloping gable roof but slips back and falls on one knee. "The roof is covered with ice," he says, regaining his footing.

"Get that portable ladder up here," the officer tells the other firefighter, "but, before you do, drive that axe into the roof for a foothold."

The firefighter swings the axe blade into the roof, midway up the slope, and embeds it deeply in the roof surface. The flathead part of the axe stuck in the roof is now used as a brace for the foot of the firefighter starting the roof cut. Facing the front of the building, with his left foot braced above the axe head and his right leg bent at the knee, he starts the power saw; the first roof cut is parallel and close to the ridge rafter. As he begins his second cut, the portable ladder is put in place and the vent cut is completed while he is standing on the rungs of the roof ladder. The roof deck is pushed down into the occupancy below, with the back of the axe. Smoke billows out of the opening, and flames are visible inside at the floor level.

"Captain, have your men go to the rear and vent some windows from above with the pike pole. The engine is having trouble moving in."

"10-4, Chief," replies the truck officer. He orders one of the firefighters, "Replace the portable ladder to the side of the building again, for our exit, and stay by the ladder. Warn us if the fire coming out of the roof vent starts to take off."

Unable to pass along the narrow walking space, the firefighter, followed by the captain, advances along the parapet roof, slowly and carefully, to the back of the wall. The slush on the roof gets deeper. Approaching the rear, they see that firefighters in the alley below are venting by laying a portable ladder tip into the glass windows.

"It's okay, Ladder 7!" a firefighter shouts from the alley. "We got the windows!"

Smoke billows up as the captain tells the firefighter, "C'mon, we're going back. They got it."

The two turn around and cautiously walk back toward the front of the building along the parapet wall. The darkness on the wall is interrupted by flashing red lights from the apparatus in the street. The roar of the pumping engines competes with the crashing of glass, and the smell of smoke saturates the cold, wet night air. As the captain approaches the firefighter who has been waiting by the ladder, the roof suddenly appears to explode upward into fire. Part of the roof near the parapet wall drops down like a swinging trap door, and a column of flame immediately shoots straight up out of the opening, some 10 feet into the dark night.

The firefighter standing by the ladder sees the column of fire shooting up where the roof has collapsed. He realizes that the captain has plunged through the collapsed roof and quickly climbs down the ladder to the street. "Chief!" he shouts. "The roof just collapsed! The officer fell inside the building and a firefighter is trapped on the roof!"

Peaked Roof Collapse

Operating on a peaked roof is more dangerous than operating on a flat roof. Here is why. Unlike a flat roof with an adjoining building, to get to a peaked roof, a firefighter has to climb a ladder and transfer tools from one ladder to another ladder without falling. Also unlike a flat roof, on a sloping roofs, firefighters must do acrobatic acts of climbing and balancing while cutting a roof (fig. 7–1). And on a peaked roof, unlike a flat roof, if a firefighter loses balance and falls there is no parapet to keep from sliding down the roof to the ground below.

In additon to the falling hazards, a peaked roof's structural framing, roof deck, slate or tile roof shingles are all collapse hazards. During extinguishment in the initial states of attack and during overhauling in the later stages of a fire, a firefighter must sometimes operate above, below, or around sloping roofs. Roof venting , chimney fire operations, protecting wood roofs from wild fire, and overhauling concealed pockets of roof charring require firefighters to operate on sloping peaked roofs. The three most common types of sloping roofs are the gable roof, the hip roof, and the gambrel roof. (The other, less common types are shed and mansard.) The gable roof has sides sloping up from two walls; the hip roof has sides sloping up from four walls; the gambrel roof has two slopes on each of two sides, with the lower slope steeper than the upper. The mansard roof has two slopes, one each of four sides with the lower slope steeper than the upper.

Fig. 7–1. A ridge rafter peaked roof collapse

Structural Frame Collapse

A fire occurred in a two-story, brick-and-joist residence building with a sloping hip roof in the Bronx, New York. Fire originated in the top-floor, rear rooms of this 20-by-60-foot structure and quickly spread to the attic space and roof. Firefighters utilized a hand line to extinguish the interior fire and an exterior master stream at the rear for the roof fire. The roof was badly damaged at the rear, where the two hip rafters extended from the ridge rafter to the rear wall. (Rafters are sometimes called roof beams or roof joists.) Flames burned through the asphalt roof deck and shingles at this point, and pockets of fire still smoldered at the under side of the roof, inside the attic.

Ladder 44 firefighter, Michael Brady, entered the attic space and a hoseline was passed to him through a 3-by-3-foot access opening from the top floor. Meanwhile, three firefighters were working on top of the hip roof: one cutting the roof deck above the smoldering hot spots with a power saw; another acting as his guide, as he stepped backward with the spinning, highspeed saw blade; and the third standing by ready to pull up the cut roof deck with a Halligan tool. Suddenly the roof started to collapse (fig. 7–2). The deck and shingles near the two remaining front hip rafters ripped apart. The peak of the roof fell, sliding down around a masonry chimney, and the eaves pushed out beyond the bearing walls on both sides. Several rows of bricks at the top of the side bearing walls fell into an alley. One firefighter on the roof leaped across a 3-foot alley to the roof of an adjoining building; the other two rode the roof down. Brady, inside the attic, sensing the collapse above him, dove headfirst for the opening in the attic floor but the collapsing roof pinned the lower half of his body before he could escape. Crying out for help, the firefighter was caught with his head and shoulders hanging out of the access opening and his hips and legs pinned between the underside of the heavy collapsed roof rafters and the attic floor. A firefighter standing on the ladder leading to the access opening saw what had happened and quickly wedged the point of his Halligan tool between a roof rafter and the attic floor, where Brady was caught, holding the roof up several inches above the floor. Although the trapped firefighter could not free himself, this action temporarily kept the full weight of the roof from crushing his lower torso. Rescuers carefully cut away the roof rafters round the trapped firefighter, releasing him from beneath the collapsed hip roof. Brady survived and returned to active fire duty.

Fig. 7–2. This hip roof collapsed and trapped a firefighter working in the attic space.

An investigation of the building collapse revealed several contributing causes:

1. There was fire destruction of the roof deck at the rear hip rafters.
2. There were no collar beams connecting rafters of both sloping sides. (Collar beams tie opposing sloping roof rafters together and help resist the outward thrust of the roof rafters at the eaves.)
3. The wood plates on top of each side wall, supporting the roof rafters at the eaves, were not anchored into the top of the masonry bearing walls with bolts.
4. The weight and vibrations of the three working firefighters on the roof overloaded the fire-damaged structure.

In a similar incident Fire Captain Francis Federici of Bridgeport, Connecticut, was killed in a fire when a low-pitched hip roof collapsed on his company, operating on the top floor of a two-story structure. There was no attic floor to stop the falling hip roof.

Primary Structural Elements

A primary structural element is a structure which supports another structural member. The collapse of a primary structural member may lead to the failure of another part of the building (fig. 7–3). Header and trimmer beams are primary structural members in wood floor construction; ridge rafters, hip rafters, and bearing walls are primary structural members found in sloping roof construction.

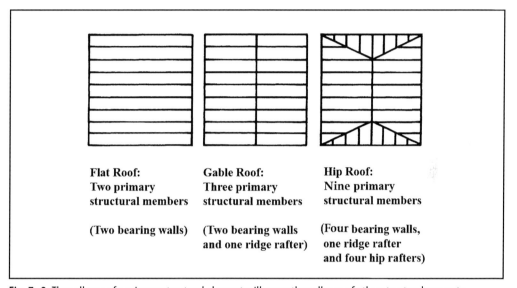

Fig. 7–3. The collapse of a primary structural element will cause the collapse of other structural supports.

Roof Deck Collapse

The fire described at the beginning of this chapter was an actual incident which took the life of veteran fire officer Harry Korwatch of the Yonkers, New York, Fire Department. A preliminary investigation of the fire building revealed that the roof of the structure was a sloping gable roof supported by heavy timber trusses, spaced 15 feet on center. On each sloping side of the roof, there were timber purlins running the length of the 105-foot structure, perpendicular to the trusses. Outside of these few structural elements, the sole support of the firefighters on the roof was a 1-inch tongue-and-groove board. A purlin is a timber laid horizontally and perpendicular to support the common rafters of a roof (fig 7–4).

Approximately 80 percent of the roof area was unsupported, 1-inch thick wood decking. Unlike a rafter roof, whose structural roof supports are close together, large areas of this roof were thin tongue-and-groove roof board. (Yonkers firefighters who walked on it described it as "bouncy" or "springy." The roof deck did not feel "spongy," which would indicate char or fire weakening; a roof or floor which has structural supports spaced several feet apart creates a springboard effect.) Flames from the fire below charred and weakened the 1-inch roof deck. When the officer stepped on this area, he plunged through the roof deck area between the widely spaced trusses and purlins. The firefighter trapped on the roof was rescued from the roof at the rear of the building.

Fig. 7–4. Lieutenant Harry Korwatch of the Yonkers, NY, FD was killed when the roof deck collapsed.

Sloping Roof Construction

The three most common types of wood construction used for sloping roofs are timber truss, plank-and-beam, and rafter construction (fig. 7–5). When a firefighter climbs on top of a sloping truss roof, the stability of the roof deck depends on the number of purlins running perpendicular to the trusses and on the truss spacing. In general, of the three construction methods, the truss roof has the largest area of unsupported roof deck. The plank-and-beam roof has the next largest area, and the rafter roof the least amount, making it the safest roof to walk on. Some older plank-and-beam sloping roofs, however, do not even have a roof deck beneath roof shingles. Plank-and beam construction found on barns or farm storage buildings in rural areas may have a lattice or grill of 1-by-3-inch boards spaced every 6 inches, instead of a solid roof deck. Poor construction of this type will not support the weight of a firefighter.

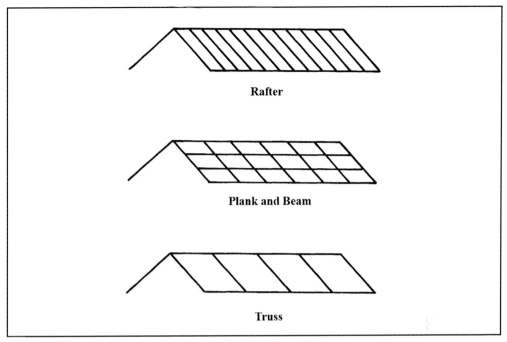

Fig. 7–5. The roof supports determine the stability of the deck during a fire.

Roof Rotting

In the North, accumulations of snow or ice on a sloping roof caused by improper drainage can result in areas of rotted or decayed roof deck. At these points, the roof deck can collapse beneath the weight of a firefighter. Roof rotting is often found at roof edges, where roofs change slope and where a sloping roof abuts a vertical plane. Firefighters should avoid walking on these areas whenever possible. If a small portion of the rotted roof gives way a firefighter could lose balance and fall from the roof.

Slate and Tile Shingle Collapse

Roof coverings can be divided into two categories: built-up roof coverings and prepared roof coverings. Built-up roof coverings consist of several layers of materials applied to a roof; for example, a tar and gravel roof. Prepared roof coverings have materials nailed to a roof deck, such as slate, tile, asbestos, wood, asphalt, and sheet metal. Slate and tile shingles present a collapse hazard to firefighters. Slate is a rock naturally found in smooth-surfaced layers; tile shingles are pieces of unglazed, fired clay, stone, or concrete (fig. 7–6). One slate or tile shingle can be 2 inches thick and weigh up to 10 pounds. Ten or twenty broken, razor-sharp shingles falling several stories from a roof can be extremely dangerous to firefighters working below.

Fig. 7–6. Razor sharp slate shingles falling from a roof can seriously injure a firefighter.

At a fire in a four-story, 20-by-60-foot brownstone mansion in upper Manhattan, slate shingles collapsed from a sloping roof and severely injured FDNY Lieutenant Robert O'Connell. Flames had raced up an open stairway from the first floor to the roof, which was covered by slate shingles and sloped toward the front of the building. As fire cut off the interior stairs, a trapped occupant was being removed from a second-floor window into the bucket of an aerial platform. Flames lapped out of the top-floor window and blew back against the sloping roof; fire was burning in the attic space and through the roof. Firefighters directed an exterior hose stream at the roof fire from an adjoining building, and advanced hoseline up the flaming open stairway to effect an interior attack. Suddenly, a number of slate shingles collapsed off the roof. Sharp, jagged pieces of slate crashed into the bucket of the tower ladder on top of the victim and the fire officer. One large piece of razor-sharp slate struck the arm of the officer, shattered his wrist, and severed arteries, tendons, and nerves in his arm. After the injury, the officer never regained full use of his hand and retired because of this disability.

There are several causes of slate and tile shingle collapse:

- The wooden roof deck to which the shingles are nailed can be destroyed by fire.

- All stone is adversely affected by sudden changes in temperature—it cracks and breaks when suddenly heated by flame.
- A powerful hose stream striking a sloping roof can drive slate or tile from the rooftop into the air, causing broken pieces to fall around the perimeter of a burning building where firefighters are working.

Firefighters stretching hoselines, climbing ladders, or donning breathing equipment near fire buildings have been killed and seriously injured by falling objects. Window air conditioners, broken glass, building fragments, slate shingles, tools dropped accidentally, large pieces of smoldering furniture, and even people jumping from windows have fallen on firefighters below (fig. 7–7). Another danger of slate and tile roofs is that they conceal a fire-weakened roof structure. Slate and tile, considered fire-retardant roof coverings, will protect a wooden roof deck from sparks or airborne burning embers which land on a roof. This fire-retarding effect, however, also slows any fire and heat in an attic from traveling upward. The roof deck and roof rafters below a slate roof can be almost completely destroyed by fire, and yet the roof shingles can look stable and undamaged. There will be no scorch marks or blistering on the slate from the fire below. Firefighters have been killed by falling through collapsing slate and tile roofs which appeared sound from above, but which had the wood supports and rafters below charred and burned away. Slate and tile roofs have the same dangerous effect as a tile or terrazzo floor above a serious fire: they conceal the amount of heat, smoke, and flame below and then collapse in one large section without any warning signs.

Sloping Roof Pitches

The chances of a firefighter losing his balance and falling off a sloping rooftop are far greater than those of a firefighter being injured in a sloping roof collapse (fig. 7–8). Even on a low-pitched roof, ice or snow can make the roof too slippery to walk upon. In warm climates, slate and tile roofs can become slick from rain or hose streams. Shingles can also crack and slide out when walked upon. In areas near water, the humidity sometimes causes a mildew or a fungus to grow on a rooftop, creating the same effect as ice. Even asphalt shingles, which offer fairly good traction, can melt and become slippery when heated by a fire or the sun. A sloping roof is measured by its pitch. A pitch of "4 in 12" indicates there are four units of roof rise to twelve units of rafter span. A "2 in 12" is a low pitch, while a "15 in 12" is an extremely high-pitched roof. A low-pitched roof, such as that on a ranch-type house, may be walked on, depending upon the type of shingle and the weather conditions; a medium-pitched roof (5 or 6 in 12) should have a roof ladder secured at the ridge, or be butted by a parapet wall at the base to ensure footing for walking or operating. A high-pitched roof, like that of an "A" frame or an English Tudor roof, cannot be walked on safely, even with the assistance of a roof ladder. A firefighter must operate from an extended aerial ladder or while standing in the bucket of a tower ladder.

Fig. 7–7. Falling shingles and other debris can injure firefighters below.

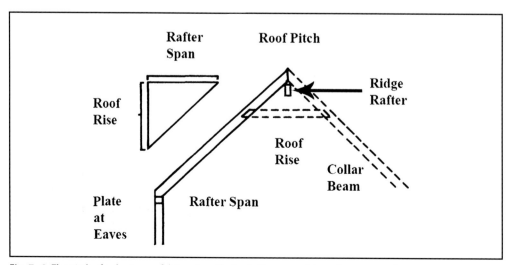

Fig. 7–8. The pitch of a sloping roof determines if a firefighter can safely walk upon a roof.

Lessons Learned

There are two ways firefighters can protect themselves when working on a peaked roof. They can use a roof ladder that is hooked on to the ridge rafter or they can operate from an aerial or tower ladder. A roof ladder will protect firefighters from a roof deck collapse and falling from the roof. However a roof ladder will not protect a firefighter from a roof rafter collapse. If the rafters collapse, the roof ladder will collapse into the burning attic with the firefighters (fig. 7–9). Only standing and operating from an aerial ladder or aerial platform can protect a firefighter from a rafter collapse. If the roof rafters collapse, the firefighter will be independently supported by the aerial or platform when the entire roof fails. If a fire department has a standard operating procedure that requires firefighters to operate and vent peaked roofs the department must have an aerial ladder to do this safely.

Fig. 7–9. Using a roof ladder can protect a firefighter from a roof deck collapse.

When firefighters are operating at a private home with a sloping roof, it is more effective to remove smoke from the structure by venting top-floor windows than by cutting a vent opening. When preparing a pre-fire plan for a hazardous building, firefighters should identify the type of roof construction and include it in the plan. In rural areas, outside city limits or zoned areas, roof construction may not conform to the local building code. Also older structures, built before a local building code was enacted, need not conform to code requirements for roof construction. These structures may have substandard roof construction that will not support firefighters. All fire departments should develop standard operating procedures (SOPs) for operations involving sloping roofs. These should be based upon life safety, fire containment, and property protection—in that order of importance. Because a sloping roof sheds water and snow when properly pitched, it may be designed to support less of a live load than a flat roof. Such a sloping roof will support fewer firefighters than a flat roof will. Firefighters should understand that, when they walk upon the roof of a burning building, they risk the possibility of plunging through a fire-weakened structure into a dark, smoky, flaming, cage-like enclosure of four walls from which they will not escape alive.

Note: For more information about peaked roof collapse, search the Web for: *Firefighter Fatality Investigation F2005-09*.

8

TIMBER TRUSS ROOF COLLAPSE

An alarm is received at dawn for an industrial area in the outskirts of town. The area consists mostly of large one-story factory buildings, junkyards, and empty lots littered with illegally dumped rubbish. The captain of the first-responding ladder company climbs into the front seat of the truck, checks to see that his crew is aboard, and signals the driver to roll. It has been a busy night tour, and this run hurts the most. After nothing but false alarms and rubbish, it seems to the captain that his company serves only as a plaything for children and a supervisor of burning garbage. But as the apparatus approaches the location of the alarm, it appears that this run might be different. In the dark gray, early morning light a cloud of smoke can be seen drifting above some distant buildings.

As the ladder company turns into the street behind the first-due engine, the officer directs the driver to place the truck in front of a one-story 75×100-foot brick-and-joist machine shop, below an almost motionless cloud of smoke. The engine company has already stretched a hoseline to the front of the building. The truck captain determines this to be the fire building and directs his "forcible entry team" to force open the normal entrance door. If necessary, several large windows and a truck entrance door could be ventilated later by the driver. Both exposures 2 and 4 are attached, similar buildings. A ground ladder is raised to the roof of exposure 2. This will enable the "roof firefighter" to get to the roof of the fire building quickly and safely and begin venting the skylights and scuttle covers.

The driver raises the aerial ladder to the roof of exposure 4. The outside-vent firefighter takes a power saw up the aerial ladder to the roof, for it looks as though the roof firefighter could use it. The outside-vent firefighter will then check conditions at the rear of the fire building and report to the chief whether a heavy stream can be positioned there. This firefighter will also attempt to vent the rear window of the fire building from the roof. (In this part of town, the rear doors and windows are often bricked up and sealed tight.) These tactics cover venting entry search and rescue for the entire structure.

Below, firefighters enter the factory's dark interior. There is heat and smoke drifting near the ceiling, just above their heads. Since the area they search contains only machines, the firefighters assume the fire is small. Near the corner on the right side of the factory, a small fire sends up flames into the smoke banking down below the ceiling. Firefighters vent the rear windows from inside the building. The hoseline is ordered charged. It seems like a routine operation until they hear a distant sound of breaking glass. The firefighter on the roof radios the ladder captain by Handy-Talkie: "Say, Captain, I just vented the skylight glass. This is a bowstring truss roof, and there is heavy fire in the roof."

He starts up the power saw and begins to cut the roof. The engine and ladder officers look at each other in disbelief as they shine their handlights up into the smoke overhead; they can see neither fire nor truss—just a ceiling of smoke floating 2 feet above their heads. Before he entered the fire building, the captain remembered to look above the roof for a roof curve, but none was visible. "This is a bad situation," he thinks to himself. He has experienced two other truss roof collapses at fires and knows that the actions taken in the next few minutes will mean the difference between life and death for some firefighters.

The chief arrives on the scene and the captain notifies him of the truss roof. Just then the firefighter on the roof radios the captain. "I just vented another skylight and the frame collapsed into the building." The smoke begins to rise. Three huge timber trusses 30 feet above the machine shop floor become visible. Dark red flames mixed with smoke swirl wildly between the truss webs. Burning pieces of skylight framework fall through the smoke. The chief enters the structure, looks up at the ceiling, and orders everyone out. Via Handy-Talkie, the truck captain orders his men on the roof to get to the adjoining building immediately. A collapse is imminent.

Last to leave the building, the captain sees firefighters setting up a portable deluge nozzle near the truck entrance door, which is now open. Other firefighters stand on the sidewalk awaiting orders. He shouts for everyone to back off the sidewalk. As he pushes some stragglers back with his outstretched arms, the entire 7,500-square-foot timber truss roof crashes down and instantly the front masonry wall collapses out into the street. Clunks of brick and mortar tumble behind the heels of the firefighters running from the collapse.

Timber Truss Roof Collapse

"A tragic, timber truss trilogy" is a term used to identify accounts of the three most deadly timber truss roof collapses in the history of the fire service—three timber truss roof collapses in the 1960s, 1970s, and 1980s that killed 16 firefighters. The first part of the deadly trilogy is the Cardinal bowling alley timber truss collapse. Five firefighters in the Ridgefield, New Jersey, fire department were killed responding on a mutual aid call to a fire in Cliffside Park, New Jersey. The truss roof of a bowling alley collapsed and pushed out a concrete wall, burying the five firefighters beneath tons of concrete cinder blocks. The firefighters were operating outside the burning building. This deadly blaze occurred on October 15, 1967. The second part of the story occurred in Brooklyn, New York. A supermarket timber truss collapsed and six FDNY firefighters fell through the roof into a raging fire (fig. 8–1). The six firefighters were working on the roof of the burning building. This fire happened on August 2, 1978. The third part of this tragic story took place in Hackensack, New Jersey. On July 1, 1988, five firefighters operating inside a burning Ford auto dealership were caught and trapped by tons of falling wood truss beams and concrete ceiling.

Chapter 8 | **Timber Truss Roof Collapse**

Fig 8–1. The truss roof collapse at the Waldbaum's supermarket in Brooklyn, NY, in 1978

The Cliffside Park, New Jersey, Cardinal bowling alley building was one story, 106×120 Type III ordinary construction with a timber truss roof and cinder block wall. The fire on arrival involved a large area of the bowling alley. Soon after arrival there was an explosion and the interior attack strategy was changed to exterior operation and roof venting ordered. When the roof collapsed it pushed out an enclosing wall. The falling wall killed five firefighters from Ridgefield, New Jersey, operating a hoseline outside the building. There could have been more deaths. When the roof collapsed, one firefighter slid down the sloping roof and was hanging on a hoseline. Another firefighter lowered himself, while hanging on to a broken edge of the roof, and extended his leg to the trapped firefighter, who then scrambled up his leg and body to safety. The truss roof collapsed 30 minutes after arrival of the firefighters (fig. 8–2).

Fig. 8–2. This truss roof collapse in Cliffside Park, NJ, killed five firefighters from Ridgefield, NJ, who responded on mutual aid.

The Brooklyn supermarket timber truss collapse occurred in a one-story ordinary constructed timber truss roof and brick walled building, 120×120 feet with a suspended ceiling 18 inches below an ornamental tin ceiling attached to the bottom chords of the truss. The fire, on arrival, was in a mezzanine—a small intermediate landing between the first floor and the roof. An interior firefighting strategy was launched and the visible fire in the mezzanine was quickly extinguished. However, unknown to the firefighters, concealed fire traveled up in to the truss roof space area and spread outward. Twelve firefighters fell into the burning crevice after the roof failed. Six fell into the store and survived and six other firefighters were killed in the explosive roof fire after the collapse. The truss roof collapsed 32 minutes after arrival of the firefighters.

The Hackensack, New Jersey, building was a one story brick and timber truss roof 224×175 foot auto dealership of Type III ordinary construction with a wire-and-lath and ½-inch cement ceiling. The fire, on arrival, was in the attic roof space created by the bowstring timber truss section. Firefighters were attempting to extinguish the fire from inside the building when one burning 80-foot timber truss crashed through the ceiling and trapped six firefighters in the building. One firefighter was able to escape, two other firefighters retreated to a dead end storage room, where they were asphyxiated, and three firefighters were pinned beneath the falling roof and ceiling. The truss roof collapsed 36 minutes after arrival of the firefighters (fig. 8–3).

Chapter 8 | **Timber Truss Roof Collapse**

Fig. 8–3. Five firefighters from the Hackensack, NJ, FD were killed in this bowstring truss roof collapse.

The "tragic timber truss trilogy" shows the three ways firefighters can die when operating at fires involving timber truss construction:

- Firefighters can be killed operating outside a burning timber truss roof building. When trusses collapse they can push out a masonry wall. The falling roof can cause a secondary wall collapse.
- Firefighters operating on the roof above a burning timber truss can fall through the collapsing roof into a fire.
- Firefighters operating inside a burning building can be crushed and burned to death when the collapsing truss roof falls on top of them.

The timber truss roof has a history of death and destruction in the Northeast. The buildings are old and in some instances suffer from neglect and improper alterations. The truss buildings in the South and West are new; however when they age they will become a similar collapse danger. On March 8, 1998, Los Angeles Fire Captain Joseph Dupee was killed by a timber truss roof collapse.

What is a timber truss? *Timber* is wooden construction larger than 2×4 inches but not large enough to be classified as heavy timber or mill construction. The *truss* is a structural composition of large wood members joined together in a group of triangles and arranged in a single plane so that loads applied at points of intersecting members cause only direct stress (tension or compression in the members). The timbers in a truss are joined together by bolts which pass through the center of metal connectors. The most common connector is the split-ring metal connector, which is embedded in prepared depressions on the face of the timber. Its purpose is to relieve the bolts of shearing stress.

A timber truss roof can be built in a variety of shapes. There is an inclined plane truss (gable shape roof) a parallel chord truss (flat roof) and a bowstring truss (arch roof). The bowstring timber truss is the most common design. Its curved top chord creates an arched roof; its bottom chord is horizontal timber. Both chords outline a bow with a string attached to each end. The wooden web members connecting the top and bottom chords are of smaller dimensions; however, they are critical to the overall strength of the truss section during a fire (fig. 8–4). When attacked by flames, the entire truss section may fail as soon as the smallest web member weakens. In other words, under fire conditions, the truss is only as strong as its weakest member (fig 8–5).

Fig. 8–4. A timber truss is a structural composition of large wood members joined together in a group of triangles and arranged in a single plane so that loads applied at intersecting members will cause only direct stresses, tension, or compression.

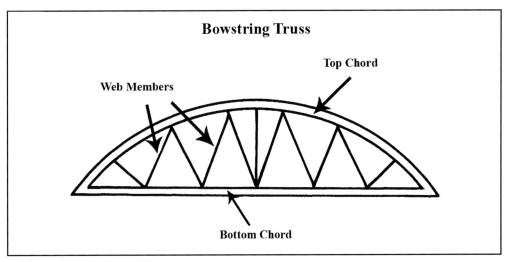

Fig. 8–5. The strength and stability of a truss depend upon the smallest web members.

Truss construction is the most dangerous roof system that a firefighter will encounter. It is known to collapse during the early stages of a fire, and it will often cause the subsequent collapse of the front or rear masonry enclosure wall of the structure (fig. 8–6).

Fig. 8–6. A timber truss roof that caused the collapse of a front wall

In the Bronx, New York City, two bowstring truss roofs collapsed. At each fire, when the truss roof fell, it caused the collapse of the front masonry wall. A close examination of both collapsed structures revealed that the front wall, which had failed, had received in recessed pockets, the ends of roof joists, which sloped down from the top chord of the front truss. These joist ends did not have self-releasing fire cuts. As the ends of the roof joists resting on the upper chord of the front truss went down with the collapsing roof, the ends of the joists rigidly held in the front wall were forced up and outward. These joists acted as levers, toppling the upper portion of the front wall out into the street. A bowstring truss roof has four bearing walls. The ends of the truss sections are supported by the side walls. The front and rear walls support the sloping hip rafter or roof joists extending from the front and rear truss sections.

The question arises: Why don't the rear walls collapse as the front walls do? One reason the rear walls did not collapse at the Bronx fires is that they were more stable and resisted the lever action of the falling roof joists. The rear walls had only a few small openings in them, while the front walls had many openings, making them less able to resist any movement of the rigidly encased, fixed, sloping roof joists. There were large openings used for trucks, double-sized windows, and a normal entrance door for employees. Openings in a wall weaken its load-bearing capabilities and permit fire spread.

"Size-Up" Errors with Timber Truss Roofs

A *size-up* is not something a firefighter does for only a short period before beginning operations. A firefighter is continually sizing up the fire. Firefighter take in information they see, hear, touch, and smell, consciously and subconsciously, during the entire fire. They rely on this information for safety and for firefighting effectiveness. A building with a bowstring truss roof is different from most other structures, however, and can send a firefighter dangerous, misleading size-up information that can have a disastrous effect. Most textbooks and post-fire analyses of bowstring truss roof collapses state that firefighters and chiefs can easily and quickly identify the bowstring truss roof by its curved roof. That's not always true. A high, decorative front parapet wall can conceal the roof shape from all operating members, except those positioned on the roof.

During a fire, early identification of a truss is key to a safe operation. When the truss is identified early, serious injury can be avoided by a defensive strategy. A firefighter who discovers any type of truss in a building should immediately relay this information to the officer in command of the fire. The firefighter should not assume that others are aware of it, no matter how obvious it may appear. When entering a burning room or area, if you are able to walk in an upright position (not forced to crouch because of heat banked down below the ceiling), you may subconsciously assume the fire is small. This assumption can be a deadly error in judgment in a building with a bowstring truss roof or in any room with a high ceiling. The large space created by the concave, arched underside of the bowstring truss roof acts as a "heat sink"—that is, it allows large amounts of heat and smoke to flow upward from the fire floor below.

Upward flow delays the build-up of heat and smoke at the floor—which firefighters normally use to sense a fire's severity. If a fire occurs in an enclosed building with a flat ceiling and burns for a period of time, the smoke and heat accumulate inside and bank down to the floor. These combustion products can prevent a firefighter from moving more than several feet beyond the building's front door. The same serious fire in a building with a bowstring truss roof might allow firefighters to enter the structure, walk upright to the rear of the occupancy, and become trapped under a collapsing truss roof.

Chief officers who have directed fire operations in buildings with bowstring timber truss roofs tell of receiving conflicting reports from those firefighters operating above and below a fire. The firefighters on the roof state that the fire is severe and beyond control of the hose team operating below. The firefighters within the building state that the fire is small and that quick extinguishment will he forthcoming. The chiefs who have operated at such fires successfully are guided by the report from above, and they withdraw the men working above and below the fire (fig. 8–7).

Fig. 8–7. The roof size-up is more accurate than the size-up report given by interior operating forces. When a chief officer receives conflicting reports the chief should rely on the roof report over the interior report.

When firefighters first enter a bowstring truss-roofed building with a very low fire load (such as a machine shop), they may be misled into thinking there is little to burn. Smoke or a ceiling may conceal a tremendous structural fire load in the roof system. Timber trusses, roof beams, purlins, and the underside of a wooden roof deck will exist high above any content at floor level. In addition, over the years an accumulation of paint or flammable vapors from a spray process at floor level may coat the timber trusses, encouraging flame spread and smoke generation. In a timber truss building, the main fire will be in the roof structure, not in the content below.

One of the lessons learned at the Brooklyn supermarket post-fire investigation was that there are three size- up indicators of a truss roof in a building: One is a large open space without columns, which indicate a long span roof support such as a truss above. Another indicator of the presence of a truss roof (bowstring truss only) is a mounded roof shape. And finally, there are certain occupancies that frequently use truss construction in the roof, such as supermarkets, bowling alleys, garages, theaters, places of worship, auto dealerships, piers, and armories.

Fire Strategy for Timber Truss Roofs

Bowstring timber truss roofs are most often found in prewar industrial buildings, garages, lumberyards, piers, bowling alleys, and supermarkets. Often the property is used for manufacturing and storage. Generally, the public does not enter the truss portion, and employees are the only occupants. Since interior decoration is not a factor, the timber truss roof supports are often visible from the floor level below; the trusses are not concealed by a ceiling. This absence of a ceiling greatly assists firefighters. It allows identification of the trusses from inside the structure if smoke is not excessive and gives the interior hose stream an open path to extinguish any fire in a timber truss quickly. A truss roof concealed by a ceiling is much more dangerous to firefighters than a truss roof without a ceiling. (Six career and two volunteer firefighter in New York State were killed in two fires in buildings that had timber truss roofs hidden by ceilings.)

Firefighting strategy for timber trusses concealed by ceiling

The three buildings with truss roofs that collapsed in the timber truss trilogy of fires, killing 16 firefighters, had trusses concealed by ceilings. When a building has a timber truss concealed by a ceiling, the access to the truss attic space may be from a remote ladder enclosed in a closet leading up to a trap door in the ceiling. The chances of finding this ladder and trap door and getting to the attic space to extinguish a serious fire is slim to none. And the chances of opening a ceiling below a timber truss and extinguishing it with hose streams from below is also rare.

For example, the ceiling enclosing the timber trusses in the Brooklyn supermarket was ornamental tin attached to purlins—the bottom chords of the timber truss roof. Below the tin ceiling there was another suspended panel ceiling. To gain access to the fire in the 10-foot-high truss attic, firefighters tried to open the suspended ceiling and then open the ornamental tin ceiling. Above this second ceiling were 3×8-inch wood beams, 16 inches on center (purlins), that extended from the bottom chord of one truss to the bottom chord of the adjoining truss. A hose stream directed back and forth from below into the truss attic would be broken up by the purlins, rendering the extinguishment effect minimum.

At this fire, the eventual strategy was to place a hose stream on the roof and attempt to cut the roof open to gain access to the fire and extinguish with hoselines on the roof. When the truss collapsed, firefighters fell into the fire. The lesson learned was there can be no extinguishing a fire in a truss space from above.

At the Hackensack fire, the strategy was different. The strategy was to extinguish the fire in the timber truss space from below. Firefighters with hose streams operated through a ceiling trap door standing on vertical ladders. When the timber truss and ceiling collapsed on top of them, they were trapped inside the burning building. The lesson learned at this fire was that there can be no extinguishment from below the burning truss through a trap door in the ceiling. When a fire extends to a timber truss attic space, prepare for a defensive outside attack and protect exposures. Order all firefighters above and below the burning timber truss attic space out of the building. The concealed fire will spread throughout the large truss attic space, collapse the roof with one or more walls, and create a major fire plume, radiating heat and spreading fires to adjoining buildings.

Firefighting strategy for timber trusses not concealed by a ceiling

A typical fire inside a structure with a bowstring truss roof that does not have a ceiling may originate in combustible contents at floor level. As the fire grows in size, flames will extend upward, igniting a timber truss or the underside of the roof decking. Heat and smoke will then travel along the curved underside of the roof, pass through the open webbing of several trusses, and stratify at the highest point of the curved roof some distance away from the original fire. A secondary fire at this high point, hidden by smoke, may feed on the roof deck. It may burn undetected by firefighters who are extinguishing the original fire at floor level below.

The fire strategy of a first-arriving engine company at a timber truss roof building without a ceiling should be to attack the fire directly with a large-diameter hoseline. A powerful water stream, capable of reaching a distance of 50 feet, will be needed to extinguish fire in the upper portions of the trusses. This first hoseline should attack the main body of fire. If this first stream does not control the flames within the first few seconds of water discharge, and it appears that the fire will increase, interior firefighting should be discontinued and firefighters withdrawn. It is difficult to justify a long-duration, defensive firefighting operation inside a structure with a timber truss roof where there is no life hazard. The fire must show immediate signs of extinguishment by the first hose stream, or an outside attack should be ordered for the safety of the firefighters.

After all units are withdrawn, firefighters should anticipate an early collapse of the roof and a subsequent failure of one of the enclosing masonry walls. In some timber truss roof systems, the web members are enclosed by fire-retarding materials such as drywall or plasterboard on one side. This material will cause the truss sections to appear like curtain boards, so fire and smoke will be confined to the bay between those two truss sections directly above the fire. If collapse of this compartmented roof occurs, timber trusses will fail one at a time. If the web members are not covered on one side with fire retardant, then fire, heat, and smoke will pass through the open spaces between the web members, and the entire truss roof will weaken and fail at once, with all the timber trusses failing at the same time. If the first hoseline is successful and controls the fire at floor level or in a single timber truss, and it appears that the fire will not spread, a back-up, large-diameter hoseline should be stretched into the building. This second hoseline should sweep the underside of the roof, if necessary. This technique will extinguish any possible secondary fire hidden behind the smoke and heat at ceiling level.

Roof venting

Since smoke is the main killer at fires, roof ventilation is a life-saving tactic in an occupied building. Roof ventilation is also necessary to improve a fire environment so firefighters can approach a fire with a hoseline for extinguishment. At an unoccupied bowstring truss roof building, however, roof ventilation may be too dangerous to risk the lives of firefighters. In the early stages of a fire, where an interior attack has a chance for success and the first hoseline quickly controls the fire, smoke is not a factor in preventing the firefighters from making a close approach. Any smoke generated by the fire will be diluted by the large air spaces created by the concave roof spaces between the bowstring trusses. Smoke will rise to the top of the roof space, and firefighters will be able to walk upright into the fire area.

When roof stability is assured and roof venting is needed, firefighters can sometimes open the skylights or scuttle covers quickly and leave the roof immediately. Also, in a bowstring truss-roofed building which does not have the web members enclosed with plasterboard, vent openings may be made in the front and rear sloping portions of the roof, thereby providing horizontal cross-ventilation of the entire roof space (fig. 8–8). At some fires, the firefighters assigned to the roof will be the first people to identify the roof as truss construction. The officer in command must be notified of this by firefighters on the roof. This truss identification will only observed from the roof shape when the truss is a bowstring truss roof. A parallel chord or gable truss will not be identified by the roof shape.

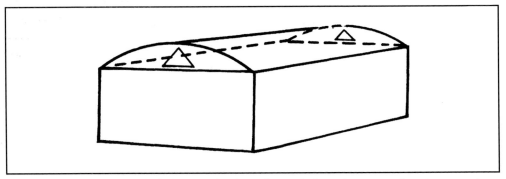

Fig. 8–8. To quickly determine if a fire has spread to the roof space of a bowstring truss, a triangular cut may be made in the sloping front or rear of the roof from the safety of an aerial platform.

Attic space quick access

If a small fire, such as in an overheated motor or electric wiring, exists in a bowstring truss-roofed attic space that is concealed by a ceiling, and the access ladder to the truss attic space can not be found, quick access can be obtained into the concealed roof space by making a roof cut opening at the sloping ends of the roof deck. A hose stream can be directed into a truss roof area through this roof cut opening if the small, normal access opening cannot be located.

Evacuating a Truss Roof

In a bowstring truss-roofed building where the web areas of each truss are enclosed with fire stopping and a serious fire burns in one bay between trusses like in the Brooklyn supermarket fire, one truss is likely to fail first. In a building 100 by 100 feet with truss sections 20 feet on center, one truss failure will collapse a 4,000-square-foot section of roof (a 40×100-foot area). The direction in which a firefighter proceeds from the roof center above the failing truss is critical. If he goes 20 feet in a direction perpendicular to the failing truss and reaches the adjoining truss, he may be temporarily safe (fig. 8–9). If a firefighter decides to head toward the nearest parapet wall, and travels 50 feet parallel to the truss, he or she may still go down with the collapsing roof. The same principle would apply to firefighters inside the building, trying to get out of the path of the falling truss. Firefighters attempting the initial hoseline attack on a fire should be positioned behind a truss section next to the one involved in fire. The reach of a powerful hose stream will allow them to be out of the collapse zone of a single truss. A firefighter's knowledge of building construction not only assists him in fire extinguishment—it increases his chances of survival.

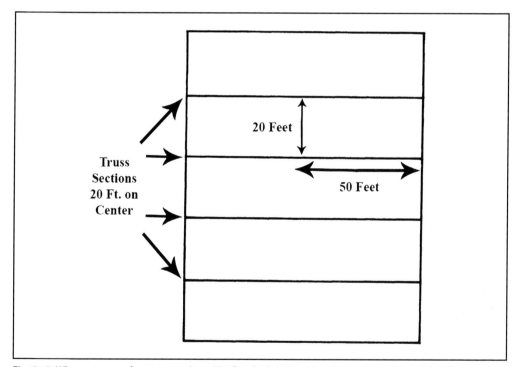

Fig. 8–9. When a truss roof appears weakened by fire, firefighters ordered to evacuate the roof should retreat in a direction perpendicular to the roof trusses.

Lessons Learned

The key to safety when operating at a fire in a truss constructed building is communications. When a truss of any kind is discovered, firefighters must notify the incident commander. Only when the incident commander *knows* it is a truss building can he or she take precautionary action. Teamwork of all ranks at a fire prevents injuries and deaths. If any one of the following three actions had not been taken in the scenario described in the beginning of this chapter, the outcome of the collapse would have been tragic. First, the firefighter venting the roof alerted all members inside the fire building of the unseen truss hazard. Second, the chief, entering the structure for a firsthand view of the fire area, made a quick, correct decision to order the men out. Finally, the captain commanded all firefighters to leave the sidewalk in front of the building. Without this third follow-up order, there would have been serious injuries and loss of life despite the other two excellent, professional firefighting actions.

Note: For more information on timber truss roof collapse, search the Web for: *Firefighter Fatality Investigation 98F007*.

Place of Worship

When we examine timber truss roof collapse dangers, we must consider places of worship. The large open space required in a place of worship usually means a timber truss roof. From an incident commander's point of view, a burning church, synagogue, mosque, temple, or other house of worship is an ordinary Type III collapse danger. A church, synagogue, mosque, or temple is a poorly designed and dangerous structure. Pre-fire plans should be prepared for a house of worship, identifying the type of long span roof system (usually a timber truss) over the building, and fire chiefs should recommend a defensive outside attack with emphasis on protecting nearby exposure and preparing for collapse. A defensive preplan firefighting strategy should consider a collapse of the steeple, the bell tower, the ceiling, and the timber truss roof, followed by the side walls.

A preplan may reveal that a 100-year-old building has undergone many renovations. The walls and columns may look like stone but are usually stone imitations with plaster over wood lath. The large open area or small concealed spaces behind imitation stone walls, hollow columns, or in the attic can allow fire to spread to timber truss roof supports. If a church, synagogue, mosque, or temple was not considered a sacred house of worship most would be shut down for violations of the fire and building codes. And all of these structures would be immediately ordered to install a fully automatic extinguishing system throughout the entire building.

Flames spread

While attending a service at the church of the FDNY chaplain in Saint Francis Church on West 31st Street in Manhattan, I noticed the stone mosaic figures depicted on the walls and ceiling of the church looked very shiny and new compared to previous years. The gold and red and blue color mosaic pieces were bright and glistening. After the service I asked the chaplain, "Did they install new mosaic walls in the church.

He said, "No, they just cleaned the entire inside of the church. There was a hundred years of wax vapor built up on every thing. Vapor given off from the burning candles, 24 hours a day 7 days a week for several decades, built a film of wax on everything."

This is one reason an exposed timber truss roof of a place of worship will rapidly burn and collapse—it has a wax coating on the surface.

Case studies of truss collapse in places of worship

In 1979, two Valley Stream, New York, firefighters John Tate and Michael Moran, were killed in a synagogue fire. After knocking down a fire in a small room next to the main altar, heat and flame were quickly reduced and firefighters assumed the blaze was extinguished. But unnoticed in the smoke, flame had spread up to the attic space through a small spiral staircase leading to the attic. In the attic, concealed by a recently installed ceiling, truss roof supports were burning. As firefighters were leaving to refill air masks, the ceiling and timber truss roof collapsed. The collapse and ensuing fire trapped several firefighters, killing two (fig. 8–10).

Fig. 8–10. Two Valley Stream, NY, firefighters were killed when this inclined plane timber truss roof collapsed.

In 1992, James Hill and Joseph Boswell, two Memphis, Tennessee, firefighters knew they had fire in a church attic. A hoseline was stretched in to the church and a portable ladder placed near a side wall and the ceiling was being opened with pike poles to expose the fire in the attic. Without warning, the ceiling and lightweight trusses collapsed into the church, trapping several firefighters. Two firefighters were killed.

In 1999, Phillip Dean, Gary Sanders, and Brian Collins, all Lake Worth, Texas, firefighters, died when the truss roof of the Precious Faith Temple church collapsed.

In 2007, Richard Stefanakis and Charles Brace of the Pittsburgh fire department died when a bell tower collapsed into a church and caused a secondary collapse of the truss roof beams. In addition to the fatalities, 29 firefighters injured.

Collapse dangers place of worship structure

The unique design of a house of worship makes it a collapse danger. When assessing a city or town's ability to withstand earthquakes, structural engineers analyze the stability of different parts of a building. They want to determine which parts of a building fail and which parts of a building withstand the ground shaking during an earthquake. Studies have found the most unstable structures in a community are often places of worship. And the part of the place of worship that is most unstable is the tower and steeple. A massive stone tower has a wooden interior stairs or ladder, and several platforms support one or more heavy church bells suspended by ropes or cables. These church towers and steeples are considered less stable and more likely to collapse first during an earthquake.

Other unstable parts of a building are chimneys on roof tops, parapet walls over stores, nonbearing walls, and bearing walls (a bearing wall is a wall that supports a weight other than its own weight) and floors of buildings. When small-scale replicas of buildings are constructed in earthquake-testing laboratories, scientists observe how they react during ground-moving effects of an earthquake. The laboratory test for an earthquake starts out with a minor ground shaking and with each test, the strength of the earthquake gets stronger. As the platform moves and simulates a mild earthquake, the church steeples collapse first, next a stronger quake causes the chimneys to crumble, and then with more shock, the freestanding parapet walls topple. Finally as the shaking increases to maximum, the test building's nonbearing walls cave in and finally the bearing walls collapse and the roof and floors collapse with them. Scientists and engineers rate parts of a building as to their susceptibility to collapse during an earthquake. The church tower and steeple are among the most unstable and are among the first to fail. This lesson is meaningful to the fire service. When a smoke explosion or partial collapse occurs during a fire, it shakes the structure too.

Aggressive interior attack on a place of worship

Despite the dangers noted, firefighters are optimists. We must believe we can succeed and extinguish any fire in a place of worship. This optimism is confirmed by the statistic that 95 percent of all fires are extinguished by the first attack hoseline. However that can-do attitude and an aggressive interior attack strategy are most effective on residence building fires, rather than on places of worship fires. At a church fire, upon arrival, the first officer should locate and size up the fire and determine if the blaze can be extinguished by a hoseline. If the fire is too big, an exterior defensive attack should be the initial strategy. If the fire is extinguishable, the first line is taken into the place of worship to the seat of the fire and the fire extinguished. The largest hose should be stretched to give the firefighter the most water power with the greatest reach. You only get one chance to extinguish church fire. A second large-diameter hoseline should immediately back up the first line. As soon as the flames are knocked down, the walls and attic should be checked for concealed fire spread. If two hoselines do not extinguish a fire in a place of worship, firefighters should withdraw. You do not conduct a defensive attack from the inside of a church as you may do inside a multiple dwelling or high-rise building. If two lines do not do the job, withdraw, and fight the fire from the outside.

Defensive Operations at a Church or Temple Fire

When the incident commander arrives at a fire in place of worship and the first and second hoselines have been stretched and supplied with water, the strategy is to prepare for a defensive master stream. As long as you have sufficient resources to advance the interior attack lines, start positioning apparatus for the defensive operations. The interior attack may or may not be successful, but this should not stop the incident commander from being proactive and planning for the defensive operation in case the hose attack fails. If the house of worship has a rose window—a large stained-glass window at the upper reaches of the front of the church or temple—this is where the first aerial ladder should be positioned to operate. An aerial master stream should be positioned in a corner safe area (an area at the exposure "A/B or A/D side of the building) and the aerial platform should be supplied with a large diameter hose and raised to the rose window. This window should be vented by a firefighter in the raised bucket. The entire window should be vented. This is the one window in a church or temple that is high enough and big enough to vent smoke and heat at the upper reaches of a burning place of worship. It can be vented faster than a firefighter cutting a roof opening. The primary venting of this rose window may also allow an aerial master stream to discharge a heavy caliber stream into the upper portions of the church. The aerial master shooting 50 to 100 feet through a rose window, unimpeded, can sometimes break up fire and heat waves at the ceiling peak of a church or temple where the inside hose streams can not reach. In some instances, a choir loft structure or pipe organ at the front of the church will block the aerial master stream coming through the rose window.

In many urban areas the rose window and side stained-glass windows are covered with unbreakable clear plastic. This creates major venting problems and allows smoke and heat buildup and a subsequent backdraft smoke explosion to occur. Firefighters who are venting should notify the incident commander when unbreakable plastic covering is discovered over stained-glass windows. However, if there is no blockage over the rose window, a single master stream directed through the rose window may extinguish the fire at the upper peak near the ceiling of the church or temple. When inspecting a place of worship, examine the rose window and its possible effective use for venting and as access for an aerial master stream. After this aerial stream is in position, other master streams should be positioned around the structure in parking lots and on roofs of adjoining buildings to be used to protect exposures.

Venting a place of worship

After the rose window is vented, the decision must be made to vent the side stained-glass window of the place of worship. The first two hoselines in a place of worship must extinguish the fire or they must be ordered to back out, so a defensive outside attack can be started. During the initial hoseline attack, venting may not be necessary because the smoke and heat will rise up to the underside of the high ceiling, and so smoke will not prevent firefighters from seeing or approaching the fire during the initial attack. If the fire is extinguished by interior hoselines, the side stained-glass window may not have to be vented. However if the interior must be vented for the advance of the first two interior hoselines, then the stained-glass windows nearest to the fire on both sides of the building should be vented first. If after this venting, the interior attack fails and the strategy changes from interior to exterior attack the decision must be made by the incident commander to vent all the side stained-glass window.

At many churches and temples, the side windows are not stained glass and not valuable. But in some places of worship, the side stained-glass windows are extremely valuable. Unfortunately stained-glass windows must be broken. However this creates a public relations problem when the reason for the glass venting is not understood by the public or parishioners. Parishioners sometime donate the money for each window and have their names on the windows. One reason for window venting is, if you do not vent the side stained-glass windows and the building remains closed up, the church may suffer a smoke explosion that could raise the roof and push out the walls. Also, if you do not vent the side windows, the fire may flash over when the firefighters are inside operating the hoselines. And if you do not vent the side widows, the smoke and heat may bank down on the firefighters searching and operating inside. So during the initial stages, portable ladders should be raised to the side stained-glass windows near the fire and where the firefighters are operating hoseline inside. The first venting should be limited to the windows on both sides of the church near the fire inside the building. The opposite window should create cross ventilate.

To vent tall windows, portable ladders, or aerial platforms, should be placed against the church sidewall, upwind of the stained-glass windows and a firefighter on a ladder with a pike pole should break the windows. Window venting should start near the top of the window and work down, because the smoke and heat will be near the top of the buildings interior. Care should be taken because the smoke and superheated gases flowing out of the vented windows may suddenly ignite when they reaches the fresh air. Firefighters should try to be upwind of the smoke and heat.

When you analyze the top of the side stained-glass windows of most churches or synagogues and then look at the height of the ceiling, you quickly see if you vent a side stained-glass window, this will not vent the smoke that is stratifying at the upper reaches of the interior near the ceiling. When you look at the interior of a church or temple, you will see the underside of the high ceiling is higher than the highest point of the side stained-glass window. Even if all of the side windows are vented, the fire will stratify up at the underside of the high ceiling, above the top of the window and spread to wood timber trusses.

Interior fire spread

When comparing a *defensive* strategy for a place of worship fire, you think of a Type IV heavy timber mill building. But when comparing a place of worship fire for an *offensive* attack, you think of ordinary Type III construction. The inside walls, columns, and ceiling of a place of worship may look like stone. However they may actually be plaster over wood lath, covering concealed spaces and poke-through holes. Do not be fooled by this imitation stone veneer surface. As one firefighter, who survived the Ebenezer Baptist church collapse in Pittsburgh that killed two firefighters said, "Beware of the house of worship stone veneer. This veneer makes the structure appear to be constructed of something other than it is. The church had a false appearance of solid 1×3-foot limestone blocks. This was only a façade and gave a false sense of security to the structure. We were under the impression the building was a lot stronger than it was; and there was not way it could collapse." Behind a stone veneer could be lath and plaster concealing a hidden space (fig. 8–11 and fig. 8–12).

Fig. 8–11. This church steeple and roof prior to collapse during overhaul and killed two Pittsburgh, PA, firefighters.

Fig. 8–12. The scene after the steeple and roof collapsed in Pittsburgh, PA

If the interior attack is successful and the fire is extinguished by the interior attack hose teams, firefighters should immediately try to open up the walls and ceilings near the smoldering fire. Check the concealed spaces for fire. If fire spreads to the concealed spaces, it may spread up to the large attic space. Flames may spread to an attic through the hidden voids behind the side walls, via hollow imitation stone columns, and through small holes in the ceiling level around chandelier lights. In addition to opening up the concealed spaces after a fire is extinguished, firefighters should quickly gain access to the attic space and check to see whether fire has already spread to the large concealed space. Finding the stair that leads up to the attic may take some time and climbing a narrow spiral stair may also slow the firefighters, however this is an important action. If the fire is in the attic, the ceiling could collapse or the truss roof beams could also fall and trap firefighters below. If fire is allowed to spread in the attic unnoticed, there could be a collapse on firefighters performing salvage, washing down burned content, and overhauling.

Collapse Danger of a Place of Worship

Structural engineers have identified a church or temple tower or steeple as one of the most unstable features of a structure during an earthquake. The tower is the square structure rising above the church roof. Sometimes there is a steeple constructed atop of the tower. The steeple is the tapered pointed structure on top of church tower that has a cross at its tip. A temple may have a dome at its top. The steeple on a tower is more unstable than a domed tower. When the steeple is located at the front of the structure, this exposure A wall must be considered a collapse danger hazard.

Wall collapse

The roof of a church with a peak is supported by the sidewalls. Sidewalls are running parallel with the peak ridge are the bearing walls. In a church, the weight supported by the walls is the roof. These side-bearing walls are primary structural members. In a church or synagogue, the roof is the structural member supported by the sidewalls. What does all this mean to an incident commander? It means that if the roof burns and starts to collapse because it is interconnected, it could push out the sidewalls. Conversely, if a wall fails the roof would lose its support and collapse into the church or temple floor. Because of the church steeple and the interconnection of roof and side bearing walls, the exposure A, B, and D sides of a burning place of worship are most dangerous areas during a fire.

Ceiling collapse

If the fire spreads to the attic of the church, there is plenty to burn. In an attic of a place of worship that has a Gothic plaster ceiling beneath a peaked roof like Saint Patrick's Cathedral in New York City, there are tons of wood. In the attic are 2-foot-thick timber truss beams. There is the plank wood underside of the roof deck. There a wood-lath bent covering curved to the shape of the plaster Gothic arch ceiling below. There is a wooden walkway from the back to the front of the attic space. If fire reaches the attic spaces of most places of worship, it cannot be extinguished with handheld hoselines. Access to the attic space is through one small door and there is no possibility to vent. Firefighters will have to be withdrawn and a defensive attack, using master streams strategy, used. Before the roof collapses, the ceiling may collapse. If a large part of the ceiling collapses, it will create an explosion like eruption inside the church that will blow out windows and knock firefighters off ladders. As a large church ceiling collapses, it causes compression below the falling ceiling of superheated air, smoke, and flame inside the church. This compression can blast out windows. The collapsing ceiling will also create a vacuum above the ceiling. This instant vacuum sucks air into the now-open attic space, creating a flashover of superheated smoke and fire gases that had accumulated in the attic before the collapse.

Finally, an added danger of fighting a fire in a place of worship is the emotional factor. Many firefighters are religious churchgoers and some firefighters at the scene may even attend the place of worship that is burning. When a church or temple burns, it usually attracts the local parishioners and they are watching their holy place of worship being destroyed by fire. Inside the burning house of worship there are sacred books, scrolls, gold and silver tabernacles, and other sacred objects. All this sometimes leads incident commanders, sector officers, fire officers, and firefighters to take unusual risks that might not be taken at an ordinary public residence or commercial building fires. Again, even at a fire in a place of worship the priorities of firefighter must be adhered to. The first priority is the life hazard that includes the firefighters, second priority is incident stabilization, and property protection is the last priority, even in a place of worship.

Lessons Learned

- **Destination of first attack hoseline.** The initial attack hoseline is taken in a front or side door and attacks the seat of the fire. This hoseline must be the largest diameter hose available. Maneuverability of hose is not factor. Large amounts of water and a high-pressure stream with maximum reach are the necessary hoseline requirements at a fire in a large place of worship. A second hoseline must be taken in the place of worship to back up the first attack hose team. This second backup line may be needed to assist the first line in extinguishing a large body of fire. Because there are few partitions in a church or synagogue and because there may be a delay in discovering and reporting a fire, a large fire may be present upon arrival of the first responders.

- **Primary venting.** Primary venting at a place of assembly should be the rose window at the front of the building. The rose window may the one window that can vent smoke and pent-up combustible fire gases at the upper reaches of a place of assembly. Primary venting also includes the stained-glass windows near the fire on both sides of the structure. These sidewall windows should be vented to create cross venting.

- **Exterior fire spread.** The burning church or temple may have the interior fire spread fire problems of Type III ordinary construction, but the exterior fire spread problems of a place of worship are similar to a Type IV, heavy timber mill building. The exterior fire spread problem at a burning place of assembly will be radiated heat coming through the open stained-glass window, and a burning sky filled with airborne flying embers.

- **Collapse hazard.** The engineers studying earthquakes have identified the collapse dangers of a burning place of assembly. They are the church or temple bell towers, the steeple on top of the tower, the ceiling, the sidewalls, and the roof. Protect firefighters from all these collapse dangers by using aerial master streams inside corner safe areas, outside the collapse danger zone, directed from above the unstable wall and from flanking positions on each side of the danger.

Glossary of Terms for Places of Worship

APSE. Part of the church that is a semicircle or U shape wall

BUTTRESS. Masonry built against a wall to give additional support

CHANCEL. A space reserved for clergy that includes the altar and the front choir area

NAVE. A main seating area of a church

ROSE WINDOW. A large round window at front of Gothic church

TRIFORIUM. A middle story of a church (side balconies)

TRANSEPT. A space the runs at right angle to nave and chancel

DOME. A hemispherical roof on a circle tower or base

GOTHIC. Church architecture of 12th century that features a pointed arch

Note: For more information about church collapse, search the Web for: *Firefighter Fatality Investigation NIOSH 2004-17*.

# **9**

FLAT ROOF COLLAPSE

On a hot summer afternoon, fire is raging throughout a top-floor apartment in a large, four-story, 50×75-foot brick-and-joist apartment house. The first attack hoseline cannot advance because of extreme heat and flames shooting out of the burning apartment doorway. The actual fire is located in a bedroom several rooms back. The flames shooting out of the doorway into the path of the firefighters are being driven by a strong wind blowing through the top-floor residence. Fire broke one window in the rear bedroom, allowing a strong cross-current through the apartment. "Vent the top floor," the chief orders the captain of the second-arriving ladder company, Ladder 6.

The captain and four firefighters, equipped with roof venting tools, climb the aerial ladder of the first-arriving ladder company, which has already been raised up to the roof of the burning building. When they reach the roof, they climb off the aerial ladder on to the flat rooftop of the fire building. The windblown smoke on the roof engulfs them. A screeching power saw can be heard in the distance. Unable to see anything in the smoke, they proceed cautiously toward the sound of the power saw. With body weight always on the rear leg, each firefighter slides one foot forward, feeling first for an object in front of him, such as a low parapet wall. When it meets no obstructions, the forward foot is pressed down to test the roof deck's stability and support. Only when stability is assured does the rear supporting leg move up with the forward leg.

Suddenly, the wind and the smoke direction change. The roof of the fire building becomes visible . Two firefighters are at the rear of the roof, working feverishly. One is bending over at the waist, moving backward, cutting a second vent roof cut. Several feet away the other firefighter is pulling up roof boards of the first roof vent with the adz end of the Halligan tool. Smoke is rising everywhere around the place where the firefighters are working. Smoke pushes out of the vent cut where the firefighter is pulling up roof boards. Smoke rises up around the base of several soil pipes. Smoke flows out of the top of 4-foot-high structures housing old, sealed-up dumbwaiter shafts. Smoke completely fills the open doorway of the bulkhead structure over the interior stairway. Smoke is coming from the top of the stair enclosure where the stair skylight was apparently broken. Smoke is also climbing up over the roof-edge parapet from the rear of the building, indicating that some of the rear windows have been vented.

As the captain moves forward to assist the two firefighters of the first-in ladder company ventilating the flat roof, he shouts above the noise of the screeching power saw for his firefighters to start the cutting saw. As they walk forward on the roof, it feels solid. The captain looks at the tools carried by his firefighters. Unable to be heard above the noise of two power saws, he taps on the shoulder the firefighter who is carrying the axe and Halligan tool and points toward the firefighter struggling with the roof boards. He next taps on the shoulder of the firefighter who is carrying two 6-foot pike poles and points toward the side parapet wall, indicating that he should vent any remaining top-floor windows by reaching over the roof edge, breaking the windows from above with the pike pole.

The officer and the two remaining firefighters proceed to a point near the other firefighter. As they walk forward, the roof begins to feel soft and spongy beneath their boots. The captain signals the firefighters to stop. He moves forward, cautiously feeling the roof's stability. The solid roof beam supporting the roof deck can be sensed by the officer. He proceeds forward, walking on the roof beam portion of the roof deck, attempting to find a spot as close as possible over the fire below. Although firefighting procedures state that a roof vent should be made directly over the fire, the officer knows from experience that in many instances the roof is so weakened by fire that it is too dangerous to operate directly over the fire. At such a fire, the roof vent must be cut as close as possible to the point directly above the fire.

The captain is moving close to the original roof cut out of which flames are now beginning to shoot up, mixed with the smoke. With his body weight on the leg placed above the roof beam, the officer steps down on to the roof deck between joists with his other foot. The charred roof deck below the tar paper feels too soft to support a firefighter. He retreats several feet to where the roof deck feels solid, turns to the firefighter carrying the power saw, and, with the point of the lock puller tool he is carrying, the officer draws an imaginary 4×4-foot outline on the roof where he wants the roof vent. As the firefighter lays the spinning blade into the roof, the screeching noise of the two saws can be heard by the firefighters in the street below.

The captain signals the remaining firefighter to act as a safety man for the roof-cutting firefighter. The safety man places one hand on the back of the firefighter, who is bent over, moving backward with the power saw. At the same time he visually checks the area behind the firefighter cutting the roof to ensure there are no tools, holes, roof objects, or parapet walls which could cause him to trip and lose his balance while holding the high-speed cutting blade. The continuous contact of that hand on the back of the firefighter who is moving backward with the saw signals to him that the area behind is clear of obstructions.

After Establishment of the Roof Operations

With the roof vent operation established, the captain walks over to the bulkhead door opening, and crouching down, reaches his arm and lock puller tool into the smoke- and heat-filled doorway. He slides the tool across the floor of the bulkhead landing, checking to be certain that no person has collapsed on the floor just inside that smoke- and heat-filled doorway. He then proceeds to the side parapet of the roof. As he does so, he checks the progress of the roof venting. Flames shoot straight up out of the first completed roof opening. Firefighters have pulled up the roof boards of the second opening and are pushing down the ceiling beneath the roof deck with pike poles. Smoke mixed with flames is rising out of this vent. A third roof cut is being started by the two firefighters with saws working together. Looking over the side parapet, the officer sees a gooseneck ladder leading from a top-floor fire-escape landing. He thinks to himself, "When they advance the hoseline through the apartment, this will be an avenue to launch a search for victims into the top-floor apartment from the roof."

Flames are now blowing out of the windows beneath him. He pulls his head away to avoid the heat. Looking down the side of the building, he sees the assigned firefighter leaning over the edge of the parapet swinging the hook with one hand, breaking the last window of the top floor. Flame and smoke are coming out of every broken window. The captain does not hear the hose stream attack team advancing below him; there is no sound of a hose stream striking the underside of the roof and no sign of a water stream shooting out of any of the top-floor windows. "There must be some problem," he thinks to himself.

Back with the roof-cutting firefighters, he sees three good-sized roof vent openings. Flames are now coming out of each one. The wind again changes direction. Smoke again covers the entire roof area. The firefighters, with their tools, are ready to leave the roof after completing the assignment. The captain assembles them all at the side parapet wall and radios the chief-in-command at the street level. "Ladder 6 to Battalion 1."

"Battalion 1 to Ladder 6: go ahead with your message."

"Ladder 6 to Battalion 1: the roof and top floor venting has been completed."

"Battalion 1: 10-4. Get your men off the roof and report to me in the street. A second alarm has been transmitted for this box."

"Ladder 6, 10-4." The captain points to the tip of the aerial ladder sticking up above the roof parapet and says, "Let's go."

Each firefighter walks along the parapet, tools in one hand, the other hand holding on to the parapet wall. The officer is the last in line as they proceed to safety along the parapet. A smell of gas drifts up on to the roof from one of the vented windows. "A gas meter in one of the apartments must have ruptured," thinks the captain. The firefighter directly in front of the captain moves cautiously along the wall. Smoke from a window below that is engulfing the firefighter suddenly explodes into flame. The firefighter staggers away from the parapet. The captain tries to grab him, but fails. The roof collapses beneath the firefighter. Flames and smoke shoot up out of the hole. "Urgent . . . Urgent! "Ladder 6 calling! We have a roof collapse! A firefighter fell inside the top-floor apartment!"

Flat roof collapse

The National Fire Protection Association's 10-year study (1990 to 1999) of 56 firefighters who were killed by building collapse showed that 19 died as a result of roof collapses. Working on a flat roof of an ordinary construction building (fig. 9–1), with a fire raging below, is a hazardous operation. If the roof collapses, firefighters will plunge through the roof into a dark smoky flaming cage-like enclosure of four walls from which they will not escape. Any injuries from a fall will be secondary to burns received from the fire. Roof venting operations during fires must be carried out every day throughout the nation. And to survive this dangerous assignment, firefighters working on roofs of burning buildings must know how to size up a fire below.

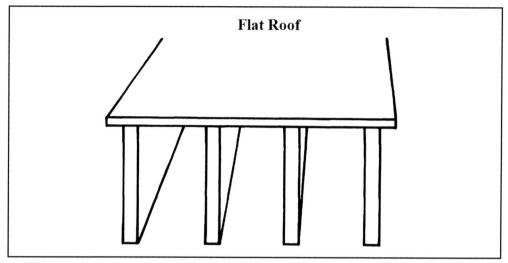

Fig. 9–1. A tar and wooden roof deck attached directly to joists spaces 16 inches on center is a common type of roof found in ordinary constructed buildings.

This is easier said than done. Size-up of a fire below is difficult because a watertight, wooden roof deck, covered with many layers of paper and tar, acts as insulation, concealing the heat of the fire and roof instability (fig. 9–2). Some textbooks tell firefighters to use their senses of sight, hearing, and touch to determine roof stability and that they should always look for signs of structural failure, or listen for noises which precede collapse. However, firefighters' senses are considerably reduced at the scene of a fire. Smoke or darkness often limits vision; the screeching of power saws and the rumble of hose streams below can drown out most sounds; and moving firefighters can keep a roof bouncing. Firefighters need more than their senses to evaluate a roof's stability. They must know roof construction.

At a fire, the more knowledge firefighters have of a structure, the more safe and effective our operations will be. In the past firefighting experience helped us to evaluate the risks involved in operating on a roof when fighting a fire. However today, roof support systems are changing and our experience is not effective. Today roof construction is mostly lightweight materials: steel bar joist, wood parallel lightweight truss, wood I-beams, and metal C-beams. Modern roof construction is becoming more deadly to firefighters operating on roofs. The past experience we have gained when operating on conventional solid wood beam roofs does not apply to lightweight roof construction. The fire service must conduct pre-fire plan inspections of all new construction and study the roof support systems. Lightweight construction like unprotected steel bar joist construction that can fail in 5 or 10 minutes of fire exposure must be documented and programmed into fire dispatch systems. First responders must be notified when responding to fires. Firefighters should be trained in defensive strategies for operating at fires in buildings with lightweight roof construction. A fire officer must realize the roof collapse potential in new and renovated buildings using lightweight roof construction has increased. A veteran firefighter's size-up experience of older brick and solid wood joist roofs will not be effective when operating on roofs of burning buildings constructed with lightweight construction.

Fig. 9–2. A firefighter must know a building's roof system to operate safely.

To determine the risks of a flat roof collapse, we need to know three facts about any type of roof construction:

- What type of support system has been used? Is it unprotected steel bar joist construction, lightweight wood truss, wood I-beam, sheet metal C-beams, or ordinary solid wood joist construction (fig. 9–3)?
- What method was used to connect the joists to the supporting wall? Is it a "simple" supported joist system, or a "fixed" joist system?
- Is there a built-up structure above the main roof supports (sometimes termed an inverted, or raised roof), or a rain roof covering the old roof, constructed to promote roof drainage or is the roof deck nailed directly to the roof support system?

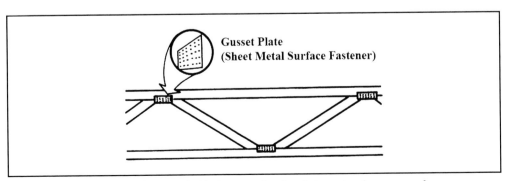

Fig. 9–3. A parallel chord wood truss roof with sheet metal surface fasteners is a dangerous roof.

These facts of roof construction may not always be available to members who are operating on a roof during a fire. However, fire officers must realize that without this information, the evaluation of a roof's stability will not be accurate.

Roof Support Systems

The floors and roof of the World Trade Center building were lightweight construction. The floors and roof were supported by long-span, steel bar joist construction. Instead of a structural steel support of columns girders, so the floors were unsupported. They were steel bar joists 60 feet long, unsupported by any columns or girders. The National Institute of Standards and Technology (NIST) determined these steel bar joist supports were one of the first structural elements to fail during the fire.

In the United States, new and renovated one- and two-story commercial structures are now using steel bar joists in place of wooden joists to support roofs. These steel bar joists are actually metal trusses, which have the top and bottom chords equal in size and parallel to each other. They are used to support the roof deck by spanning the area between bearing walls or steel girders. Examining a steel bar joist roof, we will find a fluted metal steel decking that is welded to the top chord of the bar joist. Above the steel decking will be several inches of an insulating fiber, topped by a thinner layer of a tar and gravel mixture.

The steel bar joist system may be unprotected, or covered with spray-on fire-retarding insulation, or have fire retarding provided by a membrane panel ceiling that is suspended below it. However, the ceiling barrier system will only provide fire protection if it is installed correctly. It must present a tight barrier against any fire below, and have no panels removed or damaged or improperly set. If fire penetrates the fire-resistant ceiling and heats the steel bar joists, the roof will collapse rapidly. In fact, to quote the NFPA's *Fire Protection Handbook*, 19th edition, section 12-78, "An exposed steel joist system may collapse after only five or ten minutes of exposure." The handbook further states, "Experience has shown that either because of inadequate installation procedures or because of required maintenance after a ceiling has been installed, membrane ceilings will not resist the effects of a fire well."

Recently NIST, investigating the 9/11 World Trade Center collapse, stated that there is no scientific basis for deciding the necessary thickness of fluffy spray necessary to fire protect open web steel joists for one or two hours of fire resistance. It appears the construction industry has been guessing at the thickness of the spray-on material required to protect thin bar steel for one or two hours.

Many one-story commercial buildings, housing stores (strip stores) have been renovated and steel bar joist trusses have replaced the conventional solid wood roof beam. These renovated *taxpayers* give no indication of their steel bar joist roof systems from the exterior or the roof of the building. When the roof is being cut, sparks from metal cutting metal will appear. However, unless the roof is cut or firefighters operating below discover the metal trusses after opening the membrane ceiling when examining for fire, it is almost impossible to identify this type of roof support system.

The key to safe fire operations in a building with any type of truss, steel or wood, is early identification of the truss construction by firefighters. At the first sign of any type of truss construction, firefighters must notify the officer in command.

At one fire, units responded to a low-temperature fire that generated large volumes of smoke. The fire was in a dry-cleaning store located in the Bronx. In this instance, roof ventilation was successfully carried out, even though the roof was of steel bar joist construction. Firefighters did not know the construction until the fire was extinguished. A post-fire investigation revealed the fire involved only (content) clothing, and did not develop into large-scale active combustion. Had the fire burned for a longer period before the arrival of the fire department, and had high heat and visible flaming been present, the roof might have failed while members were operating on top of it. Noncombustible construction (Type II) uses steel bar joists for roof and floor construction. This type of construction is designed for low hazard occupancy such as dry cleaning. However if the occupancy changes to high hazard (content) occupancy such as storage, wood working, or flammable material, which can create a rapid hot fire, this type of steel truss roof construction will fail rapidly.

When a fire building has a steel bar joist system supporting the roof, horizontal ventilation of windows and doors, in advance of hoselines, is preferred to vertical ventilation, which requires firefighters to operate on the roof above the fire.

Lightweight wood truss is another dangerous roof construction. Fire testing firefighting experience show this construction can fail within 5 to 10 minutes of fire exposure. This construction, widely used in private dwellings and stores, is a contributing cause of firefighter deaths. On average throughout America, one firefighter every two years is killed in collapse of lightweight wood truss construction. The wood I-beam is another deadly type of lightweight construction.

Wood Joists

The conventional solid wooden joist is the most common flat roof supporting member used in older ordinary constructed buildings. This joist provides greater stability to a roof during a fire than its lightweight roof joist counterpart. For one thing, wood joists are spaced close enough together so that if one becomes weakened by fire, the adjoining joists assume the roof load. In addition, the wood joist is thick enough to allow fire to consume its outer portion, and still retain some of strength. An important feature of a wood joist supporting system is that it does not collapse suddenly when weakened by fire; it will generally give some warning. In large part, this is due to the fact that a wood joist roof has a wooden roof deck, which is nailed to the top of the joist and covered with roofing paper and tar. The roof deck will soften first, which serves as a warning of an impending, serious joist failure. A firefighter walking on a fire-weakened roof deck may plunge his foot through the roof deck and, because of its closely spaced joist system, in most instances can grab on to a beam and avoid falling into the fire below. That is not the case with many of the new lightweight roof supports. They are spaced farther apart and will allow a firefighter to fall between them into a fire. In addition, the steel bar joist system and the lightweight wood truss, unlike the wood joist system, may weaken and fail before the roof deck shows an indication of collapsing. When comparing both systems, the wood joist system is far superior in terms of withstanding higher temperatures for longer periods before it fails.

Connections

The second roof construction fact that helps us to evaluate roof stability is knowing whether the roof joists are rigidly held in position at both ends in a brick cavity—a restrained beam is one restrained at bother ends. An unrestrained beam has the ends resting on a girder or brick corbel shelf, not rigidly held in position.

Brick cavity

In brick cavity construction, the solid wood beams are placed within a recessed portion of a brick wall and cemented into position (fig. 9–4). This is a restrained beam. Generally speaking, in such construction, a beam can support more weight near its encased end, and will not support as much weight at its unsupported center. Wood beam ends which are recessed and cemented within a brick wall cavity, will sometimes continue to provide support to a roof near the supporting brick wall, even when the center of the roof is weakened by fire. If the center of the roof is burned away, the beam end encased in the brick cavity may temporarily provide support as a cantilevered beam. This is the basis for advising firefighters to walk close to the parapet wall on a fire-weakened roof.

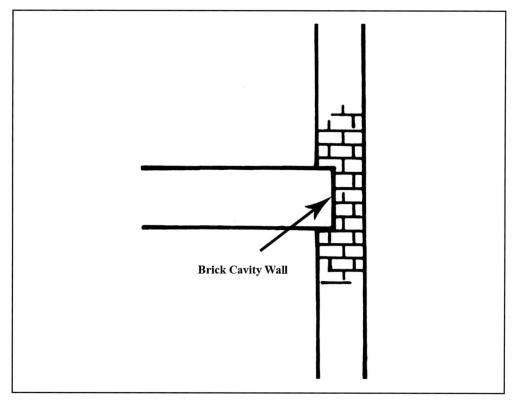

Fig. 9–4. A brick cavity roof beam connection is a restrained connection.

Corbel shelf

In corbel shelf construction, the wood joists are not rigidly encased in a brick wall, but are simply resting on a girder or a brick shelf (fig. 9–5). This type of construction is, of course, less stable under fire conditions than brick cavity construction. In corbel shelf construction, the beam ends may rotate off their resting position when the center of the beam is burned away. A beam end that is simply resting on a corbel shelf (fig. 9–6), or a girder (fig. 9–7), functions as a *fire cut* beam. That is, it is a self-releasing or unrestrained beam. As the roof beams burn away in the center, any weight on the roof will cause the beam ends to rotate off the supports and slant downward toward the center of the collapse area. Most H type apartment roof joists are resting on a corbel shelf.

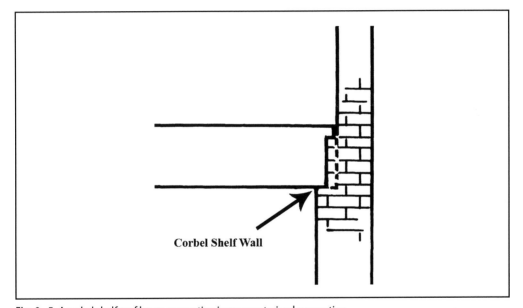

Fig. 9–5. A corbel shelf roof beam connection is an unrestrained connection.

Several years ago Firefighter John McKenna of Ladder 34 died as he was evacuating a roof. He was forced away from the parapet wall to escape a sudden blast of fire. The fire-weakened roof collapsed under his weight and the beam ends rotated off the corbel shelf. An investigation into the cause of the blast of fire that caused McKenna to move away from the relative safety of the parapet wall revealed a burned apartment gas meter. It is possible that the fire blast which forced the firefighter away from the wall, toward the center of the fire-weakened roof, was caused by the sudden ignition of accumulated natural gas which had leaked from the fire damaged gas meter and piping located in the fire apartment.

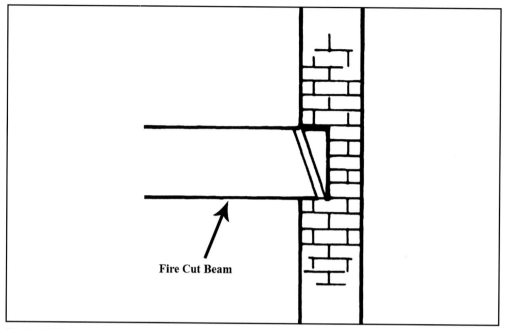

Fig. 9–6. A fire cut beam is an unrestrained connection.

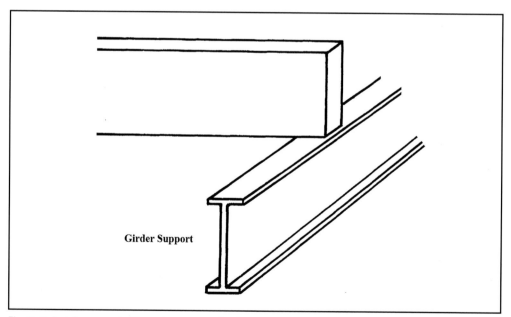

Fig. 9–7. A girder-supported beam is an unrestrained connection.

Inverted or raised or rain roof

The third fact necessary to determine roof stability is for firefighters to know whether the roof deck of the wood beam roof is directly attached to the roof beams, or whether it has an intervening space so that the roof deck is resting on a smaller wood framework that was built up above the main roof supports (fig. 9–8). On some flat roofs, the roof deck is pitched toward a drainage area in order to avoid shallow pools of accumulated rain water. Instead of varying the size of the large main roof beams, a smaller built-up structure above the main roof is created which can more easily be varied in size to create the needed progressive slope of the roof deck.

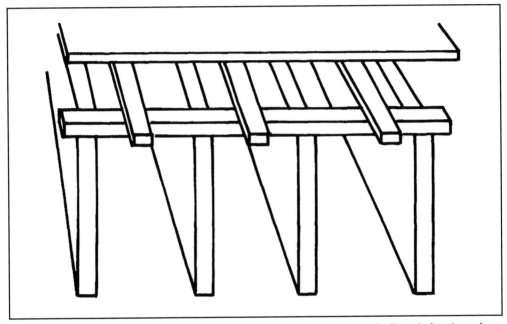

Fig. 9–8. This so-called "raised roof" or "inverted roof" may allow the deck to sag and collapse before the roof joists below are weakened by fire.

A fire that enters the roof space created by built-up framework will first damage and weaken the smaller wood framework supporting the roof deck. The roof may seem unsafe and the roof decking may sag in spots, yet the main roof beams can still be structurally sound. A softening of the inverted roof surface can cause members of a fire company, thinking the roof was about to collapse, to prematurely evacuate their position. However, if the members operating on a roof know that it is only the inverted roof supports which are weakening, and that the main supports are still stable, the intended roof operations can be completed before evacuation becomes necessary.

An inverted roof can present other dangers to firefighters. While a member is attempting to cut a roof vent opening, the smaller framework supporting the inverted, or raised, roof deck could burn away and collapse. The roof deck edge, dropping suddenly into the opening, could cause a firefighter who is operating the saw to lose balance. Additionally, the inverted roof's 2×4-inch framework is spaced farther apart than the roof beams, so there is less of a chance that a member will be stepping on a roof joist. Both feet could plunge through a fire weakened roof deck and a firefighter's body could become wedged in the roof deck and suffer serious burns before he could extricate himself (fig. 9-9).

Fig. 9–9. The roof edge around a vent cut may suddenly collapse causing firefighters to lose their balance and fall with a power saw.

One of the contributing factors of the August 2, 1978, Waldbaum's Supermarket fire, was a *rain roof*. Because the supermarket's bowstring roof was sagging and rainwater was accumulating at its depressed center, a second roof or rain roof had been built over just the center of the original roof to cover the depression. The rain roof recreated the curves at the highest points of the roof and enabled water to roll to the sides of the trusses.

Roof Hazards

In addition to having knowledge of roof construction for the purpose of determining roof stability, a firefighter must also be aware of other hazards that are found on the rooftops of buildings, and which can cause serious, if not fatal, injuries. The following are some of the dangers that can be encountered when operating on a roof.

The firefighter's trap—repairs

A collapse hazard can be created when a temporary repair to a roof is made by laying a thin layer of roofing paper over an opening without replacing the roof decking or beams. A firefighter, walking on this improperly reconstructed roof section, can fall into the concealed opening. If the supporting beams are also absent, or if they are widely spaced, as in a lightweight constructed roof, the firefighter could fall through to the floor below. During daylight hours, when a freshly repaired roof section is encountered, particularly in a partly vacant building, or one that has experienced a previous fire, be extremely cautious. Avoid walking on any repaired sections. Although this repair may have innocently been intended to temporarily keep out the elements, it sets the stage for disaster.

Scuttle covers

A potential area of roof failure is presented by the common scuttle cover. The scuttle cover, which is intended to cover a roof opening, is not designed to support the weight of a firefighter. It is, in fact, a roof deck without underlying roof beams. Quite often, the scuttle cover has been damaged and, when repaired, only a thin sheet of plywood or fiberboard may have been used beneath the roofing paper. Often what appears to be a scuttle cover is a covered-over skylight opening. When the cost of repairing a glass and metal skylight becomes prohibitive to an owner, he simply removes the glass fixture, and covers the opening with a thin wood sheet and roofing paper. An owner may reason that since the original skylight was not designed to support a person, the replacing cover need not be reinforced. Stepping on any such type of roof covering can also lead to disaster.

Disorientation

A firefighter sometimes becomes disoriented by smoke or darkness when operating on a roof. Once disoriented, there are no guides that firefighters can use to reorient themselves on a large open rooftop. As may be expected in the early stages of the fire, there may be no hoselines on the roof which could be followed back to an area of safety. And, there will be no partition. A firefighter disoriented on a roof should not walk upright but instead get down on hands and knees and crawl, with his or her hands feeling for the presence of the roof deck and its stability, before moving forward.

Shafts

Some attached buildings are connected at the front half by a party wall, which can easily be scaled by a firefighter seeking to gain access to the roof of the fire building from an adjoining building. At the rear half of the building, however, there is a shaft that separates the buildings. The parapet wall at the rear half is the same height and of the same construction as the party wall at the front half. And, as is often the case, a firefighter may cross over the front party wall to the fire building to perform roof operations. However, after completing roof operations, if he did not notice the shaft in the darkness or smoke, and attempts to quickly evacuate the roof at the rear by scaling the parapet wall, he will fall into the shaft.

When straddling an apparent party wall, and not able to see the roof surface due to smoke or darkness, the experienced and knowledgeable firefighter will drop a tool or other object, and listen for it to strike the roof he intends to step down on before he crosses that party wall. There are many dangers confronting a disoriented firefighter on a roof during a fire operation. A confused firefighter could fall into an open shaft or walk off the edge of a roof which has a low parapet wall or no wall at all. Vent openings or trench cuts, made during the fire, in addition to skylights and scuttle covers which have been removed for ventilation purposes, all present hazards to the disoriented firefighter. When momentarily blinded by smoke or darkness, and there is no immediate danger, it may be best for a firefighter to remain in place until the wind changes and the smoke clears, or light is provided and he can see again. If roof evacuation is urgent, and a firefighter is disoriented, the safest course he can take is to crouch down and move ahead slowly, using a tool or an outstretched leg to feel for the presence of the roof deck and its stability, before moving all of his body weight forward.

Evacuation

Firefighters operating on a roof should continuously be aware of an emergency escape route. During the entire fire operation, this escape route should be available to them. Never allow fire to get between your escape route and your roof position. Be aware that when cutting a roof vent opening over a top floor fire, or making a trench cut, it is possible that fire exploding out of the roof cut could block your exit. If you're ordered to evacuate a roof, and there are two paths of escape, take the safest one. In some cases, the quickest way off a roof would mean walking directly over the fire on a weakened roof; not a wise course of action. When receiving a "direct" order to evacuate the roof area immediately, do not take this path if another means of escape is available. Take the safest way off the roof, even if it's the longest way around.

Lessons Learned

There can be no question that operating on a roof during a fire is hazardous duty—many dangers exist on a rooftop. And, in the early stages of a fire, remember that firefighters may sometimes operate without the benefit of a supervising officer. For these reasons, it is considered good practice to assign experienced and knowledgeable members to this duty. Comprehensive training in the hazards presented by the roof environment during fire conditions should be the standard for all firefighters. Finally, no firefighter responding to an alarm expects to fall into a fire through a collapsing roof. Yet, it happens every year. The difference between the firefighter that survives and the firefighter that doesn't could depend on the amount of protective clothing he wears. Remember, the use of a helmet, gloves, hood, bunker type turnout gear and an operable self-contained-breathing-apparatus is the last line of defense for a firefighter who is trapped in a fiery environment.

Note: For more information of flat roof collapse, search the Web for: *Firefighter Fatality Investigation F2002-05*.

10
LIGHTWEIGHT STEEL ROOF AND FLOOR COLLAPSE

Near midnight, two engines, a ladder, and a battalion chief are operating at a fire involving a one-story, 200×200-foot noncombustible-construction garage. Flames engulf the front half of the building, and large volumes of smoke are pushing out of the windows in the rear. As first-in ladder company members try to force the front doors, the first engine company, assisted by the second unit, stretches a hoseline to the fire. The battalion chief orders the ladder company to vent the roof. Two firefighters, a 20-year veteran and a rookie, carrying out his command, start to scale the 20-foot ladder that has been placed against the structure. The older firefighter carries a gasoline-powered circular saw in a canvas sling on his back; the younger firefighter, an axe and a 6-foot pike pole. Reaching the top of the ladder, the firefighter in the lead swings the gasoline-powered circular saw over the parapet wall, removes the sling from his shoulder, and steps onto the roof. As he picks up the cutting tool and walks toward the center of the 40,000-square-foot flat roof, he stops at the point he believes is over the fire and lays the saw on the roof deck. There are no skylights or scuttle covers nearby.

Picking up the saw by the carrying and control handles, the firefighter vigorously shakes the tool to mix the gasoline and oil fuel tank mixture and then replaces it on the roof deck. Crouched down in the darkness, he turns on the starter switch, places the other lever to full choke position, locks the trigger handle to full throttle, and, rising up slightly while holding the saw steady with one foot and one hand, he sharply pulls the starter cord. The saw engine coughs twice. He again pulls the cord, and this time the gasoline engine roars to life. The firefighter flips down his eyeshield, adjusts the choke, unlocks the trigger handle, and picks up the saw. Increasing the blade revolutions by squeezing the trigger up to full speed, he leans forward and lays the cutting blade into the roof, letting the high-speed carbide blade slice through the gravel and tarpaper roof deck. Its high-pitched shriek drowns out the roar of the engine. As the firefighter leans forward with his feet spread apart, a horizontal shower of blue-white sparks shoots between his legs, peppering the inside of his right boot with hot metal fragments. A steel roof deck, he thinks, completing the first cut. Carefully lifting the high-speed blade out of the asphalt and steel roof deck, the firefighter then lowers it into the roof again to cut a short 45-degree-angle "pull cut" across the end of the first larger cut. This small cut will provide an opening for his partner to hook a pike pole and pull up the cut portion of the roof. He starts the second cut at a right angle, working back toward the first cut. When the cutting blade does not meet the resistance of any supporting joist below the roof deck, he surmises that it must be a lightweight steel open-web bar joist roof, with the joist spacing farther then 4 feet apart (fig. 10–1).

Fig. 10–1. A lightweight Type II constructed building with an open-web bar joist roof presents a collapse danger to firefighters. As a precaution, the use of horizontal ventilation is recommended.

The two roof cuts meet, and the firefighter cautiously lifts the blade out of the roof, walks around the roof cut, and starts the third side of the roof vent cut. Keeping both feet placed firmly on the roof side of the vent cuts, he guides the saw out of the roof cuts. The section of roof deck he already cut starts to sag downward several inches; the firefighter knows that if he straddles the cut and the roof suddenly hinges down into the fire, he could be thrown into the vent hole with the spinning saw blade. As he continues working, the blade noise changes—the shrieking stops, the gasoline engine begins to labor. The blade is binding! He lifts the blade up out of the roof slightly, then lowers it again. The shrieking of the blade cutting the steel roof resumes.

Halfway to completion of the final cut, the area of roof suddenly hinges down into the fire. The saw kicks upward violently. A blast of flame, heat, and smoke erupts out of the vent hole. Holding the spinning saw steady in front of him, the firefighter slowly backs away from the roof opening until he is out of the blinding smoke and he can see his partner. Flames shoot several feet out of the roof vent. "Go to the parapet and tell the chief the roof is vented!" he shouts to the rookie firefighter.

Without warning a 40-square-foot section of roof deck around the flaming vent opening begins to collapse. The roof slowly sinks, as the supporting steel joists and girders below the roof sag and bend, causing the flat roof to slope downward into a 45-degree-angle concave depression. The firefighter with the saw loses his balance. As he starts to slide into the sinking roof, he releases the saw. It rolls down into the bottom of the sinking roof, disappearing into the flames and smoke. Attempting to climb out of the sinking hole, the firefighter keeps slipping on the melted asphalt and gravel roof covering. Unable to crawl up out of the roof depression, he lies flat on the side of the sloping roof and shouts to his partner, now out of sight, "Help! Get me out of here! I can't climb up!"

Flames start to spread up the asphalt roof covering toward him. Engulfed by dense black smoke, he tries not to slide farther down the sagging roof. The young firefighter now appears near the edge of the sunken roof carrying a portable ladder. He swings it into the sinking roof area and lowers the tip down into the smoke, rung by rung. Seeing the descending ladder, the trapped firefighter reaches out and grabs a rung, then scrambles on to the ladder and climbs out of the roof hole. Now standing on the flat portion of the roof, the two firefighters pull up the ladder and quickly carry it to the parapet wall. They lower one end to the sidewalk below and climb down to safety, just as the remaining portion of the roof sags down into the fire.

Steel Bar Joist and C-beam Roof Construction

Steel does not melt during a fire. Heated steel expands, bends, twists, sags, and buckles when heated by the flames of a fire, but it doesn't melt. Steel starts to distort and lose strength during a typical structure fire where flames reach 1,000 to 1,100 degrees F. It requires more that 2,400 degrees F to melt steel, and this temperature is rarely reached at a structure fire. However, steel, when heated, may expand and push out a masonry wall; steel may buckle and suddenly collapse a floor; steel may warp and twist, allowing roof beams to cave in; and steel may sag and bend causing a concrete floor to buckle, crack, and fail. Steel, when it distorts from the heat of a fire, can cause a buildings collapse.

Steel is an alloy of iron and carbon used as a structural material. Since World War II the use of steel in buildings has dramatically increased. Over the years, steel is replacing wood and masonry in structural systems. Columns, girders, beams, and trusses in buildings are now steel supports, instead of wood or masonry. Today there are even all-steel buildings. Everything, inside and outside—walls, roof, floors, columns, girders, beams, and trusses—are steel. Steel buildings are inexpensive and quickly constructed. However during a fire they can collapse fast. When steel distorts and moves, the building can collapse. The trend over the past half-century is toward lightweight buildings, and that means more steel buildings and less concrete.

Steel has greater tensile and compressive strength than wood or masonry and steel can be shaped more easily. Steel can also be bent and twisted into shape to increase its strength. This is called *shaped steel*. Shaping the steel gives it greater load-bearing capacity and uses less material. This shaping allows for lighter weight steel buildings. Thin shaped steel can be used instead of thick heavy steel structural elements. For example, manufacturers can bend thin bars of steel to make truss construction. The bent shape can give the metal greater strength while using less material. Two of the most common structural elements formed by bending steel are the bar joist truss and the C-beam. The steel bar joist truss is is a long steel bar (web) bent at 90-degree angles and welded to angle irons (chords) at the top and bottom of the bar bends. The steel C-beam is a 16-inch-wide-by-20-foot-long strip of $^3/_8$-inch of sheet steel bent into the shape of the letter C to create a 20-foot roof or floor beam. The bar joist truss and the C-beam are replacing the thick solid I-shaped beam in modern lightweight construction.

Other common examples of shaping steel to increase load-bearing capability are: hollow tube steel columns, steel partition wall studs, and fluted floor decking used to support a masonry floor. The World Trade Center, one of the tallest high-rise buildings, was constructed mostly of shaped steel. The twin towers had tubular steel bearing walls, fluted metal steel deck floor supports, and bent bar steel truss floor supports. This steel was shaped to increase its load-bearing capability, in order to reduce the amount of steel and to lessen the weight in the structure. Remove mass and you can build higher at less cost.

The failure temperature of steel is acknowledged to be in the range of 1,000 to 1,100 degrees Fahrenheit or 538 to 593 degrees Celsius. At this temperature the steel distorts, it does not melt. The National Fire Protection Association (NFPA) conducts tests to evaluate protection methods of insulating steel from heat effects of fire. They use a so-called "standard time temperature fire" (fig. 10–2) The standard time temperature test is a compilation of many test fires conducted in ordinary type construction buildings in 1918. This temperature curve used during a test fire is considered an average fire. This test is used in most fire protection testing laboratories today. This standard test fire documents steel failure (distortion) when the average temperature of the steel is between 1,000 to 1,100 degrees Fahrenheit (F) or 538 to 593 degrees Celsius (C). The standard furnace fire tests use only small samples of floor sections or wall areas to determine a structural elements reaction under fire conditions. And so it gives results only for small test sections of steel. A floor test sample need only be 18 feet by 14 feet.

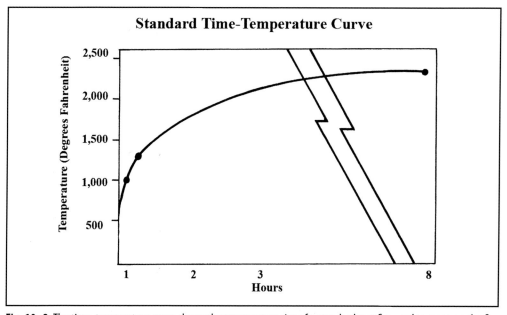

Fig. 10–2. The time-temperature curve shows the temperature rise of a standard test fire used to measure the fire performance of construction assemblies.

Experience has shown that large sections of steel exposed to fire may fail sooner at a real life fire. There is no full scale testing that simulates an actual building in an actual fire. For example, the National Institute of Standards and Technology (NIST), 9/11 study conducted tests where a 17-foot piece steel and concrete floor passed the required fire resistance testing; but a 35-foot piece of the same steel and concrete failed a duplicate test. Fire resistance test fire results cannot accurately predict how a larger building section will perform during an actual fire. Longer sections of steel beam collapsed before short span steel sections when using the same test. This century-old time temperature furnace test fire used by testing laboratories to rate steel fire resistance is questionable in the opinion of many experts.

There is a difference between temperatures in a structure fire and a test fire such as the time temperature test fire. It takes 2,400 degrees F of heat to actually melt steel. A typical structure fire can reach a fire temperature of 1,700 degrees F. And the standard time-temperature-curve test fire which is used as the fire temperature to determine fire protection for steel reaches 1,000 degrees F in five minutes; 1,300 degrees F in ten minutes; 1,700 degrees F. after one hour; 2,000 degrees F after four hours; and 2,400 degrees F is the highest temperature of the test fire is reached after eight hours.

Thin steel such as the open-web steel bar joist and steel C-beams can reach the critical temperature of 1,100 degrees F (593 degrees C) much faster than thick solid steel I-beam sections. Large solid steel can absorb more heat before reaching critical temperature for collapse and so takes more time to reach the critical distortion temperature. The critical failure temperature of steel failure depends on many factors besides the heat of the fire. Steel failure depends on: the load it carries, the dimensions of the steel and the span of the steel beam, and the fire protection insulating covering over the steel surface.

Fire Protection Methods

Three common methods to protect steel from the heat of fire are: encasement of the steel with concrete; the installation of a fire-retarding ceiling between the occupancy and the steel above; and finally the most popular, but least effective, method, applying a spray-on fire retarding (SFRM) material over the entire surface of the steel.

Concrete encasement

The best method to protect steel, from a firefighter's point of view, is to encase it in concrete. During a fire, the concrete covering adheres to the surface of the steel and makes the noncombustible steel fire resistive. A good covering of steel with one to two inches of concrete has proven effective in the past. In buildings constructed before World War II, the steel is rarely affected by fire when encased in concrete, so there is little distortion to the concrete covered steel. The one drawback is chunks of concrete encasement can collapse during a fire. Small chunks of concrete covering can fall from the steel when heat by fire. This is caused by *spalling*. Spalling is a term used to define the expansion of moisture in the concrete into steam. This expansion takes place inside the concrete during a fire. Moisture is naturally found in concrete. During a fire, the heat makes the moisture expand into steam and this expansion causes chunks of the concrete encasement to fall on firefighters advancing in a hoseline or searching for victims. The small pieces of falling masonry are an injury hazard; however, this is better than having the roof or floor above collapse down on firefighters because the steel failed.

Membrane ceiling

Another method of protecting steel is the use of a ceiling to separate the steel building supports from a fire in the occupancy. The use of a fire-retarding ceiling to separate any room and content fire from the steel roof or floor above is less expensive, but not as effective as the concrete encasement method. This so-called "membrane ceiling" method of protecting steel uses a suspended ceiling system of lightweight steel frame and rods holding one-hour, fire-retarding ceiling panels, This sealed barrier placed between the occupancy where a fire may occur and the steel support system above is supposed to stop fire and heat from exposing the steel above. Often it does not for several reasons The membrane ceiling method of fire protection is only effective if the ceiling panels are tight and in proper position within the metal frame. There may be missing panels; or panels may not be properly in place. There also can be penetrations in the membrane ceiling, such as openings for fluorescent lights and air conditioning ducts that are not fitted with approved assemblies to restrict fire spread. These openings will render the membrane ceiling useless.

There is also a collapse danger with a membrane ceiling. If the suspension system holding the ceiling is not properly constructed the ceiling can collapse on firefighters. This collapse is more serious than the concrete spalling because the entire metal framing may fall like a net over firefighters. If the membrane ceiling collapses the framing and wire hanger straps can entangle and trap firefighters in a burning room. The metal frame holding the panels and the hanger rods, not the ceiling tiles, are the collapse hazard to firefighters.

Spray-on fire-retarding material (SFRM)

The third, and least effective, method of protecting steel from heat of a fire is the spray on application of a fire-retarding material over the steel surface (fig. 10–3). This spray on fire-retarding material (SFRM) is applied over the entire surface of a steel column, girder, and beam. The fire retarding is also sprayed over the entire underside of steel-fluted floor deck and is supposed to insulate and protect the steel from the heat of a fire. Depending on the thickness of the sprayed material, the steel may receive a one-, two-, or three-hour fire-resistive rating. It rarely protects steel for the designated time at a serious fire. Post-fire investigations have shown this spray-on material to be dissipated by the fire and the steel to be distorted by heat. Despite these findings, this fire-retarding method for steel continues to be the preferred method to protect steel in new buildings because of its cost effectiveness when compared to the other methods. This so-called "direct application" of fire protecting structural steel is less expensive and a faster method than encasement or use of a membrane ceiling. In the past, an asbestos base liquid was sprayed on steel for fire protection, but today it is outlawed. The material sprayed on steel today for fire protection is a slurry of vermiculite, volcanic rock, or mineral fiber mixed.

Chapter 10 | **Lightweight Steel Roof and Floor Collapse**

Fig. 10–3. Spray-on fire retarding material to protect steel

The fire service has known this method of spray-on fire protection was ineffective since its introduction in the 1970s. For example, after every serious fire in New York City fire-resistive, high-rise office buildings, the spray-on material was missing in the fire area and the steel floors and ceilings protected by this spray-on fire retardant were seriously buckled, warped, and sagging. The concrete floors in the fire area were cracked, sagging, and heaved upward. The floor above the fire had become too dangerous for firefighters because of this distortion. In Type I resistive and Type II noncombustible construction, steel-fluted floor systems protected by spray on fire retarding material often supports 2 or 3 inches of concrete flooring above. The spray-on fire retardant fails and steel floor beams and girders are visually sagging and warped. When spray-on fire retarding material (SFRM) fails to protect steel from the heat of a fire it is because

- The steel is not prepared properly (cleaned) to allow the spray-on material to stick properly. Tests have shown applications of primer paint on steel reduce adhesion of spray on fire retarding effectiveness by one third to one half.
- The spray-on slurry is often not mixed properly.
- Workers do not apply the spray-on material evenly.
- Other workers doing subsequent work nearby easily remove the critically important fire protection.
- Even if applied properly, over several years, vibrations and HVAC air movement in the plenums causes the fire retarding spray-on to blow off and leave the steel unprotected.

The exact thickness of the spray-on fire-retarding material used to protect steel was brought into question by NIST during its investigation of the World Trade Center (WTC) collapse. During the construction of the WTC, the thickness of the spray-on was first three quarters of an inch. This was supposed to give the steel floors a two-hour rating. Then after inspections and tests in the 1980s indicated that this was inadequate, the thickness of the WTC spray-on insulation was increased to 1½ inches thick. The thickness of the 1½-inch spray-on was, once again, said to provide fire-retarding protection of the steel bar joists supporting the floors for two hours. During an investigation after 9/11, no fire testing documentation was found to justify the thickness of either ¾-inch or 1½-inch spray-on fire retarding material to provide a two-hour fire rating for the steel floor trusses. Another problem revealed after the collapse of the World Trade Center towers was that the spray-on material had difficulty adhering to the steel round bar of the truss. Spray-on fire retarding material adheres better to a web member that is angle-iron shaped, rather than round bar shape. Spray material flakes off the thin bar that forms the web members of steel trusses. As result of these finding, the New York City fire department has prohibited the use of open web-steel joists in nonresidential high-rise buildings until appropriate fireproofing standards are developed and promulgated.

Today almost 40 years after being used to protect steel there is no standard thickness of spray-on fire retarding agreed upon by the construction industry. It was a guessing game. Only now after the WTC collapse, the construction industry is reevaluating the standards for use of spray-on fireproofing, to protect steel in high- and low-rise buildings.

Adhesion, consistency, and thickness are three criteria that the construction industry now determines are necessary criteria for the SFRM effectiveness on steel during a fire. Adhesion has been found to be a more important factor for spray-on fire retarding method of protecting steel than previously understood. Investigations of spray-on fire protection effectiveness have found air movement in a plenum area created by a central air system blows the spray-on fire-retarding material off the steel. Even if it is applied over the steel properly, the air movement and the vibrations can cause insulation to flake off the steel.

Steel In Renovated Buildings

Many renovated buildings replace wood and masonry structural elements with steel. The replacement of heavy wooden roof and floor joists with the lighter steel open-web bar truss joists is a common renovation change (fig. 10–4). A lightweight steel open-web bar truss joist supported the floors of the World Trade Center and the roof of the Charleston, South Carolina, Sofa Super Store. The open-web bar joist is a long steel bar (web) bent at 90-degree angles and welded to angle irons (chords) at the top and bottom of the bar bends. This type of construction is used in new commercial buildings classified as Type II noncombustible construction. However, it is also used to replace old wooden joists in renovated ordinary type, brick-and-joist, buildings. Buildings of noncombustible construction have masonry or steel nonbearing enclosure walls and steel column-and-girder framework and use lightweight steel open-web bar joists as roof and floor supports. Noncombustible construction, using steel in place of wood for interior structural framework, is well suited for large-area, commercial buildings with low hazard content.

Hazards of lightweight steel bar joists

The lightweight steel bar joist must be viewed by firefighters as extremely hazardous in roof and floor construction for two reasons: the failure characteristics of unprotected steel and the joist spacing (fig. 10–5). The fire service has known for decades, when unprotected, the steel bar joist may collapse after five or ten minutes of fire exposure. Advancing a hoseline inside, or operating on a roof of a burning building, built with steel bar joist trusses is a high risk because of early collapse. If the average response time of an urban fire company is five minutes; a suburban or rural fire company is somewhat longer. If a serious fire occurs in an unoccupied building with a roof supported by unprotected lightweight steel bar joists or today's new steel C-beam is so severe that the stability of the roof is in question, then it must be considered too dangerous to send firefighters on the roof for ventilation and for interior attack with a hoseline. In an unprotected Type II steel building, fireground commanders should consider a strategy of horizontal ventilation and defensive exterior attack.

Fig. 10–4. Lightweight steel bar joist roof construction

Fig. 10–5. Unprotected lightweight open-web steel bar joist can collapse within 5 to 10 minutes of exposure to fire.

Ventilation tactics

Some one-story, noncombustible buildings using lightweight steel joists are constructed with large rectangular windows at the upper portion of the masonry or corrugated steel enclosure walls. These windows are effective for smoke venting (refer back to fig. 10–1). Located at the top portion of the walls, where heated smoke would accumulate, these windows have a horizontal length that is greater than the vertical depth of the window area. Ventilating several of these windows can be accomplished faster and more safely than making a 4×4-foot roof cut. In the 20th century, roof ventilation has been a very effective and relatively safe fire department operation when carried out on a wood joist roof support system spaced 16 to 24 inches on center.

This is not the case in the 21st century. Roof construction that varies from this solid wood joist design is more dangerous. When a different roof construction, such as lightweight steel bar joists, or C-beam is introduced into a community in which standard wood construction has predominated, the different collapse characteristics present a new safety hazard to firefighters. The steel bar joist and C-beam will collapse more quickly. An experienced firefighter who has advanced hoselines into burning ordinary constructed buildings or who cut vent openings in a roof of solid wood joists cannot transfer his or her sense of safe operating time to a building with a lightweight steel roof. Firefighters have developed a sense of how long they can work, inside a burning building or operate on a roof over a fire, based on the type of roof construction used in ordinary construction, This experience can not be applied in Type II noncombustible construction with steel trusses. Over the years fire officers develop an ability to size up a fire. This size-up is correct 99 percent of the time; however, the 1 percent of the time when this size-up is incorrect, a roof or floor collapses, killing or seriously injuring a firefighter. Today this incorrect size-up is occurring more often because of a change in noncombustible Type II construction throughout America. Unprotected lightweight construction using steel bar trusses or steel C-beams, which fail during the early stages of a fire are often the cause of the size-up misjudgment (fig. 10–6).

Fig. 10–6. This roof supported by lightweight steel bar joist collapsed during the early stages of the fire.

Spacing of lightweight steel bar joists

An important design difference between a wood joist and a steel bar joist and C-beam roof support system is the spacing of joists. Open-web steel bar joists can be spaced up to 8 feet apart, depending on the size of the steel used and the roof load. This wide spacing creates several dangers to a firefighter cutting an opening in a roof deck. First, when the outline of the roof cut is near completion, and if the roof deck is not directly above one of the widely spaced steel bar joists, the cut roof deck may suddenly bend or hinge downward into the fire. If a firefighter has one foot placed inside the roof cut opening, he or she risks losing balance and falling, with the saw, into the fire below. The wide spacing of the steel bar joists allows a firefighter to fall through a roof opening when visibility is poor because of darkness or smoke.

In the past, the wood joists spaced 16 inches on center could prevent a firefighter from falling through the roof openings if he lost his balance. The mask cylinder and back plate could snag on the closely spaced joists, or the joist could be grabbed by the falling firefighter. When several large roof vent cuts are planned over a fire, a firefighter cutting the roof will over-cut the initial roof vent opening. On a fluted-steel roof deck with steel joists spaced several feet apart, the roof deck area around the initial roof vent hole near these over-cuts is extremely unstable; the edge of the roof deck may bend downward and drop a firefighter through the roof opening. The blade of a power saw cutting a roof deck sinks several inches into the roof. If the blade is drawn across the steel bar joist, it can slice the top chord completely in two. In some instances, the top chord is the main load-carrying member of the steel joist. Over the past 100 years, training for roof operations has been carried out under the assumption that a firefighter was working on a wood-joist roof 2×10 or 2×12 inches in size and spaced 16 to 24 inches on center. This assumption is no longer true. A firefighter must know the type of roof construction and know its collapse potential.

Causes of Failure of Unprotected Steel

Four factors determine the speed with which unprotected steel will fail during a fire: temperature of the fire, the load stress, the steel thickness, and the fire size.

Temperature of the fire

Fire load, the amount of combustible material which can burn, includes combustible content and combustible structure. Unprotected steel used in noncombustible building are safe in buildings with low combustible content rating. If the noncombustible steel building contains content in the amount, or combustibility, considered as high fuel load, fire protection systems such as automatic sprinklers and fire partitions must be added to the building; sprinklers with the back up of compartmentation in case the sprinklers fail. In some buildings, such as those of heavy timber construction, more fuel may be supplied to a fire by the wood structure (timber columns and wood floors) than by the content. In a building classified as noncombustible, which feature lightweight steel bar joists instead of wood joists, a considerable amount of fire loading is eliminated from the structure. Unprotected steel has no fire resistance, a steel building can be quickly destroyed by fire. It will collapse when heated to temperatures that are easily attained at a fire. When heated, steel will bend, sag, warp, and twist unless it is encased, cut off, or covered in some type of insulating material.

The principal danger to a firefighter in a burning noncombustible structure that contains unprotected steel is its potential for early collapse. The fire service considers 1,100 degrees F (593 degrees C) to be the failure temperature of steel, for at this temperature, steel will lose almost half (40 percent) of its load-carrying capacity. This temperature, however, is not the temperature of flame and heat of the fire. The temperature within the steel itself, not just that of the surrounding temperature, must be raised to 1,100 degrees F (593 degrees C) before it will fail. Because steel is a good conductor of heat, there is a time lag between the time the fire area reaches 1,100 degrees F (593 degrees C) and the time the steel itself reaches this temperature.

The high temperature of a fire can also cause steel to expand. Masonry walls have been moved to the point of collapse by expanding steel girders and beams during fires. A 50-foot-long steel beam that is heated uniformly over its length from 72 degrees F to 972 degrees F, the average increase in length can be 3 feet 9 inches. This increase in size will either push out a enclosing wall or cause the steel beam to buckle.

Firefighters should have some idea of the heat which can be generated by a fire. The standard time-temperature test fire gives us some idea of how rapidly temperature can rise during a fire. The two important facts of the time-temperature curve are:

1. Within the first 5 minutes, the temperature of the fire will rise to 1,000 degrees F.

2. At 10 minutes, the temperature reaches over one-half the total temperature rise (1,300 degrees F) attained after 8 hours.

Load stress of steel

The second factor which determines steel failure during a fire is the load supported by the structural member. The greater the supported load, the faster a structural steel member can fail. In modern noncombustible buildings, roofs are not designed to support the same load as the floors below. A floor must be capable of supporting contents and people, but a roof is designed to support only the weight of rain or snow accumulation. (In the South, a roof may be designed only to resist a wind load.) In older brick-and-joist buildings, the flat roofs could often support the same load as the floors below. This result was achieved by accident, not by design, because the builders were not as cost-conscious as they are today. The roof joists were the same size and spaced the same distance apart as the floor joists below.

Today, when designing a roof support system, builders do not consider the weight of firefighters and their equipment on a roof or inside the building when it is burning. In a noncombustible building, the open-web steel bar joists or C-beams used in the roof construction will either be spaced farther apart than the floor bar joists, or the roof joists will be of steel of a smaller dimension. A roof designed only to support a snow load may have a load capacity of 20 pounds per square foot.

The roof deck is also a load factor which can influence the collapse of a steel joist. The heavier the supported roof deck, the faster the collapse of heated steel roof beams. There are two common types of roof decks used above the fluted-steel deck of a bar joist roof or C-beam support system: a lightweight insulation or a precast concrete plank. The heavier concrete plank roof deck may reduce a firefighter's safe operating time on top of or below the unprotected steel roof of a structure.

Thickness of the steel

The size of the structural element is another factor which determines steel failure. A heavy, thick section of steel has greater resistance to fire than a lightweight section. A large, solid steel I-beam can absorb heat and take a relatively long time to reach its failure temperature, while a lightweight steel beam, such as an open-web bar joist, or C-beam can be heated to its failure temperature much faster. By increasing the mass of a steel structural element, we can actually increase its fire resistance to a limited degree. An unprotected, built-up steel column of sufficient mass could even be given a one-hour fire resistance rating if tested in a furnace by means of the time/temperature curve and if the unprotected steel absorbed sufficient heat to prevent the column's cross-section area from reaching an average of 1,100 degrees F (593 degrees C) within that time.

Unfortunately, the trend is toward lightweight steel construction, and the costly "overbuilding" found in older steel buildings has ended. The heat-sink property of large-size steel is its capacity to absorb heat from a fire and its ability to conduct or transfer heat of a localized fire away from the point of flame contact to the cooler interior regions of the steel. This heat-sink property can lengthen the time required for the temperature of the steel to reach the collapse temperature; however, even large-size steel will eventually collapse in a fire. There is little "heat sink" capacity in thin lightweight steel structural supports.

Fire size

The size of the fire is the final factor that can affect steel failure. If a small-area fire comes in contact with a portion of a large steel beam, the steel will absorb heat and transfer it away from the flaming area to cooler parts of the structural element. A fire could burn for some period of time before it heats the entire steel beam to its failure temperature. On the other hand, a large-area fire in which flames involve much of the steel beam in a short period of time will heat the steel beam to its critical temperature more quickly (fig. 10–7). A so-called "flash fire," suddenly involving a large area with flame, can heat steel rapidly to its failure temperature. Type II noncombustible unprotected steel buildings are designed for low hazard occupancies like schools, offices, and retail shops. This unprotected steel construction should not house high hazard occupancies like wood-working shops, furniture storage, flammable liquids, and furniture refinishing plants unless deluge-type sprinklers and smoke detectors are installed.

Fig. 10–7. This large steel girder buckled and pulled away from the wall and collapsed during a large area fire in a stationary store.

Lessons Learned

There is a difference between the terms *noncombustible* and *fire resistance*. Steel is noncombustible, not fire resistive. Noncombustible steel will not add fuel to a fire, but steel can not resist fire. It will collapse from the effects of fire. The heat of a fire destroys the load-bearing qualities of steel. If you want to make steel fire resistive you must protect it. You can make steel fire resistive by covering it with insulation. A heat-resistant material—brick, terra cotta, concrete, plaster board and, least effective, spray on insulation—can give steel fire resistance for one, two, three, or fours hours depending on the type of insulation; You can also separate steel from a potential fire area with a membrane ceiling.

Firefighters must remember unprotected steel bar joist trusses and steel C-beams are not fire resistive. They can collapse after 5 to 10 minutes of fire exposure.

Note: For more information on unprotected steel bar joist roof collapse, search the Web for *Firefighter Fatality Investigations F2003-18 and F 2007-18.*

11

LIGHTWEIGHT WOOD TRUSS COLLAPSE

An engine company responds to a fire in a row of two-story town houses under construction. Arriving at the scene, the company sees smoke coming out around the second-floor eaves of several of the new wood-frame row houses. A construction worker runs up to the arriving pumper and excitedly tells the officer, "There is smoke coming from the ceiling of the top floor of that building!"

The captain of the engine company turns to order the hoseline stretched but sees a firefighter already pulling the nozzle and preconnected 200 feet of 1½-inch hose off the apparatus. The captain runs back to the radio and gives a preliminary status report. "Engine 8 to Communication Center. We have smoke showing in a row of town houses under construction. The attached structures are two-story, 20 feet wide by 40 feet deep, wood-frame construction. Engine 8 is on the scene. Transmit an alarm for a working fire. Send two more engines and two ladder companies."

The officer jumps out of the cab of the engine and catches up with the firefighter who is stretching the hoseline. The officer helps the firefighter assemble excess hose in the second-floor hallway for an advance into the fire. He then opens the door to the apartment slightly; dark brown smoke and the smell of burning wood blow out of the opening. The officer quickly closes the door and looks around; construction work is almost finished in this building. The plaster board ceilings and walls are in place, but there are many open voids awaiting utility finishing touches to be made to the town house.

After sufficient excess hose for an advance is in place, the officer calls the pump operator on the Handy-Talkie, "Engine 8 to pump operator, start the water."

"Engine 8 pump operator to Engine 8 command, O.K., here it comes, Captain."

The officer looks at the firefighter with the nozzle, who is crouched down on one knee, and says, "Let me know when you are ready."

The firefighter, moving quickly, drops gloves and helmet on the floor, pulls the face mask straps over the top of his head, replaces helmet and gloves, picks up the nozzle, closes it fully, then cracks it open slightly to bleed the oncoming rush of air from the hose, looks up, and nods to the officer. The captain pushes the door open all the way; smoke billows out. The mask-equipped firefighter and officer disappear inside the smoke-filled doorway. Smoke is banked down to the floor. The heat is stratified just above their heads.

As they advance, the nozzle is ready but not yet opened because there is no red glow visible through their face mask lenses, just dense smoke and the heat above. Cautiously moving forward, they encounter no fire. They crawl through several rooms of smoke and still see no fire. A crackling sound of fire can be heard, but no flame can be seen. "We must be in the wrong apartment. Back out to the hallway." The muffled voice of the officer speaking through the mask is heard in the smoke.

The two begin to back the charged hoseline out of the apartment, when suddenly the ceiling in front of them collapses downward, showering the room with sparks and a flaming wood truss roof beam. The burning crisscross of the truss web members is clearly visible. One end of the flaming truss has fallen to the floor; the other remains held up above the ceiling. The firefighter quickly opens the nozzle and directs the stream on the flaming truss. Another burning truss crashes through the ceiling. This one collapses behind them, blocking the path back to the apartment door with a maze of burning and broken truss web members. Sparks are now raining down into the entire apartment through the broken ceiling.

With the nozzle open, the stream of water blasts away at the fire, but the heat and flame continue to flow from the ceiling. The captain and the firefighter must lie on the floor to escape the superheated gases. Another wood truss collapses through the ceiling on top of the firefighter. The room bursts into flame; he drops the nozzle and attempts to make a dash for the exit doorway through the trusses blocking his path. Punching, kicking, and ducking through one truss, he trips, gets up, and crashes through the next maze of web members blocking his escape. There is a blast of heat, and another collapsing truss, falling from the ceiling, knocks him to the ground. He slowly staggers up from the floor but falls, seriously burned, into another collapsed burning truss. Entangled with broken pieces of wood and overtaken by flames and heat, the firefighter stops moving. The captain, fighting his way to the rear of the flaming apartment, crashes through a glass window and falls from the second-floor window. He lands in a high pile of sand and rolls down into construction rubble. Staggering up on one knee, he radios, "Urgent! Urgent! Engine 8, we have a firefighter trapped inside the second-floor apartment!"

Lightweight Wood Truss Construction

There are two types of wood truss systems used in building construction: heavy timber truss systems and lightweight wood truss systems. Most building codes define heavy timber roof construction as that in which the wood dimension is at least 4 inches wide and 6 inches deep. Connections and fastenings used to connect web members and chords of heavy timber trusses are made of steel bolts and plates. The lightweight wood truss, on the other hand, incorporates wooden members which can be as small as 2 inches wide and 4 inches deep—and these wood pieces are connected by sheet metal surface fasteners called *gusset plates* or gang nails. This connector, which is critical to the integrity of the lightweight truss during a fire, is a piece of sheet metal with many V-shaped points punched through it. These V-shaped nailing points fasten only the surface of the 2×4-inch wood truss. Three wood truss members are sometimes held together by sheet metal points that penetrate the wood surface to a depth of only ¼ inch to ½ inch. This type of lightweight truss system is being used at an increasing rate in the construction of homes and apartment houses throughout this nation (fig. 11–1). By engineering calculations and practical firefighting experience, lightweight trussed rafters may be expected to collapse after about ten minutes in a fully developed fire. The fire services of this nation are alarmed and disturbed by this new lightweight structural element.

Chapter 11 | **Lightweight Wood Truss Collapse**

Fig. 11–1. There is nothing larger that 2×4-inch lumber in this lightweight truss roof and floor constructed building.

The sheet metal surface fastener

The main concern of the fire service is the sheet metal surface fastener used to connect the truss members together. The surface fastener, which only connects the outer ½ inch of wood truss members, is a deficient structural connection from a fire protection point of view. The design of the truss can be defended from an engineering viewpoint, but no architect, engineer, building construction contractor, or code official can defend the sheet metal surface fastener. This device is a dangerous structural connection (fig. 11–2).

Fig. 11–2. The fire service should seek to outlaw the sheet metal surface fastener.

In any structural element, the critical area subject to failure during fire is the point of connection. At a fire, the points of connection are the first to fail. The most serious defect in the surface fastener connecting lightweight wood trusses is the insufficient depth of penetration of the nailing points. As the name indicates, the metal surface fastener only fastens the outer ½-inch surface of the truss. The V-shaped nailing points enter the wood to a depth of only ½ inch. During a fire, when the outer layers of wood char, the surface fastener loosens more quickly than would a nail or steel bolt, which penetrates the entire thicknesses of truss members. In addition, even if the fire is not of sufficient intensity to char the wood, heat from the flames can warp the thin sheet metal surface fastener and cause it to curl up and pull away from the wood truss.

The lightweight wood trusses are prefabricated at a factory and shipped to the construction site, where they are stored until needed. If these trusses are improperly transported or stored at the site, or if they are dropped or handled roughly, the metal surface fastener can pull away from the wood surface or become loosened. In this instance, the truss has been weakened even before it is installed in the building. Still another problem with the metal surface fastener is corrosion. When installed in an enclosed, well-insulated roof space, where moisture becomes trapped, the sheet metal surface fasteners will probably rust and weaken. The treatment of wood trusses with certain fire-retarding chemicals or wood preservatives may also lead to corrosion of these fasteners.

Fire Spread

A flat roof or floor supported by a lightweight wood truss will allow fire to spread more quickly throughout the concealed space than one supported by a solid wood beam. Unlike the new lightweight wood truss, a solid wood beam has some minor fire containment value. For example, when a fire travels up into a concealed ceiling space, flames may travel in the space between the solid beams, but fire spread in a direction perpendicular to the solid beam will be blocked. There is no such temporary perpendicular fire blocking with an open-web lightweight truss. When fire enters the concealed space of a parallel chord lightweight truss, it spreads, simultaneously, quickly between the length of the truss and perpendicularly through the web members. If fire enters any attic or concealed roof space with a lightweight truss, it will spread rapidly and quickly involve the entire area (fig. 11–3).

Fig. 11–3. When fire enters the concealed spaces there is a 100% faster fire spread. Flames spread through web members.

The rate of fire spread inside a building's concealed space will be 100 percent faster than in concealed spaces of a building with conventional solid wood beam construction. Condominiums, town houses, and private homes throughout this nation are being built with lightweight wood truss construction held together with sheet metal surface fasteners. These roof and floor supports can be expected to fail more rapidly than solid beam constructions, which have nails penetrating several inches into the wood at connecting points.

Lessons Learned

It is not always possible during a fire operation to identify a building as having lightweight wood trusses. So, to safeguard the firefighters at a structure fire from collapse of any type of truss, everyone must be aware of the presence of the truss in the building. There is a trend in the fire service of requesting laws requiring the truss building to be marked. Hackensack, New Jersey, has truss buildings marked with a triangle. Cheasapeake, Virginia requires truss buildings to have a letter T and New York has two vertical lines to identify truss construction. However laws requiring truss marking, identifying truss construction in most cases only applies to commercial buildings. Laws requiring marking of building identifying truss construction, in most cases, exempt residence buildings. One notable difference are local marking ordinances being enacted in Bergen County, New Jersey. Unfortunately the problem of lightweight truss roof and floor collapse is in residence buildings.

So what is the fire service to do? The fire service can not depend on these laws to safeguard firefighters. Each fire department must identify truss buildings in their community, program them into the dispatch system, and notify fire responders of the truss danger before they arrive at the scene. Preplanned visits, close-up inspection of the truss design and the sheet metal surface fasteners, and, finally, the development of a defensive standard operating procedure based upon collapse potential are necessary safeguards for firefighting operations. For example, during inspections the truss constructed building address must be identified. This information is recorded and programmed into the dispatch system and when a fire call comes in for the address of the building, the first responders must be notified by radio while responding of the presence of truss construction.

Once on the scene of a fire in a truss constructed building, the incident commander must have a firefighting strategy for combating various types of fire inside. The recommended strategy for fighting a fire in a building with lightweight wood truss roof or floors is as follows:

- **Content Fire:** If fire is burning content such as a couch or mattress, standard operating procedures should be followed. Stretch an interior hoseline and extinguish the fire.
- **Structure Fire:** If fire involves the structure and is burning throughout the concealed roof or floor spaces, the people should be removed and exterior attack should be the strategy.

James Pressnall of the Irving, Texas, Fire Department is the first known firefighter to die after being trapped by the collapse of a lightweight wood truss, inside a burning two-story apartment house on February 27, 1984. This firefighter's death was the first of many more to come as lightweight wood truss construction spreads throughout the nation.

Since Firefighter James Pressnall died in 1984 there have been many other firefighter deaths due to collapse of lightweight truss construction:

- James Pressnall, Irving, Texas, 1984
- Todd Aldridge, Orange County, Florida, 1988
- Mark Benge, Orange County, Florida, 1988
- Alan Michelson, Gillette, Wyoming, 1990
- James Hill, Memphis, Tennessee, 1993
- Joseph Boswell, Memphis, Tennessee, 1993
- Strawn Nutter, Louisville, Kentucky,1994
- John Hudgins, Chesapeake, Virginia, 1996
- Frank Young, Chesapeake, Virginia, 1996
- Edward Ramos, Branford, Connecticut, 1996
- Brant Chesney, Forsythe County, Georgia, 1996
- Gary Sanders, Lake Worth, Texas, 1999
- Brian Collins, Lake Worth, Texas, 1999
- Phillip Dean, Lake Worth, Texas, 1999
- Lewis Mayo, Houston, Texas, 2000
- Kimberly Smith, Houston, Texas, 2000
- John Ginocchetti and Tim Lynch, Manlius, N.Y., 2002 (fig. 11-4)
- Cyril Fyfe and Kevin Olson, Yellow knife Canada, 2005
- Arnie Wolfe, Green Bay, Wisconsin, 2006

Fig. 11–4. Firefighters Tim Lynch and John Ginocchetti died when the floor collapsed in this private dwelling.

The trend toward use of truss construction is increasing. Collapse during fires in truss buildings that kill and injure firefighters will be increasing. This sad fact will not change, so the fire service must change. We must change the way we fight fires. The defensive firefighting strategy described above must be used for fires in truss constructed buildings. A veteran fire officer said, "Chief, we cannot use defensive firefighting when a fire involves a truss structure as you recommend."

My question for him was, "What would you say to me if you were the chief, and a firefighter in your department was killed by a collapsing truss?"

He did not have an answer. I did.

Note: For more information about truss construction, search the Web for: *Preventing Injuries and Death to Firefighters Due to Truss Construction NIOSH 2005-132.*

12

CEILING COLLAPSE

The restaurant manager is about to close up his store for the night when he sees smoke rising at the ceiling near the grease vent duct over the stove. He immediately notifies the fire department. The time is just after 1 a.m. Minutes later, the first-in company arrives at the fire building, a one-story brick-and-joist structure 60×75 feet in area, consisting of the restaurant, a hardware store, and a meat market. The first-arriving ladder captain and the forcible entry team are directed to the rear kitchen area of the restaurant by the manager, who meets them at the front door. The first-arriving engine company stretches to the front door of the restaurant, not committing the hose placement until the exact location of the fire is confirmed.

Inside the restaurant near the stove, the captain sees light smoke drifting down from a crack between an ornamental stamped tin ceiling and the vent exhaust duct. The sheet metal duct is designed to exhaust cooking vapor from a hood over the restaurant stove to the outer air. Penetrating the ceiling and roof, the duct might have conducted heat from the cooking vapor to a nearby combustible roof or a ceiling wood beam. The kitchen ceiling is an old, ornamental tin ceiling made of 2×4-foot strips of painted, decorative tin nailed together. The captain immediately orders a firefighter to open up the ceiling with a 10-foot pike pole. Since the lights are still on in the kitchen, the firefighter can determine the tin ceiling seam nearest the exhaust duct. Using the point of the pike pole, he punches up the tin ceiling's 2×4-foot sheet near the edge of the section he does not want to pull down, creating a split in the seam, and places the hook of the pike pole in the small opening in the seam. Catching the top of the tin ceiling section he intends to remove, he pulls the hook down with a short, sharp motion. One end of the tin ceiling section falls away from the roof beams and framework.

The captain glances up into the opening and sees bright orange flames vigorously fanned by the air movement in the cockloft. "Ladder 6 to Battalion 1," the captain calls on his radio.

"Go ahead, Ladder 6," the chief responds.

"Chief, we have heavy fire in the cockloft of the restaurant. Tell Engine 8 to bring the line in here. One of my men is going into the adjoining store to check the ceiling; we'll need another line in the store."

"Battalion 1, ten-four."

The captain radios the firefighter assigned to vent the roof, "Ladder 6 to roof, what are the conditions up there?"

"Roof to Ladder 6. Captain, there is fire coming through the roof around the exhaust vent over the restaurant. I am going to cut a vent opening in the roof deck around the vent."

"Ten-four," says the captain. "Ladder 6 to Ladder 6 chauffeur."

"Ladder 6 Chauffeur. Go ahead, Captain."

"Force entry into the adjoining store and pull the ceiling in the rear; I believe fire has extended above the ceiling over the store."

"Ten-four, Captain, will do," responds the chauffeur.

Meanwhile, inside the restaurant kitchen, with the engine company's hoseline charged and in position to operate, the firefighters begin the dangerous and difficult job of pulling down the tin ceiling. As they pull vigorously, razor-sharp sections of ceiling, long pointed nails, ceiling dust, and broken pieces of wood furring strips fall into the restaurant kitchen. The more ceiling area the firefighters pull down, the more flame they expose.

Outside, the chauffeur of the ladder company and another firefighter try the doorknob of the adjoining store. It is locked. The chauffeur examines the lock and sees that it is a typical cylinder type. He drops his hook to the floor and places a lock puller around the cylinder containing the keyhole. With the assistance of the other firefighter, he pulls out the cylinder and then uses a key tool to release the door lock. He enters the store and quickly walks to the rear of the butcher shop, where he starts to open up the ceiling near the partition wall separating this store from the adjoining restaurant. It is a 15-by-60-foot suspended ceiling. Vertical wood hanger strips, measuring 1-by-½ inch, connect a wood-grid framework ceiling to the roof joists. The ceiling sheathing is composed of acoustical tile squares glued to gypsum board. Flames already fill the concealed space above the ceiling. Suddenly, the entire suspended ceiling and framework crash down in one section and strike the chauffeur on the back of the head and shoulders. He is driven to the floor and pinned face down with his arms and legs outstretched. "What happened?" he wonders, trying to clear his head. It is pitch dark. "The ceiling collapsed on top of me!" he realizes.

There are no sounds of rescue, and the firefighter begins to panic. "They may not know I'm trapped here! The fire above the ceiling will start to increase in size now that the ceiling is down."

"Help! Help! I'm trapped! Get me out of here!" There is no sound. "I've got to get myself out of here," he says aloud. The firefighter attempts to push himself up using his hands and knees but he cannot budge the heavy ceiling above him. His right foot is wedged tightly beneath a large fluorescent light fixture attached to the fallen ceiling; his left leg has some space to move. There seems to be more space toward his left side, and he squirms sideways in that direction. He yanks his right leg, pulling his foot out of his boot. It is so dark he cannot even see the floor in front of him. Now he starts to smell smoke and hear the fire crackling above the ceiling. His helmet has been knocked off his head by the ceiling collapse. He has to get out from beneath the ceiling quickly, he thinks, or the flames will roast him alive. He crawls to the left but does not know if he is going in the right direction, toward the front of the store. He might be crawling toward the rear of the store, away from his rescuers. The space beneath the ceiling is getting larger (fig. 12–1). Rising up on one knee, the firefighter punches the ceiling with his fist, breaking open some fiber tile and plaster. He sticks his head through the hole in the sheathing and sees, through the smoke, the flashing red lights of the fire trucks out in the street. He ducks down again because of the heat and knows that the store is about to flash over. Quickly crawling beneath the fallen ceiling toward the direction of the store front, the firefighter strikes his head on pieces of broken ceiling tile and a light fixture.

Chapter 12 | **Ceiling Collapse**

Fig. 12–1. Voids allowing firefighters to escape are created near centers and shelves when a ceiling collapses.

    As he moves forward, the void beneath the ceiling grows smaller. Finally he can go no farther; he becomes wedged between the ceiling and the floor. Once again he tries to raise himself on his hands and knees and push the heavy ceiling up off his back. He cannot. He is trapped. "God, I'm finished!" he thinks. "I'm going to burn to death beneath this ceiling." He hears fire roaring above his head. "It flashed over!" Now he feels heat on his back, heat conduction through the fallen ceiling. Backing up, he moves to the rear of the store. The space beneath the ceiling here is larger and cooler, but he knows he is going away from his only avenue of escape.

    Suddenly he hears firefighters' muffled voices coming from the front of the store. "Quick, get that hoseline in here! Hit that fire! There's a firefighter trapped in there!"

    "Help! Help!" the chauffeur calls out, but his cries are drowned out by the rumble of the hose stream striking the under side of the roof joist above him. Again he crawls forward into the heated space, but hot water is dripping down through the ceiling. Suddenly he feels a hard metal object pushed beneath the ceiling on his right side. It is a portable ladder. Now he hears voices above the ceiling, shouting, "Okay guys, lift up this end of the ladder! One! Two! Three! Lift!" The ceiling next to the trapped firefighter begins to rise up and break apart as the far end of the ladder is raised by the firefighters. Quickly the chauffeur crawls out from under the ceiling and staggers free.

Seeing him, a firefighter shouts, "Hey, Captain, here he is! We've got him!" The rescuer grabs the freed firefighter as the latter stumbles and falls to one knee. "Are you all right?" he asks. The chauffeur nods his head. He has never felt more alive in his entire life!

Ceiling Collapse

The danger of ceiling collapse during firefighting is misunderstood. Because a ceiling is a non-load-bearing structural element and supports only its own weight, when it collapses during a fire it does not cause the collapse of other parts of the structure and is therefore placed low on a scale of structural importance in a building's framework. A ceiling collapse, however, should be placed high on a scale of collapse hazards that can cause death and serious injury to firefighters, but it is not. The danger of ceiling collapse is underestimated. Why? One reason is that, when a ceiling fails during a fire, firefighters below often are not killed or injured by the falling ceiling but by the ensuing fire or smoke. The fact that a ceiling collapses and traps them or prevents their escape somehow becomes lost in the investigation. So from a medical examiner's point of view, the exact cause of death is burns or smoke inhalation, from a firefighter's point of view, the cause of death is a ceiling collapse. Like a heavy wood or steel net, the ceiling drops over the firefighters and traps them inside the burning building.

Types of Ceilings

Ceilings can be divided into two broad categories: those that are directly affixed to the floor or roof joists above and those that are suspended several inches or feet below the joists by vertical strips of wood, steel hangers, or thin wires. A directly affixed ceiling, as found in older buildings, is a lath-and-plaster ceiling, which has strips of wood or wire lath nailed to the under side of the joists and several coats of wet gypsum plaster applied to the underside of the lath. In newer buildings, a 4×8 square foot dry gypsum board that is $1/4$- to $3/8$-inch thick, is nailed directly to the underside of the joists.

There are other ceiling types directly attached to the joists, such as ornamental tin and wood paneling, but wet or dry gypsum board is the most common (fig. 12–2). During firefighting operations, a directly affixed ceiling usually collapses in small pieces rather than in one large section. This type of collapse does not usually cause serious injury, since only the plaster sheathing falls, not the wood or metal lath above it. The collapse of a directly affixed ceiling is often caused by water absorption into the plaster from a hose stream. As firefighters move through an involved structure, they direct their stream at the upper levels of the fire room to cool the atmosphere before advancing. While some of the plaster is broken and knocked down by the force of the stream, much of the water is absorbed by the remaining plaster ceiling. As the hose team continues advancing, the water-soaked hot plaster ceiling sometimes drops away suddenly from the lath, striking the firefighters on the head and shoulders.

Chapter 12 | **Ceiling Collapse**

Fig. 12–2. A ceiling attached directly to the floor above

When the plaster is only 1 or 2 inches thick and when the firefighter is equipped with a properly fitted helmet, there is usually no serious injury. But, in some older structures, where plaster ceilings are 3 or 4 inches thick and 15 to 20 feet in height, the collapse of a large area of thick, heavy plaster can knock a firefighter unconscious or cause a serious head or neck injury. Firefighters struck by large, heavy sections of ceiling plaster, unaccounted for after searching or venting operations, have been found unconscious, lying face down in several inches of water. The degree of injury caused by a local collapse of a plaster ceiling is determined by the thickness of the falling plaster, the height of the ceiling from which it falls, and the condition of the cushioning system inside the firefighter's helmet.

Suspended ceilings

The most dangerous ceiling is a suspended ceiling. Suspended ceilings are most often found in three types of occupancies: stores, top floors of multiple dwellings, and renovated buildings. These suspended ceilings sometimes called "dropped ceilings" or "hanging ceilings" are constructed several inches or several feet below the supporting roof or floor joists above, the ceiling hangs below the joist, connected by vertical hangers that may be strips of wood, steel, or thin wire. The reason for a suspended ceilings in stores is to cover the prior out-of-fashion old ceiling above (fig. 12–3). Instead of removing the old ceiling, a new ceiling is installed to cover the old one.

The suspended ceiling in the top floor of a multiple-floor dwelling is to create a space to insulate a top floor from sun rays. Suspended ceiling in renovated buildings can be found on every floor in every apartment. The suspended ceiling in renovated

Fig. 12–3. Suspended ceiling are found in renovated buildings on all floors.

Fig. 12–4. The grid work and framing can trap a firefighter if the ceiling collapses.

buildings is to reduce the floor height in order to conserve the energy required to heat and air cool the apartment. A suspended ceiling presents a serious collapse hazard to a firefighter when it fails during a fire, because when the hanger or vertical supports are destroyed by flame, the entire grid framework, in addition to the ceiling sheathing, collapses. This ceiling grid system can weigh several thousand pounds and trap a firefighter beneath the ceiling or inside a void. A firefighter caught below a collapsing suspended ceiling may be lucky and become trapped in a void created by furnishings and crawl out from under the ceiling. Or the firefighter may be unlucky and be pinned to the floor, or be trapped inside a small sealed void space at the rear of the collapse, or be pinned beneath the heavy ceiling grid. The firefighter could be strong enough to break though the collapsing ceiling framework and then become engulfed in fire (fig. 12–4).

There is often a subsequent rapid fire increase after a suspended ceiling collapse, and firefighters pinned by the ceiling or trapped inside a sealed void or breaking though the falling ceiling can be burned to death or asphyxiated by the toxic smoke. Firefighter Kevin Kane of FDNY, Ladder Company 110, was caught and trapped by a collapsing suspended ceiling in a top floor fire in a vacant multiple dwelling. While pulling ceilings to check for fire in the space above a suspended ceiling in an adjoining apartment, the ceiling and grid collapsed. He broke through the collapsing ceiling framework, was engulfed in flame, and killed (fig. 12–5).

Fig. 12–5. FDNY firefighter Kevin Kane of Ladder 110 died when a suspended ceiling on the top floor of this multiple dwelling collapsed.

There are three common types of suspended ceilings: a wood grid (fig. 12-6), a metal grid with a permanently affixed ceiling (fig. 12-7), and a lightweight metal grid system with a permanently affixed ceiling, and a metal grid system with a removable panel ceiling (fig. 12-8). The part of a suspended ceiling that traps or pins a firefighter inside a burning building is the grid system or framework behind the ceiling sheathing. The grid system is the heaviest part of the ceiling. The firefighter may break through a plaster or panel ceiling, but he will not be able to penetrate a grid system of a 2×4-inch framework of screen of wire lath or expanded metal lath. Dropped suddenly over a firefighter inside a burning store, this heavy net of wood or steel is a deadly trap.

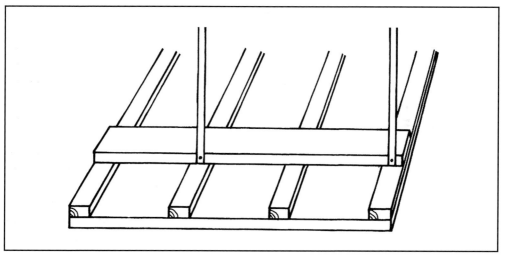

Fig. 12–6. A suspended ceiling with a wood grid system

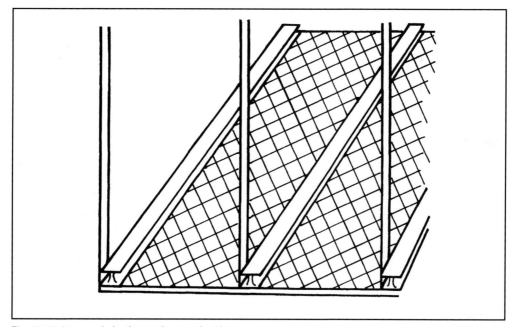

Fig. 12–7. A suspended ceiling with a metal grid system

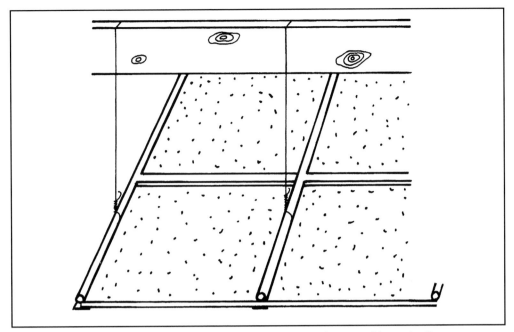

Fig. 12–8. A suspended ceiling and a lightweight metal grid system with a removable panel ceiling

Of the three types of suspended ceiling, the lightweight metal grid system with removable ceiling panels does not create a crushing or pinning hazard but can entrap a firefighter in wire and metal framing. The supporting wire and thin metal framing is much lighter because the removable acoustical ceiling panels it supports are much lighter. In Houston, Texas, the roof and ceiling of a McDonalds restaurant collapsed during a fire and killed firefighters Lewis Mayo and Kimberly Smith. When firefighter Smith's body was found, she had a pair of wire cutters in her hand. Investigators believe she was cutting her way though wire and cable that came down with the ceiling.

The weight of a suspended ceiling determines the seriousness of the entrapment hazard. The heavier the ceiling, the greater danger of entrapment, and the more difficult and time-consuming the rescue effort. Architects' manuals specifying average weights of building materials show that a lightweight metal grid panel ceiling weighs one pound per square foot, while the heavier metal grid system, permanently affixed wire lath ceiling weighs ten pounds per square foot. Such a heavy wire lath ceiling in a 20×50-foot store will weigh 10,000 pounds.

To reduce the danger of suspended ceiling collapse, firefighters must prevent fire extension into the concealed space above it by extinguishing the blaze as quickly as possible, and check the concealed space above the ceiling as soon as possible for fire spread after the fire has been extinguished.

Fire can enter above a suspended ceiling in many ways. Flames may burn through a ceiling directly above a room content fire, or burn through a poke-through hole created by a light fixture, or pipe recess. When the concealed space above a suspended ceiling is used as a plenum for heating and cooling, fire may enter the louvered openings in the ceiling that transfer air to or from the plenum and attack the ceiling's vertical supporting hangers. Or, in a suspended ceiling with holes for recessed lighting fixtures that are not fire-stopped, fire from a room below can enter the concealed space above through the recessed light fixture openings. Moreover, in a metal grid suspended ceiling with removable panels, the panels are sometimes removed for repairs to utilities located in the concealed space above the ceiling. These panels are not always replaced, or they are replaced improperly, so that when a fire occurs below the ceiling it quickly spreads upward.

To check for fire spread inside a ceiling space, firefighters should open the ceiling from below with a pike pole and examine the spaces above. In some instances, they will discover several layers of concealed spaces. When checking a ceiling space for fire spread, firefighters must pull down a section of every layer of the ceiling. Some old commercial occupancies have two ceilings and two concealed spaces: There is a concealed space behind the original ceiling of ornamental tin or lath and plaster, and a concealed space above an added suspended ceiling of a lightweight metal grid system with removable panels. This was the ceiling construction in the restaurant fire that killed Boston firefighters Warren Payne and Paul Cahill. A firefighter pulling down a section of only the lowest ceiling must check for flames spreading undetected in the concealed space created by the original ceilings. If the hangers are destroyed by flames, one or both ceilings can collapse on the unsuspecting firefighters inside the building.

When a second suspended ceiling is constructed, numerous holes are punched through the first ceiling in order to connect hanger straps from the lowest ceiling to the roof or floor joists. These holes create a "honeycomb" effect between the ceiling and the concealed spaces and are avenues for vertical fire spread through all three ceilings. Once fire enters the concealed space behind the lowest suspended ceiling, it is certain to spread to the other spaces. Ceiling alterations are not constructed according to building or fire code standards or regulations. Suspended ceilings are often built by amateurs, and therefore the construction should be suspect. Post-fire investigations have revealed several suspended ceilings constructed one below the other and the hanger supports for each ceiling connected not to the joists but, improperly, to the ceiling immediately above. As a result, vertical support hangers designed to support only the first ceiling were found supporting two or three lower suspended ceilings, dangerously overloaded. When multiple suspended ceilings are discovered, the incident commander should be notified. This ceiling construction makes fire extinguishment much more difficult (fig. 12–9).

Metal hangers and framework holding up suspended ceilings

In 2007, New York City builders wanted to substitute lightweight wire to support ceilings instead of the more sturdy black iron, ¼-inch diameter rods and 1½-inch carrying framework that was required by the New York City building code. Builders wanted to replace this heavier, more costly ceiling support system with lightweight, cheaper thin wire. The FDNY considered this as another example of the trend towards lightweight building construction, which means early collapse during fire. The fire service considers fire resistance to be directly related to *mass* (bulk) of a structure. When we remove mass in a building, in this case ceiling supports systems, we remove fire resistance. The fire department Uniformed Fire Officer's Association and the wire lather's union did not want thin wire substituted for black iron in suspended ceilings. They fought the proposed change and won. New York City ceilings are still firmly supported with black iron hanger strips.

Fig. 12–9. When fire enters a ceiling with two dropped ceilings, fire extinguishment is difficult to nearly impossible.

The current ceiling support system of using black iron ¼-inch metal rod and 1½ inch carrying channels has prevented progressive ceiling collapse in New York City over many years. Without the heavier ceiling wire and framing, a partial collapse of a ceiling using thin wire could spread from wire hanger to wire hanger, resulting in the total collapse of a ceiling or a disproportionately large part of it. The black iron rod and frame ceiling support system of the current New York City Building Code creates an important safety feature. It redistributes loads during the partial failure and the robustness of this ceiling system provides additional strength to ceiling construction during a fire. It helps handle the impact of the firefighter's hose stream impact or the weight of the water build-up from sprinklers.

Ceiling construction is important to the safety of firefighters so the Uniformed Fire Officers Association Local 854 and the Wire Lathers Union Local 46 teamed up to challenge the change and won. City hall ruled in favor of the FDNY and the Wire Lathers Union Local 46.

The discussion points brought to City Hall on behalf of firefighter safety were:

- Ceiling failure is one of the leading causes of firefighter death by collapse. This according to a 10-year study, (1990 to 1999) conducted by the National Fire Protection Association.
- Six New York City firefighters died by ceiling collapse: Earnest Marquard, Joseph Beetle, Stanley Skinner, Thomas Earl, Kevin Kane, and Peter McLaughlin. These were not the strong suspended ceilings using black iron

rod and 1½-inch channel framework currently in use that collapsed. The ceilings were wood frame ceilings.

- Strong metal ceiling supports are hailed by the fire service because firefighter's hose streams must be directed toward a ceiling where heat is highest and the impact of the hose streams could collapse a ceiling.

- Firefighter's hose streams can discharge 250 gallons nearly one ton of water per minute. At some fires, the water weight is trapped above the suspended ceiling causing the ceiling to collapse.

- Firefighting strategy requires firefighters to pull down part of a ceiling after each fire to check the concealed space above for fire extension. When firefighters must remove part of a ceiling to check above for fire spread and the ceiling supports are not robust, the entire ceiling may collapse.

- Flame and smoke often spread upward through the ceiling. Once fire enters the space above a ceiling, flame may spread unseen throughout an entire building in concealed spaces on each floor. It is important to have strong metal ceiling supports in the space above a ceiling.

- Combustible ceiling electrical cable insulation increases the chances of ceiling fires. Electricity demand can overheat wires. Also cable fires are started by short circuits and/or arcing. Millions of miles of electric cable covered with combustible rubber or plastic insulation have been introduced into the voids above suspended ceilings, in plenums and interstitial spaces over the past 20 years to service our computers, fax machines, telephones, scanners, and printers. This has increased the fuel loading inside spaces hidden above ceilings. The weight of the new electric cable requires strong ceiling supports.

- In many instances, old cable is not removed when new cable is installed above a ceiling. The old cable is cut and left behind, abandoned in the space above a ceiling. This adds to the fire load and weight burden place on ceiling supports.

- Under the current building code, the required suspended ceiling support systems of ¼-inch rod hangers and 1½-inch ceiling channel framework holding up suspended ceilings have saved firefighters' lives. The fire service does not want lightweight thin wire supporting suspended ceilings—it can fail fast and collapse ceilings on firefighters.

Lightweight thin wire in a suspended ceiling system allows early suspended collapse and introduces more wire that can entangle and trap a firefighter. Firefighters can become entangled in these the mazes of thin wires supporting the suspended ceiling. The wire catches on protruding parts of a self-contained breathing mask, such as the air bottle, the mask regulator, and the mask harness. A firefighter caught in wire after a ceiling collapse will not be able to escape a room that flashes over. Firefighters train for wire entrapment. A procedure FDNY firefighters use to prevent wire entrapment is a hand movement that simultaneously opens both buckles of a mask to let the equipment fall to the ground. Then the firefighter must escape the burning room quickly.

Rescue of Firefighters Trapped Below Ceiling Collapse

The ceiling collapse that is killing New York City firefighters is the wood grid suspended ceiling. Immediately after a wood grid suspended ceiling collapses, there is sometimes a rapid increase in fire in the space directly over the fallen ceiling. When a ceiling collapses in one large section, the air space below the ceiling is momentarily compressed and the air space above the fallen ceiling is briefly placed under a vacuum. When a fire is smoldering in the space above a suspended ceiling and the ceiling collapses, a large inrush of fresh air enters the space above the fallen ceiling, fills the vacuum, and is mixed with superheated gases and heat which were previously confined to a concealed space above the ceiling. An explosive fire increase may occur. If the grid system of the suspended ceiling is constructed of wood framework, the fire will be fueled by the fallen ceiling grid on the floor and by the joists above.

Firefighters trapped below this ceiling will have very little time to be rescued before the fire and smoke reach them. The most important first step for the rescue of firefighters trapped beneath such a collapsed ceiling during a store fire is to sweep the collapse area above the ceiling with a hoseline and knock down the flames. If a suspended ceiling at the top floor of a multiple dwelling collapses and fire increases, a firefighter may become trapped at a window calling for help. A ladder should be ready at every fire to rescue trapped victims or firefighters suddenly trapped by ceiling collapse and fire. If rescue attempts are begun without extinguishment efforts, the trapped firefighters may succumb to the products of combustion before they are reached. One fire officer, half buried and trapped beneath burning rubble, told the rescuers digging him out, "Never mind digging me out of here! First cool me off with the hose stream—I'm burning!"

Lessons Learned

A company officer can direct the following precautions to protect firefighters against the dangers of ceiling collapse:

1. Train firefighters to recognize occupancies that have dangerous suspended ceilings, such as stores, top floors of multiple dwellings and renovated buildings.

2. Do early identification of suspended ceilings. The officer can order that the ceiling be opened and examined as soon as possible. Before committing firefighters to a serious fire in a large-area occupancy, the officer should know the type of ceiling in the structure. If firefighters discover a suspended ceiling, they should immediately communicate this information to the incident commander in charge of the fire.

3. Do proper ceiling examination with multiple ceilings. When checking a concealed space above a suspended ceiling for fire spread, the firefighter should open the ceiling and be able to see the roof or floor joists above. If there are several suspended ceilings, the concealed space behind each ceiling must be examined before an accurate assessment of fire conditions can be made. If the fire is spreading, a defensive strategy should be considered.

4. Be aware of ceiling overload caused by hose streams. When a suspended ceiling's vertical hanger strips have been weakened by flames, the additional weight of water from hose streams directed into a concealed space to extinguish a spreading fire can contribute to a collapse. One gallon of water weighs about 8 pounds. A hose stream discharging 250 gallons per minute is pouring more than one ton of water a minute into a ceiling.

5. Use proper methods of opening a ceiling with a pike pole. A fire-weakened suspended ceiling can collapse if it is not opened properly. If several firefighters using pike poles pull down on the grid system instead of on the ceiling sheathing, or if the pike pole pulls down the furring strips instead of the wood lath, they can pull the entire ceiling down from its supports. This action may be necessary in some overhauling situations; however, if other firefighters are overhauling at the rear of a store and the ceiling is collapsed by firefighters near the front, the collapsed ceiling and resulting fire will trap firefighters in the rear.

6. Make use of collapse shelters. A firefighter operating at a serious building fire with the possibility of a suspended ceiling collapse should attempt to stay near large pieces of furniture or stock that can serve as shelters or voids if the ceiling falls. An open floor area or empty store represents a dangerous environment because, if the ceiling collapses, a firefighter operating below will be pinned to the floor and receive the full impact of the heavy ceiling. Serving counters extending from the front entrance to the rear of stores have saved the lives of many firefighters caught beneath collapsing ceilings.

7. Practice shoring techniques. If firefighters must work beneath a suspended ceiling that has partially collapsed or is sagging badly, they should use a portable ladder to shore up the weakened ceiling temporarily. The vibrations of overhauling could cause a partially collapsed ceiling section to pull down the entire ceiling suddenly.

8. Practice safe ceiling dismantling and rescue. When a suspended ceiling collapses, the grid system must first be cut apart with a power saw, and then the ceiling sheathing must be cut or pulled apart. Firefighters may then crawl beneath the ceiling into shelters to search for victims.

9. Tunnel beneath collapsed ceilings with portable ladders. To examine areas below collapsed ceilings quickly, firefighters can sometimes use a portable ladder as a raising lever. One end of the portable ladder is slid under a section of ceiling and the other end raised by several firefighters. This technique may allow firefighters to crawl beneath a fallen ceiling or break up a ceiling if the grid system is lightweight.

10. Use hoseline protection. To protect against an increase of fire after a ceiling collapse, firefighters should quickly position a hoseline to the area of the collapse. If rescuers can operate and there is no fire increase, those handling the hoseline should stand by; if there is an increase, the hoseline should be used to control the fire and prevent smoke generation.

11. At a fire in the top floor of a multiple dwelling or a renovated building, a firefighter should remain at the turntable of a ladder to be ready to rescue

a firefighter trapped by a ceiling collapse who comes to a window to escape the ensuing increase in fire.

12. Be aware of other avenues of rescue attempt. At a ceiling collapse in a store if firefighters are trapped in shelters or voids at the rear of a store beneath a collapsed suspended ceiling, rescue attempts should be made from adjoining stores. Partition walls, separating stores in a large one-story building, can be penetrated faster than rear masonry walls. Windows or doors the rear of a store are usually bricked up or securely locked to prevent theft. Hoselines should be positioned into adjoining stores to prevent fire spread. Entry into a store beneath a ceiling collapse can sometimes be achieved from below. Going down an outside cellar entrance may then allow access to the store via an interior cellar entrance.

13. Maintain control of the rescue attempt. After a collapse, the officer in command should try to limit the number of firefighters participating in the rescue attempt, as their combined weight can be greater than the weight of the fallen ceiling and can crush the firefighters trapped below.

12. When the stability of a suspended ceiling is in question make an inspection opening in the ceiling with a pike pole near the entrance. Check for fire in the space above the suspended ceiling before you enter the room. If the ceiling suddenly collapses you have a better chance of escape.

Note: For more information about ceiling collapse, search the Web for: *Firefighter Fatality Investigations F2003-18*.

13

STAIRWAY COLLAPSE

At 3 a.m. an FDNY engine and ladder company turns into a street lined with three-story brick-and-joist row houses. A fire has been reported in the third-floor public hallway of 108 West Fifteenth Street, but no smoke or flames are visible anywhere on the block. As the two companies slowly move down the deserted street, the engine company officer shines his spotlight on the front doorways of the similar, attached houses to locate the reported address. He spots the number and quickly dismounts from the pumper, leaving the driver of the vehicle to move it up 20 feet and give the ladder company room for aerial ladder positioning.

The engine captain enters the fire building with the ladder company lieutenant and the forcible entry team. The ladder company roof man, assigned to do vertical venting, enters an adjoining building to gain access by crossing over the party wall between buildings. From his fire inspection yesterday, the roof man knows that these buildings have a bulkhead structure and a door at the top of the interior stairway. Another firefighter, assigned the "outside ventilation" position, goes to the rear of the building. From here he will report the presence of fire spread and any trapped victims and, if a rear fire escape is present, vent windows of a rear room while the engine company moves in for extinguishment. The aerial ladder operator stays with the truck and prepares for venting, ladder removal of a fire victim, or a firefighter in trouble.

As the two officers go into the front entrance doors to the public hallway, they see a red glow reflecting down the old wooden stairs leading to the upper floors. Burning wood crackles on the floors above, and sparks are falling down the stairwell (fig. 13–1). The captain orders a hoseline stretched into the building and tells the pumper to go for a hydrant. Climbing the stairs to the top floor, the officers find the third-floor public hall in flames. This incident is not a typical hallway garbage fire. There is no smoke, the walls and floor are on fire, and the familiar smell of burning garbage is absent. From the officer's position on the wooden straight-run stair, it is not clear whether the fire has burned out of a top-floor apartment or is limited to the public hallway.

The engine captain calls to the firefighters below, "Do we have enough hose to make an apartment?"

"Yes!" is the reply from the dark hall.

"Engine 5 Command to Engine 5—start water!" calls the captain on his Handy-Talkie.

Fig. 13–1. A collapsed wooden straight-run stair

"Ten-four," responds the pump operator. Firefighters don masks as water begins rushing through the hose nozzle. While the captain supervises, Joe, the nozzleman, proceeds up the stairs quickly, sweeping the hose stream from side to side to extinguish the flames. The captain and Joe have worked together for four years; Joe was assigned to the captain's group because he possesses the right combination of experience, good judgment, and aggressive firefighter instinct. At this fire, however, the captain thinks to himself, "Joe always has to be on top of the fire. Why won't he use the reach of the hose stream to extinguish the fire from this position on the steps?" But, since it appears safe for Joe's advance, the captain does not hold him back. Together they move up to the top-floor landing, make the turn to the front of the landing, and extinguish fire as they move forward. The roof man reports that the bulkhead door was open so he did not have to force it. The fire is quickly extinguished.

An occupant of one of the apartments, hearing the noise, opens his door to investigate and then quickly shuts it to avoid the steam and the smoke. He shouts through the door that there is no fire in his apartment. Looking for hidden fire, the ladder company officer starts up the burned stairway to the roof bulkhead door, which is still smoldering. The nozzleman sprays water into a charred hole in the floor near the base of the stair. The captain recognizes the hole as a sign of arson and surmises that someone has splashed a flammable liquid down the steps from the roof and fled over the rooftops. As Joe drives the nozzle stream into the burn hole in the floor, heavy smoke hangs in the hall, making visibility poor.

Suddenly there is a crack, a loud crash, and a violent shaking of the stair landing. Everyone freezes in place, not knowing what has collapsed. When the smoke clears it becomes apparent that the entire stairway leading to the roof has crashed down upon the stairway below. The ladder company lieutenant is hanging from the bulkhead doorsill by both hands, calling for help. The roof man quickly grabs the officer by both wrists and pulls him out of the doorway on to the roof.

Shouting, "Someone might be under the collapsed stair!" the engine captain orders everyone to lift up the fallen staircase. All hands raise one end of the stairs, and it is revealed that, miraculously, no one has been crushed beneath the quarter-ton structure. As they move the fallen staircase and see the red-hot ashes fall out of the broken wire lath and plaster soffit, firefighters find that fire has also burned inside the concealed space of the fallen staircase. Inside the stair section, the wooden supports, called *carriage beams*, are glowing red. Extending from the double trimmer beams of the third floor to the double trimmer beams of the roof below the doorsill, the 3×10-inch carriage beams are burned away at the bottom and top ends. More hot ashes fall out of the broken staircase as the firefighters pull it apart for extinguishment. Had a firefighter been caught below the fallen stair, he would not only have been seriously injured by the collapsing stairway but would have been seriously burned by the red-hot carriage beams pressing down on him. The arsonist had kicked out the top riser, poured flammable liquid inside the staircase, and spilled it down the steps and the hallway.

Stairway Collapse

At the turn of the century, stairway fires in multiple-story dwellings were common. Flames burned through a wooden apartment door or a glass transom and extended out into the stairway hall on a lower floor. If the wooden stairway in the hall ignited, the flames would spread vertically up the stair enclosure and extend to each apartment, starting first at the top floor. By the time the stair was involved in fire or had collapsed, firefighters would arrive and people would be jumping out of the windows. Extension and aerial ladders would have to be raised to all floors for rescue and hoseline advancement. At row houses, where access to the rear is difficult, jumpers and fire spread in the back yards were commonly overlooked.

Today, the stair halls of most multistory dwellings have been given fire protection by building codes. The stair hall enclosure walls must have a fire-retarding rating, glass transoms over apartment doors are illegal, old wooden apartment doors have been replaced with fire-rated doors and frames, and doors are required to be self-closing so that when people run out of their burning apartment, a spring hinge on the door closes it automatically. Fire is confined inside the apartment and kept from extending into the stair hallway where other people are leaving the fire building. The stairway structure itself has also been fire-protected. The under side of the wooden staircase has been protected from flames by a fire-retarding soffit, usually made of wire lath and plaster. (Wooden soffits are illegal.) The balusters supporting the stair railing and the stringboard covering the ends of the steps are required by most building codes to be of noncombustible material.

One of the reasons that some fire departments require the initial firefighting hoseline to be positioned in the interior stairways is because building codes have protected the interior stair enclosure with fire-retarding materials. The stairway is the safest area in the wooden interior of the building and logically the best avenue for initial fire attack. Unfortunately, most building code changes and fire-protective innovations are not retroactive, so many old wooden stairs are still with us. Also, building codes do not affect private houses. The initial interior attack hose team in a private house does not have a fire-retarding stair enclosure from which to operate. This fact is one reason why rural and suburban fire departments do not always require initial hoselines to operate from the interior stairway leading to the front entrance door.

Types of Stairs

There are three basic stair types:
- A straight-run stair (fig. 13–2)
- A U-return stair (fig. 13–3)
- An L-shaped stair (fig. 13–4)

A straight-run stair is most often constructed of wood or concrete, and its steps provide a 45- or 60-degree angle with an uninterrupted passage between two floors. In multistory buildings, straight-run stairs are most often positioned directly over one another. A person walks up one flight of steps to the floor above, then reverses direction, walks along the hallway to the foot of the stairs leading above, reverses direction again, and starts up the next flight of stairs. These are "stacked" straight-run stairs. In some old multistory commercial buildings, straight-run stairs were not stacked or positioned directly over each other. Straight-run stairs in these turn-of-the-century buildings extended from the first floor to the top floor in one continuous flight of stairs. At each floor level, there was an intermediate landing. An entry door to a commercial occupancy was located off to the side of the intermediate landing. This type is a "continuous" straight-run stair. A person would walk up a flight of steps to the second-floor level, take several steps across on an intermediate landing, and continue on up the next flight of steps.

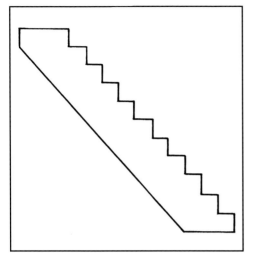

Fig. 13–2. A straight-run stair   **Fig. 13–3.** A U-return stair

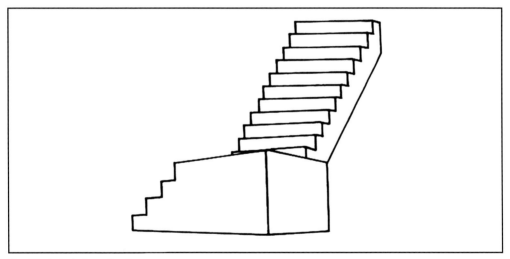

Fig. 13–4. An L-shaped stair

An L-shaped stair is most often constructed of wood and found in private homes. It is a two-section stair which changes direction at a 90-degree angle and has one section longer than the other. An L shape may be extended to a U shape. A U-return stair, most often constructed with a steel frame, is the type of stair preferred in modern building construction. It reduces the amount of floor area needed for the stair enclosure, thereby adding valuable floor space for commercial or residential use. A U-return stair breaks up the stair climb at the middle of the stair rise with an intermediate landing; here, the climber reverses direction before starting the second half of the stair rise.

Stair Construction

Firefighters often express conflicting views on the stability of stairs during a fire. Some say that a stair is the safest place for a firefighter during a building or floor collapse because, though the entire floor can crumble and fall, the stairway will remain intact. Others relate actual accounts of firefighters riding down collapsing, burned-out stairs, crashing down one on top of one another; of intermediate landings collapsing and dropping firefighters carrying tools several floors below; or of heated stair steps breaking under the weight of firefighters, causing kneecap injuries or leg burns. Which view on the stability of stairs is correct? Both are correct. Whether or not a stair will collapse with a floor depends on the stair design, Some buildings contain, in a three- or a four-sided masonry enclosure, stairs that are supported by bearing walls independent of the building's floor structure. If the floor collapses in such a building, the stairway will remain intact and the firefighters standing on it will survive.

Straight-run stairs are not enclosed by masonry walls but are supported by floor beams extending up through openings in each floor—these stairs will collapse if the floor collapses. The lower end of the stacked straight-run stair is supported by the double trimmer beams of the lower-floor opening; the upper end of the stair is supported by the upper-floor opening's double trimmer beams. As a general rule, U-return stairs found in masonry enclosures are independent of the floor structure and will not collapse during floor failure. Straight-run and L-shaped stairs are built through floor openings in the center or the side of the structure and, dependent on the stability of the floor for support, will collapse with the floor.

Stairs may be constructed of wood, steel, or masonry. A wooden, stacked, straight-run stairway will collapse in one section (as described at the beginning of this chapter) if the carriage beam supports are weakened by fire at the top or the bottom, where they are fastened to the floor's double trimmer beams. Carriage beam supports are the wood beams holding up the stairway (fig. 13–5). If the soffit or underside of the stairway is unprotected, flames from below may attack the carriage beams of the stair and a collapse of the entire stair section may occur. Also, if flammable liquid is splashed down the steps of a wooden staircase, some liquid will leak into the concealed space containing the carriage beam supports and destroy them, undetected by firefighters working nearby. Some wooden straight-run stairs may conceal their true combustible structure. Treads can be masonry, and stringboards and railings steel, but the interior stair supports can still be wood.

Steel-framed U-return stairs present a collapse danger different from wooden stairs. The most serious collapse injuries on such stairs have occurred to firefighters when the intermediate landing collapsed. The steel frames of risers, stringboards, and railings rarely collapse. When heated by fire, however, the stone masonry intermediate landings often collapse under the weight of the firefighter stepping on to them. When the falling firefighter, his tools, and the broken chunks of masonry crash down on to the landing below, their impact often collapses this lower intermediate landing. The severity of injury and the survival of the firefighter depend upon the number of consecutive intermediate landings that collapse as he falls. Firefighters have crashed through four and five stories of collapsing stone intermediate landings during a fire.

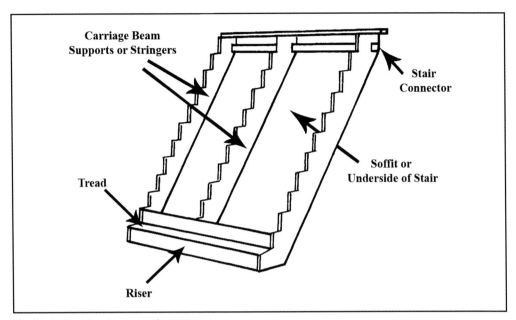

Fig. 13–5. A carriage beam wooden stair support

The steel U-return stairs that sustain most intermediate landing collapses are those which have treads and landings constructed of stone unsupported by a steel pan. The stone treads and intermediate landings are laid into a steel frame supported only by a 1-inch, angle-iron edge around the rectangular or square open area of the tread or landing area of the stair. There may be an additional strip of steel support across the center of an intermediate landing, but the structural support of the stair tread or landing is mostly the 2-inch thickness of flagstone or marble. If the stone breaks, that part of the stair collapses. When a firefighter running up a steel U-return stair to a fire floor glances up at the underside of the stair, it can be determined whether the treads or the landing may collapse. On a U-return stair, there are no soffits, so the undersides of the steps are exposed. If the firefighter sees the underside of a stone tread or landing, he or she should be alert for a possible stair collapse. Some steel stairs have the underside of a stone tread or landing completely supported by a reinforcing pan; these stairs do not collapse.

Although the intermediate landing collapse is the most serious danger to a firefighter, the failure of a stone stair tread or step is the most frequent cause of collapse of a U-return steel stair. Stone treads heated by fire crack and shatter. In some instances, they remain intact until the weight of a firefighter is placed upon them and then they suddenly fail. The firefighter's foot and leg fall through the tread opening, and the kneecap strikes the stone tread above and is broken. The steel risers above and below the fallen stone tread may be red-hot and burn the firefighter's knee or thigh. The severity of the burn will depend upon how quickly the firefighter can withdraw the shattered leg from between the hot metal sections of the steel stair.

Lessons Learned

Firefighters must use stairways during interior fire operations. The following are warning signs or safety actions firefighters can take to prevent a stair collapse injury:

- A firefighter should size up a stairway in a burning building to determine if it will be an area of refuge during a floor collapse, such as a masonry-enclosed, steel-framed U-return stair; or if it will collapse with the floor, such as a wooden straight-run stair extending through an opening in a wooden floor.
- Wooden stairs without a fire-retarding soffit are vulnerable to destruction from flames igniting the underside of the steps, resulting in an early collapse. Wooden cellar stairs are often constructed without soffits.
- Flames coming from within a wooden stair or burning on wooden steps may indicate serious, hidden, structural fire damage to the stair. Stone treads on steel-framed stairs may be cracked or shattered by heat and collapse on top of firefighters climbing below.
- When climbing a sagging, fire-weakened wooden stair, firefighters should stay close to the wall. If the outer carriage beam has been destroyed and is not load-bearing, the stair rise is supported at the wall side only. Weight close to the wall will put less stress on the stair than walking near the outer part, farthest from the wall.
- When climbing a steel-framed, U-return stair with stone treads, a firefighter should place the middle of his or her foot on the stone tread directly over the riser below and not allow the stone tread to support the weight, while avoiding stepping on the intermediate landing, if possible. The firefighter should hold onto the banister and step from a spot above the riser of the lower stair section's topmost tread to a spot above the riser of the upper section's lowest tread, and then swing around.
- When an interior stair is used as a point of hoseline attack, firefighters should avoid bunching up on the stairway. The officer and nozzle firefighter can operate a hoseline near the top of the stair while a firefighter on the floor below feeds them hose. As soon as fire conditions permit, they should move off the stair on to the fire floor.
- If a stair appears in danger of collapse, firefighters should place a ladder over the weakened stair rise and climb into the ladder rungs. They should ensure that the ladder spanning the stair rise is supported by the floor below and the floor above, so that if the stair collapses the ladder will remain in place. If a ladder cannot be used to span an interior stair rise because the ladder cannot be secured on the floor below at the base, they should gain access to the floor above by an outside ladder positioned at a window.
- When overhauling in a building that has a wooden straight-run stair extending through an opening in a wooden floor, firefighters should be aware that the wall studs of partitions below header beams or double

trimmer beams of the stair opening may be supporting these primary floor reinforcements. If these wall studs are removed or destroyed by fire, the floor above may collapse, causing a secondary stair collapse.

- Intermediate landings made of stone or marble on steel-framed U-return stairs most often collapse under the weight of firefighters descending the stair. The impact of a firefighter's weight suddenly stepping down on the landing creates more stress on the stone than his weight slowly applied when climbing up.

- Stair railings may collapse when excessive pressure is applied from the side by firefighters bunching up on a stairway. Firefighters pushing on a stair railing for support can fall from the top steps when a stair railing collapses.

14

FIRE ESCAPE DANGERS

A firefighter climbs down the gooseneck ladder from the roof to the top floor fire escape landing at the rear alley of a six-story tenement. The top floor rear apartment needs venting. Once he's on the fire escape platform, he starts to break the windows. Bracing himself on the cramped metal fire escape he winds up swinging a Halligan tool. Stepping backwards, he suddenly starts to fall through the fire escape opening. There is no fire escape landing where he planted his boot. So he falls down through the fire escape's sixth-floor opening. Dropping his tool, while falling, his body strikes the fifth-floor fire escape's waist-high outside railing. Toppling over the railing, he descends down toward the back alley, now outside the fire escape. At the fourth-floor level, his free-falling body strikes a clothes line, strung between the railing of the fire building and the back of another tenement. It breaks under his falling body weight. Falling past the third-floor fire escape, there is a telephone wire spanning the rear alley, from the fire building, to the back of a building across the alley at the second floor level. The telephone wire is reinforced by a steel wire spiked into the brick walls of both buildings. The firefighter's body hits this horizontal cable and wire at stomach level full force and he spins uncontrollably in a summersault. He crashes into the concrete back yard, hands, knees, and head. Unconscious, his helmet crushed in on one side, he lays motionless.

Later in the hospital the doctor says to the firefighter's captain and an investigating chief, "The firefighter will survive. He has a broken back and a concussion. His helmet saved his life." All three look at the helmet, held by the captain. It was crushed on the side like an Australian military hat. The brim on one side was pushed up flat against the side of the crown.

The chief asks, "How on earth did his helmet stay on during a six-story fall?

The captain responds, "Chief, this firefighter served in Special U.S. military in Vietnam. He made several combat jumps. He told me once, if you take care of your equipment, it will save your life.' He truly believes that. After each tour, he takes his gear from the apparatus: cleans it, checks it, and puts it in his locker." A good reason to wear helmet straps at all times.

Note: The firefighter who died on the 78th floor of Tower 2 on 9/11 was Ron Bucca, and I was the investigating chief.

Fire Escape Hazards

Fire escapes are rusted old metal fixtures hanging on the sides of buildings for 50 or 100 years. They are rusted, broken, missing parts, and in danger of collapse. It is amazing we have not had a firefighter killed on one of these things, one veteran fire chief told me. I agreed with him. There are many hazards of fire escapes: falls, step collapse, balcony failure, counterbalance weight collapse, drop ladder falling from guides. Fire escapes are becoming more dangerous to use during fires and emergencies. The safe condition of fire escapes are critical for occupants of a building and firefighters who must use them during fires and emergencies. Firefighters must become familiar with fire escapes in their community and the hazards they present.

Firefighters use fire escapes for many purposes: to ventilate windows, to gain access to a burning apartment in order to rescue and remove trapped or unconscious people, to force entry through windows at a minor fire, to advance a hoseline into a burning apartment, and to operate hand lines from exposed buildings at major fires.

The fire escapes on old buildings are very different from the sturdy, freshly painted fire escapes found on the drill towers of fire training centers. Some have been attached to walls of buildings a century or more and have become extremely dangerous to use because of neglect or improper maintenance. In the suburbs, fire escapes often serve attics of two-and-a-half-story houses. When the attic is occupied by a family member or a boarder, a fire escape is often attached on the outside of the structure near an attic window, providing a second exit from the attic. Multistory public schools, commercial buildings, and apartment houses are also served by fire escapes. In some neighborhoods, as a fire escape becomes more susceptible to corrosion and collapse because of age and neglect, the chances of a firefighter having to climb it during a fire become greater, for the threat of fire in the building itself becomes greater. Fire escapes can be grouped into three general classifications: exterior screened stairways, party balconies, and standard fire escapes with mechanical street ladders. Each one of these fire escape types presents different hazards and dangers to firefighters and citizens.

Exterior screened stairway

The safest type of fire escape, the exterior screened stairway used as an exit from schools and theaters is enclosed by a high metal screen or railing, and extends from the top floor of the building to the street by way of a permanent, stationary metal stair (fig. 14–1).

Unlike other fire escapes, the exterior screened stairway has no sliding "drop" ladder, gooseneck ladder, or movable counterbalance stairway from the lowest balcony to the street. It is like an interior stairway in that the rise of the step is similar, it is wide enough for two people to descend side by side, and it includes a handrail. It is enclosed with a wire screen preventing a fall outside the fire escape stairway like the one described in the introduction to this chapter. Despite these positive features, the exterior screened stairway still presents a hazard to firefighters. It is constructed of metal and becomes rusted by exposure to the elements. Weakened by corrosion, its steps can collapse suddenly under the weight of a firefighter. Firefighters are injured by fire escape step collapse. When a rusted metal fire escape screened stairway step fails, the firefighter can fall the rest of the way down the flight of steps or fall through the stairway to the stairway below. Corrosion causes the failure of the connection

Chapter 14 | **Fire Escape Dangers**

Fig. 14–1. An exterior fire escape screened-in balcony is the safest type of fire escape exit.

between the step and the stringer to which it is attached. This small space between the step and the stringer, where the connecting bolt is located, is inaccessible to normal maintenance procedures like scraping and painting. Even a visual inspection will not detect this weakness. When firefighters run up the fire escape to assist a victim down, body weight pounding on a corroded metal step can cause the step to collapse suddenly. To reduce the chances of this happening, firefighters should climb the steps smoothly and not stamp down on the tread; continuously grip some portion of the fire escape railing; place one foot on the step above and apply pressure to the step first, before putting full weight upon it; and place boots close to the side of the step. The last procedure causes less deflection of the fire escape step, thus reducing the stress on the connection between the tread and the stringer.

Party balcony fire escape

Unlike the exterior screened stairway or the standard fire escape, the party balcony fire escape has no stairway or ladder connecting intermediate balconies. There is no access to the balconies above or below, or to the street. The party balcony is strictly a horizontal emergency exit, with an exit pathway afforded through the adjoining occupancy. A person fleeing a fire opens a door or window, enters the outside party balcony, walks several feet to the adjoining occupancy, and enters it through a door or window (fig. 14–2). The real protection provided by a party balcony is the unpierced fire division between the two adjoining occupancies. Formerly constructed of metal, the party balcony today is made of wood or concrete. The hazard of the party balcony is collapse from overloading or failure of the supports fastening the balcony to the outside of the building.

Fig. 14–2. A party balcony fire escape is often overloaded with occupants fleeing a fire.

Because there are no vertical stairways between party balconies—and since most people escaping a fire do not go to the adjoining occupancies—party balcony fire escapes become quickly overcrowded during a fire. The weight of several people can exceed the load-bearing capacity of a new balcony or cause a corroded old balcony to collapse. The most tragic incident involving a party balcony fire escape occurred in Boston. During a fire in an old tenement, a metal party balcony fire escape in the rear of the structure collapsed, and a young mother and her infant daughter fell to their deaths. When arriving at a fire where people have taken refuge on any type of fire escape, firefighters should act immediately to remove them from these old, rusted, metal structures. The safest method to remove people from a party balcony is through the designed horizontal exit and into the adjoining apartment window or door. If this is not possible, and party balconies are overloaded, firefighters should use portable ladders or aerials, taking special care not to strike the weakened fire escape with excessive force when placing the tip of the ladder on the fire escape railing. If possible, the ladder should be leaned against the building next to the fire escape. In all cases, firefighters should help the victims down.

Standard fire escape

The most common type of fire escape found on residential buildings, the standard fire escape, is a series of metal balconies interconnected by narrow metal ladders (fig. 14–3). The top balcony may have a gooseneck ladder leading to the roof, and the lowest balcony will have a sliding drop ladder or a counterbalance stairway providing access to street level.

Fig. 14–3. A standard fire escape

The ladder on this fire escape is much more difficult to climb than the stair of an exterior screened stairway, because the angle of the ladder is very steep (sometimes as much as 60 or 75 degrees). The ladder step tread is very narrow and the step rise is very high. Also, only one thin bar may be available as a handrail. When firefighters see a young child or an elderly person standing out on a fire escape, they should consider it a life hazard and immediately assist the individual. The victim should be brought through the window of a safe apartment, below the fire, and taken down the interior stairway. This procedure is safer than walking down the entire fire escape. If a fire escape is used, a step might collapse or the person might lose his or her balance because of the fire escape's steep angle or unusually small steps. Falls and step collapse are the most frequent causes of injury to firefighters using standard fire escapes, but firefighters have also been seriously injured when activating the drop ladder or counterbalance stairway, which can collapse or fall apart.

Drop ladders

A sliding drop ladder is held in place on the lowest balcony of a standard fire escape by a pendulum hook. The hook holding the ladder in place is released when a firefighter standing in the street or yard places the hook of a pike pole beneath the bottom rung of the drop ladder and raises it several inches. The pendulum hook swings away, and the weight of the ladder is transferred to the firefighter's pike pole. The ladder drops straight down as the firefighter quickly removes the hook end of the pike pole from beneath the rung; however, if the drop ladder is not encased in its tracks or guides when it is released, it can fall away from the fire escape and strike the firefighter operating it (fig. 14–4). For this reason, a firefighter lowering a drop ladder

Fig. 14–4. This fire escape drop ladder is out of the tracks and will fall on a firefighter who activates it.

should stand beneath the fire escape. If the drop ladder falls out of its track, it will fall out away from the fire escape so the firefighter beneath the balcony will be protected. A safer alternative would be to use a fire department ground ladder (fig. 14–5).

Fig. 14–5. There is less danger in using a fire department ground ladder than a fire escape drop ladder.

Counterbalance stairways

On the standard fire escape, access to a street may be provided by a counterbalance stairway. These are also extremely dangerous to operate. Some of these heavy metal structures have not been tested or operated for a half a century and can collapse upon activation. Supported on a pivot, counterbalance stairways are balanced in a horizontal position by heavy cast-iron counterbalancing weights. Several hundred pounds of metal are either attached to one end of the counterbalance stairway or held up by a steel cable against the side of the building. There are many different types of counterbalance stairways. To activate a common type of counterbalance stairway, a simple manual bar that prevents the ladder from descending is moved out from beneath the stairway. A firefighter with a pike pole can do this from street level. The end of the stairway can then be lowered to the street by the pike pole or by the weight of a person walking out on it. Sometimes, however, the sudden impact of the counterbalance stairway striking the ground can cause the entire metal stairway to collapse off the fire escape lower landing; the heavy, suspended metal weights can also fall off; the cable holding the weights can snap and become a deadly whip; or the entire pulley assembly through which the cable moves can drop into the street.

When encountering people awaiting rescue on the lowest balcony of a standard fire escape with a counterbalance stairway, firefighters should use a fire department ground ladder instead of the stairs. It is safer for everyone involved. In some unusual cases, safe operating procedures cannot overcome the hazards presented by fire escapes. In Newark, New Jersey, for example, a firefighter venting windows from the fire escape's top landing of a three-story, wood-frame residence fell to his death when the entire fire escape pulled away from the building and collapsed into the rear yard. The bolts fastening the fire escape to the inside of the building were missing, and the weight of the firefighter caused the fire escape to separate from the building wall.

Lessons Learned

- Firefighters should inspect the fire escapes in their community and learn how the drop ladders and counterbalance stairs of these outside exits operate. Inspect the condition of stairs, balconies, and railings and their potential hazards. When unsafe or dangerous fire escapes are discovered, violation orders should be issued to building owners and they should be made to maintain fire escapes in safe condition.

- Size up a fire escape before operations: observe the condition and stability of the platform, check the size and location of the opening leading to the ladder and balcony below, and determine how the drop or counterbalance ladder works.

- When activating a sliding drop ladder and if the balcony is not overloaded with people and in danger of collapse, a firefighter should stand beneath the fire escape balcony. If the drop ladder falls out of the tracks the area under the fire escape will provide protection and the firefighter will not be struck by the outward falling ladder.

Chapter 14 | **Fire Escape Dangers**

- Firefighters should not stand beneath a counterbalance stairway or its cable, pulley wheel, or heavy metal balancing weights.

- When climbing a rusted fire escape, keep one hand on the stair guide railing at all times. Place one foot on the step above and apply test pressure to that step first, before putting your full weight upon it. Firefighters should remember that a missing or broken step serves as a warning: a fire escape with one defective step is likely to have another one (fig. 14–6).

Fig. 14–6. Step collapse is the most common danger when climbing a fire escape.

- When placing a boot on a fire escape step, it should be placed near the outer side of the step, where it meets the stringer. This causes less deflection of the step and reduces the amount of stress caused by the firefighter's weight.

- When ascending or descending a "gooseneck" ladder, to or from a roof, test the vertical "gooseneck" ladder before ascending or descending it. Try to pull it out from the wall to which it is attached.

- When climbing up or down a fire escape, keep one hand free to continuously grip some portion of the fire escape railing. To keep hands free, officers should attach shoulder straps to their handlights, so these can be carried in sling fashion.

- Engine company firefighters stretching hose up a fire escape should support the hoseline with a shoulder or one hand and keep the other hand on the railing.

- Ladder company firefighters should interlock their axe and Halligan tools and carry them in one hand and keep the other hand on the railing.

- When preparing to advance a hoseline into a flaming window from a fire escape balcony do not lean against the enclosing balcony rail. Test it first by pushing against it.

- When descending a weakened standard fire escape ladder, face the ladder and hold on to the guide rail. If a step suddenly fails, you will fall into the ladder, minimizing the risk of losing your balance and tumbling off the balcony.

- When the stability of a fire escape is doubtful, use a fire department ground ladder. The extra time required to obtain and place the ladder is outweighed by the dependability and safety of the fire department equipment.

15

WOOD-FRAME BUILDING COLLAPSE

All hands are heavily engaged during a major fire at an old three-story corner building. Firefighters using several outside aerial and ground master streams have confined the fire to the wood braced-frame building, keeping it from spreading to a one-story structure attached to the rear. "You wait here," says the captain to a firefighter directing a hose stream through a window of the burning building. "I'm going inside the one-story building to see if we can get a better shot at this fire." The captain enters the doorway of the one-story structure and soon reappears, waving his arm for the firefighter to come to him. The firefighter shuts down the nozzle and drags the charged hoseline through the doorway of the uninvolved structure. Inside, the captain reaches down and pulls some hose into the building. "Don't go through that archway into the fire building," he warns the firefighter. "We can hit the fire from here."

The two men crouch down in front of a large doorway between the burning structure and the attached building. The firefighter opens the nozzle and sweeps the hose stream across the flaming ceiling, extinguishing fire near the doorway entrance and playing the stream deep into the interior of the burning floor area. The reach of the pressurized hose stream enables the firefighter and the captain to stay safely inside the one-story building. "Say, Captain, you were right; we have a perfect shot at this fire," says the firefighter.

The officer nods his head. As the hose stream turns the flames at the ceiling into white steam, the three-story building suddenly starts to collapse. The walls appear to explode outward. Flaming timbers crash through the ceiling where the hose stream had extinguished fire seconds before. A blast of superheated smoke blows through the doorway into the faces of the firefighters. "Drop that line and let's get out of here!" shouts the officer. Seconds after they move back, the roof of the one-story structure crashes down on the spot where they were standing.

The floor rocks violently from the impact. Struggling to keep their balance, the firefighters stagger toward the door, while plaster dust and smoke fill the room that is crumbling around them. As they reach the concrete sidewalk, outside the half-collapsed one-story building, the front wall begins to fall. Propelled outward at a 90-degree angle, the heavy wooden structure crashes down to the sidewalk, smashing into the head and shoulders of the running captain. As the officer falls, he shoves the firefighter ahead of him. The top of the collapsing wall hits the firefighter on the back and legs, slams him into the concrete sidewalk, and pins him from the waist down under the falling structure. As the dust settles, the stricken firefighter cries out, "Help! Help! Get the wall off me!"

Wood-Frame Building Collapse

There are three ways that a wood-frame building can collapse during a fire: one wall may fall straight outward at a 90-degree angle (fig. 15–1 and fig. 15–2), the entire building may lean over and collapse on its side (fig. 15–3), or one or all four wood enclosing walls may crack apart and fall in an inward/outward collapse (fig. 15–4). A three-story braced frame structure frequently falls in an inward/outward collapse. The top two stories collapse inward, back on top of the pancaked floors; the lower story collapses outward on to the sidewalk.

Warning signs

A 90-degree-angle wall collapse is often signaled by the corners of the falling wall splitting apart from the remaining walls. The lean-over collapse is often indicated by the burning structure slowly starting to tilt or lean to one side. An inward/outward collapse may not exhibit any structural warning at all—sometimes the only indication that a collapse is imminent is a serious fire burning for a long time on the lower floor. When such a collapse occurs, firefighters report that they see no signs but that they hear a sudden, loud cracking noise, feel a hurricane-like gust of wind on their backs, and then they are engulfed in a cloud of dust as they turn to run from the falling structure. Of the three types of collapses, the inward/outward collapse is the most dangerous because it is sudden, it gives no visible warning signs prior to failure, and, unlike most other building failures, it may involve the collapse of two, three, or four walls simultaneously.

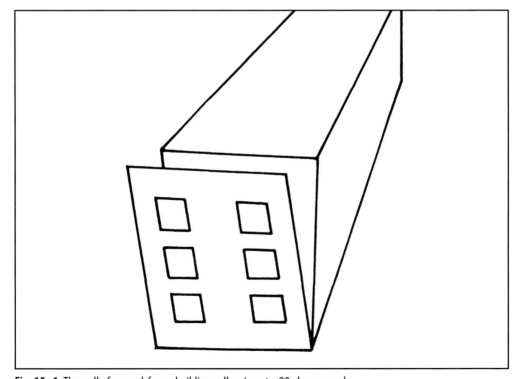

Fig. 15–1. The wall of a wood-frame building collapsing at a 90-degree angle

Chapter 15 | **Wood-Frame Building Collapse**

Fig. 15–2. A 90-degree wall collapse

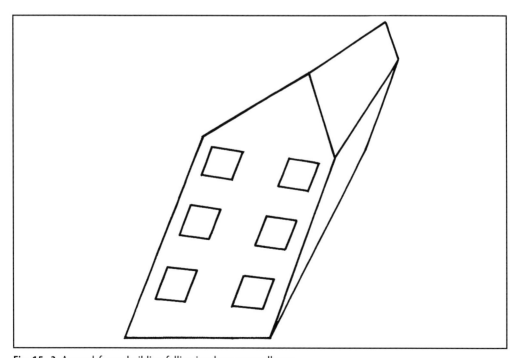
Fig. 15–3. A wood-frame building falling in a lean-over collapse

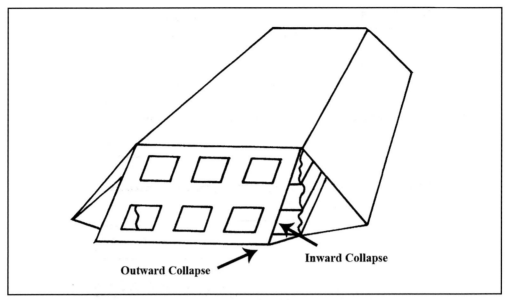

Fig. 15–4. A wood-frame building collapsed in an inward/outward configuration

During a fire in a structure with masonry walls, it is rare that more than one wall will collapse at one time (except in the case of an explosion). When a braced-frame wood building collapses, however, all four walls may collapse at one time. Only firefighters in the corner safe zones will survive a collapse. The corner safe areas are the four flanking zones around a burning wood frame building. When you look at a four-sided building from a bird's-eye view you imagine the four walls collapsing and covering the ground with bricks, you will find there are four areas at the corner of the collapse building that have fewer bricks.

Braced-Frame Wood Construction

Of the three major types of wood-frame construction in the United States, braced-frame wood-constructed buildings present the greatest firefighting danger (fig. 15–5). In a two-year period, a chief and an officer were killed and an officer and nine firefighters seriously injured in four separate collapses of braced-frame wooden buildings in New York City. Three of these structures were located on corners and one was the end building in a row of three which stood next to an open lot. As the last building was unsupported at one end, it was, in effect, the same as a corner building. All four buildings were three stories high and in each, a serious, long-burning fire had destroyed the first floor of the structure.

Corner Wood-Frame Buildings

Wooden buildings constructed side by side receive support and stability from the adjoining structure. If the lower floor of a wood building burns and one of the wood bearing walls is destroyed by fire, the structure will begin to lean to one side. Adjoining

Fig. 15–5. Braced-frame construction

structures built up against a wood building can prevent such a fire-weakened structure from collapsing. When weakened by a fire on a lower floor, however, a wood-frame corner building will collapse on its unsupported side into the street or an empty lot, The bearing walls of a wood structure, unlike those in any of the masonry construction types, are combustible and can collapse when exposed to fire. The side bearing walls on the first floor of a three-story wood building are 2×4-inch wood studs spaced 16 inches on center, the same as the bearing walls on the second and third floors. Though the bearing wall studs of the first floor support more weight than the second- and third-floor wall studs, there is no compensation for the increased deal load (unlike some multistory, masonry bearing-wall buildings, in which the lower levels of the bearing walls are thicker than the upper levels). Therefore, if a fire weakened the bearing wall studs of all three floors at the same time, the ground floor wall studs would fail first because they support more weight than the second-or third-floor bearing wall studs.

Based upon the four New York City building collapses mentioned earlier and other wood-frame building failures, it is apparent that the height of the structure affects its stability. Three-story wood-frame buildings collapse more frequently than one- or two-story wood-frame buildings. To understand how a burning wood-frame residence can collapse and how to extinguish a fire burning within a wood-frame building, a firefighter must know how a wood-frame building is constructed. The four most widely used methods of wood-frame construction over the past 200 years are:

- Braced-frame
- Balloon frame
- Platform frame and
- Lightweight wood frame construction

(Plank-and-beam and log cabin construction are also used, but are much less common.)

Braced-Frame Construction Methods

In the 18th and 19th centuries, the first large wood-frame buildings constructed along the East Coast, which still stand today, were of braced-frame construction, sometimes called "post-and-girt" construction. This type of wood-frame structure has a braced framework of vertical timbers called "posts," which are positioned at each of the four corners of the building, and horizontal timbers called "girts," which are found at each floor level. These large timbers reinforce the entire 2×4-inch wood-frame structure and are connected together by mortise-and-tenon joints. (The large timbers and the mortise-and-tenon joints are indicators of braced-frame construction.) The ends of the horizontal timbers are cut down to fit mortise openings which are cut through the vertical timbers.

Balloon Frame Construction Methods

As the population moved westward in the 19th century, the need for housing increased, and cut and finished large timbers and skilled craftsmen became scarce. A lightweight, quickly assembled wood structure, which needed no large timbers, called "balloon frame construction" replaced the Eastern braced-frame method of constructing wood structures. To erect a balloon frame structure, four wood exterior walls are constructed flat on the ground. Two-by-four-inch wood studs, extending in one piece for the full height of the wall, form the enclosing walls; the four walls are then lifted upright from the ground and connected like a box at the corners. The advantage of this type of wood construction is speed and the absence of large timbers. The drawback is a vertical void between the wall studs, which extends from the foundation sill to the attic cap and allows hidden fire and smoke that penetrate the wall space to spread vertically for two or three floors. This unobstructed opening between each stud in the exterior wall, extending from the foundation sill to the attic cap, is an indicator of balloon construction. The vertical void of a balloon constructed building will be the exterior wall. However, if an addition is added to the building, the exterior wall can become an interior wall. The attic should be quickly examined for fire spread during the early stages of a fire in a building of balloon construction.

Platform Construction Methods

Platform construction superseded balloon construction and today, is the most widely used method of wood-frame construction. The platform construction method builds a structure one level at a time. One complete level of 2×4-inch wood enclosing walls are raised and nailed together; the floor beams and deck for the next level are placed on top of these walls. The next level of 2×4-inch wood enclosing walls are constructed on top of the first, and the floor beams and deck for the next level are placed on top of these exterior walls. From a fire protection standpoint, platform construction is superior to balloon or braced-frame construction, because there are no concealed wall voids that extend for more than one floor.

Lightweight Wood Construction Methods

Lightweight wood truss construction is replacing platform construction. From a fire protection point of view, it is inferior construction. Lightweight wood truss construction suffers floor and roof collapse, not wall or "global" (total) collapse of the entire structure. The danger is early floor and roof collapse of the truss floors and roofs. The entire building may eventually collapse, but it is during the early stage of a fire, when firefighters first arrive that a floor or roof fails—especially the floors. Because floors collapse so fast, there is a saying among firefighters, "Through the door and through the floor." Tests have shown the thin metal connections used to fasten truss floors and roof beams together fail during the early stages of a fire The connectors are pieces of sheet metal that only fasten the wood surface of the trusses together. The metal fasteners only penetrate $^3/_8$ to ½ inch into the wood, so when the wood chars, the fasteners fall away. Heat of a fire can also cause the fastener to bend away from the wood truss and leave the structural element unconnected. In 2008, the Underwriters Testing Laboratory documented the lightweight wood truss and sheet metal fasteners and documented failure in less than 10 minutes of fire exposure. Because of the early roof and floor collapse and because of the large number of firefighters killed fighting fire in lightweight wood truss buildings (20 in 24 years.), it is recommended that when a fire involves truss structure such as a concealed space of the floor or roof, the strategy be to remove occupants and fight the fire defensively from the outside.

Causes of an Inward/Outward Collapse

Three factors contribute to the inward/outward collapse of a braced-frame wooden building:

- Fire destruction of bearing walls
- Failure at the mortise-and-tenon connection
- Exterior wall overload

Unlike the exterior walls of the four other basic construction types (fire-resistive, noncombustible, ordinary brick-and-joist, and heavy timber), the bearing wall of wood-frame construction *can* be destroyed by fire and can collapse when flames spread out of a window and consume the outside or inside of this load-bearing wall. Burning wood-frame buildings exhibit a rapid fire spread. When the fire department arrives on the scene, both the wooden exterior walls and the structure's interior are often involved with flame. When wood buildings are built close together or when there is a common roof space running through a row of wood houses, fire spread will be extremely rapid and will probably involve more than one structure (fig. 15–6). In addition to placing hose streams in the interior of the burning structure, firefighters will need one or more hoselines to control exterior fire spread along the outside combustible walls and to protect exposures from radiated heat.

Fig. 15–6. The common roof space is a structural defect in row buildings.

A firefighter should know which of the four enclosing walls of a burning wood building are the load-bearing walls that support the floors and roof. Because these walls are interconnected, the interior floors will collapse if the bearing walls fail during a fire. Conversely, if the interior floors collapse, they may cause bearing wall failure. In older urban neighborhoods, wood-frame buildings were built close together, with the bearing walls usually being the side walls and the non-bearing walls were the front and rear enclosing walls. This practice has changed in suburban communities. Private homes, built on large plots of land, are designed to have the larger area of the building face the street front, so the front and rear walls are load-bearing and the two side walls non-load-bearing. Condominiums and row town houses have the same design. During a fire in a suburban row of town houses, if the floors inside collapse, the front or rear walls may collapse outward. In peaked-roof buildings, the bearing walls support roof rafters and are parallel to the ridgepole. In flat-roofed wood buildings, the bearing walls are usually the walls with the greatest dimension: the non-load-bearing walls have the shortest dimension.

Mortise-and-tenon joints

The structural framework of a braced-frame wooden building that collapses inward/outward consists of vertical timber corner posts and horizontal timber girders or girts at each floor level. The corner posts and girders are connected by mortise-and-tenon joints (fig. 15–7). When a braced-frame wood timber collapses, it fails at the weakest points—often the mortise-and-tenon connection. The mortise hole has removed the center section of the corner post timber and reduced its strength; the tenon end of the girder is only a fraction of the original girder's thickness and therefore has only a fraction of its strength. In addition to this design weakness, the connection can be destroyed by fire. Furthermore, unlike concrete and steel fastenings, the wood mortise-and-tenon connection is susceptible to collapse by rotting. A vacant wooden building open to the elements can be quickly weakened by rotting structural components like the mortise-and-tenon connections.

Fig. 15–7. A wood mortise and tenon connection

Exterior wall overload

The exterior wall of a wood-frame building can be weakened by the weight of a metal fire escape landing and ladder. This heavy metal structure attached to the outside wall of a wood building is anchored to 2×4-inch wall studs behind the wood sheathing. The weight of the metal fire escape can exert a slight outward pull on the wall studs to which it is attached for support. This pull causes the wall studs to curve or bow slightly outward. The load above, supported by the curved wall studs, is no longer transmitted through the studs as an *axial* load (centered or evenly distributed), but becomes an *eccentric* load (off-centered or uneven). During a fire, the wall supporting a metal fire escape must be considered a structural danger. The weight of the fire escape will accelerate the collapse of a fire-weakened wood wall.

There are two types of masonry surfaces applied to outside walls of old wood buildings: A brick-and-mortar veneer wall can be attached to the wooden structure by thin strips of sheet metal, one strip every two square feet; or a thick stucco coating, spread on wire mesh, can be nailed to the old wooden surface of the building. These wall surfaces increase the collapse danger during a serious fire in a wood-frame building by adding considerable weight to the structure. As much as eight pounds per square foot of stucco and wire mesh have been found on a collapsed wall. Brick veneer not only overloads a wall but also hides major structural defects of the wall. It can conceal an obvious collapse warning sign, such as the wood walls splitting apart, or hide the burning of the wood bearing wall behind it. These masonry wall coverings also contain the heat and flame inside the building, thus increasing the destruction of the structural framework.

Lessons Learned

- Burning wooden buildings of three or more stories suffer global collapse more frequently than burning one- or two-story wood buildings.

- Wooden buildings located on a corner plot or standing alone are more susceptible to global collapse when exposed to fire than wood buildings in the center of a row of similar buildings.

- When a serious fire burns out the entire first floor of a three-story wood building, there is a danger of global collapse.

- Of the three types of wood-frame building collapses, the inward/outward collapse is the most dangerous. It gives no warning and can result in the simultaneous collapse of four sides of the structure.

- Large buildings three stories in height collapsing on top of smaller one-story buildings cause the smaller buildings to collapse.

- Lightweight wood truss and wood I-beam construction has made wood frame buildings more deadly. Now there is deadly inside floor and roof collapse danger added to the global outside structure failure.

- Three contributing causes of wood-frame building collapse are fire destruction of bearing walls, the mortise-and-tenon joint of a braced-frame wooden building, and the overload of an exterior wooden wall.

- Renovated wood frame buildings can create an another collapse danger. Today old buildings are renovated under so-called "performance" building codes, not "specification" building code. A performance code does not require the same type and dimension of building element be replaced in the structure. There are no specifications allowed in a performance code. The New York City performance code allows conventional construction to be replaced with lightweight materials and it allows buildings to be renovated using different designs, layout and materials. Once the building is renovated under a performance code, it is not the same. A firefighter size-up based on old construction, no longer applies. In 1998, FDNY Captain Scott LaPiedra and Lieutenant James Blackmore were killed when a second floor of a three-story braced-frame building collapsed. This building was renovated under a performance code, which allowed removal of a partition wall that was a floor support to the second floor below the collapse area. This weakened the structure and created an unprotected opening between two buildings that allowed a large area of fire to develop. The renovation sealed off the first floor rear of the building from the front entrance where the fire occurred delaying extinguishment.

- When a triple-decker wood frame building collapses in an inward/outward type all four sides may collapse simultaneously. During an inward/outward wood building collapse, where all four sides fall simultaneously, only firefighters in the corner safe zones can survive. The corner safe areas are the four flanking zones around a burning wood frame building. When you look at a four-sided building from a bird's-eye view and you imagine the four walls collapsing and covering the ground with bricks, you will find there are four areas at the corner of the collapsed building that have fewer bricks.

Note: For more information on wood frame building collapse, search the Web for: *Firefighter fatality investigation NIOSH F2002-32.*

16

COLLAPSE HAZARDS OF BUILDINGS UNDER CONSTRUCTION

"There's a fire on the top floor!" shouts the watchman as he waves down responding apparatus. An engine, a ladder, and a chief have just arrived at a late-night alarm in a 14-story, reinforced-concrete building under construction in the center of the downtown area.

"Are the standpipe and elevator in service?" asks the chief.

"The standpipe goes only to the eighth floor," replies the watchman. "The elevator power is shut off."

The chief turns to the ladder company captain. "Take two firefighters with a portable extinguisher up to the fire. The hose stretch will be delayed." The captain, with his driver and a young firefighter, walks up the dark, interior concrete stairs of the building under construction. Near the stairway they see a black iron standpipe riser that runs up through the structure.

The officer shines his light on the pipe, revealing that there is no hose but that a valve wheel has been installed at each hose outlet. "Check each valve as you go up," he tells the other firefighters, "and make sure it's closed." As they climb the stairs, they walk out of the stairway enclosure on each landing to reach the steps leading to the next level. They pass by a 10×15-foot unprotected opening and several 3×3-foot holes in the floor. "Watch out for the elevator shaft and these floor openings," he tells his crew, flashing his handlight on the hazards. The freshly poured concrete stairway ends after 13 stories, so the firefighters have to climb a handmade wooden ladder to reach the top floor.

The flimsy ladder extends up through the elevator shaft opening, with its bottom resting on a plywood platform built into the elevator shaft. There is a 130-foot fall down the shaft on either side of the temporary plywood deck. The officer steps out on the platform first, climbs the ladder to the top floor, and is followed by the other firefighters. When all three have reached the pitch-black top floor, they don their masks and spread out to look for the fire. Close up, the floor area appears to be a forest of lumber. The formwork, supporting tons of freshly poured concrete above, creates a maze of wooden timbers, and a heavy plastic curtain completely encloses the outer perimeter of the floor. In the darkness, a hellish atmosphere is created by drums of glowing red coke fires scattered all over. Heat, smoke, and coke gases drift through the air. The three men move quickly between the timbers, searching for the fire. "Hey, Captain, it's over here," shouts one of the firefighters.

The men regroup near a small fire in a pile of scrap lumber near the outer edge of the floor. Flames drift up to the plywood ceiling and a large brown scorch mark mars the plastic curtain near the fire. The young firefighter rests the portable extinguisher on the floor, removes the locking pin from its handle, points the nozzle at the base of the flames, and squeezes the handle. A stream of water is discharged. "Look around and see if there is another extinguisher on this floor. We may need it before the engine gets the hoseline up here," says the captain to the driver.

The latter disappears into the darkness on his assignment. Suddenly, the brown scorch mark ignites and burns a large hole in the curtain. A strong, steady wind blows into the floor through the expanding curtain hole and fans the small fire. Flames leap up to the plywood ceiling again and start to roar across the open floor between the forest of timbers. "The extinguisher is empty," says the firefighter, tossing it aside. He and the officer are forced back by the heat. "Captain, where is the ladder we came up on?" asks the younger man. "I can't see it."

"Never mind the ladder! Where's your partner?" says the officer. Searching, the two men scurry in and out of the wood shoring while windswept flames spread wildly across the ceiling behind them. A shower of sparks blows over their shoulders. The floor is now bright orange with dark shadows moving across it.

"Captain, look! Over there—to your right!" shouts the rookie. In a section of the floor where fire has already spread, the firefighters can see the top of the wooden ladder sticking up through the elevator shaft opening. The wooden rungs are burning, and flames are swirling around the ladder top. The firefighters move away from the flames and come to the outer edge of the structure. Pulling aside the plastic curtain they look down at the dark street. All they can see are the small, flashing red lights of the fire trucks, 130 feet below them.

Suddenly the officer and the rookie hear the voice of the chauffeur: "Captain! Over here! Quick!"

The two men move in the direction of the voice. "Where are you" calls the captain. As they weave frantically through the timber maze toward the oncoming flames, sparks and small burning embers blow into their faces.

"Over here, Captain." Now, looking downward, they spot the driver up to his waist in an opening in the concrete floor. He is about to jump down to the floor below. "I found an escape. Let's get out of here," he says. The firefighter quickly pulls himself out of the opening, and the officer shines his handlight through it. "Hurry, Captain! The fire is coming this way!"

The captain looks up and says, "Okay, go ahead—there is a hole beneath this, but if you hang and swing to one side you will make it." The driver squeezes down through the hole again, hangs by his fingers, swings his body to one side, and drops. The second firefighter does the same and is followed by the captain. Moments later, a blast of flame blows across the opening.

Collapse of High Rise Buildings Under Construction or Demolition

On August 18, 2007, two FDNY firefighters were killed in the Deutsche Bank building under demolition; on March 16, 2008, New York City, seven people died after a 19-story crane collapses from a building under construction; on May 20, 2008, 14 workmen were injured, five critically, by an explosion in a San Diego building under construction; on May 30, 2008, another crane accident in New York City killed two people.

Buildings under construction or demolition are dangerous places, and when they're on fire, extremely dangerous. Firefighters should know when running into a burning building under construction or demolition that all bets are off. It is a hazardous place. For example: temporary wooden ladders collapse under the weight of firefighters pulling hose; open elevator shafts offer a fall to eternity for unsuspecting firefighters; open stairs spread flame and smoke up or down; propane tank explosions blow firefighters off an open floor; without enclosing walls, firefighters can walk off the edge of a floor; tons of debris, tools, and building materials can rain down from upper floors.

A building, when being built or dismantled, is more deadly than an occupied and completed building (fig. 16–1). In a completed building there are many safeguards, we take for granted that do not exist in a half finished structure. For example, without enclosing walls, flame quickly spreads up the outside of the building from floor to floor at the perimeter; without interior partitions a wind-driven fire can sweep across an entire open floor instantly; without walls and doors around stairs, elevator shafts, and utility closets, fire can rapidly spread up openings at the interior; and finally, without doors and partitions, there can be no close-approach firefighting shielded behind a barrier.

Buildings under construction or demolition are unstable structures. During a fire, collapse happens fast when columns, girders, and bearing walls have not been installed or are not yet fastened together or supports are not yet protected with fire-retarding protective covering. If one section of the building fails, it can trigger a multilevel floor collapse. When weakened by fire, scaffolding, tools, and debris fall from upper floors. The major life hazard during a daytime fire will be workers trapped on upper floors, and especially the worker inside the cab of a crane. At night when a building is unoccupied, a small fire will grow unnoticed and will be a large fire on arrival. After work hours, the major life hazard will be the firefighters and the hazard will be fire and collapse of the structure.

Fig. 16–1. Buildings under construction or demolition are more dangerous than occupied buildings.

A building under construction or demolition is a strange environment. Unless firefighters inspect and become familiar with the building's interior, it will be a strange hazardous interior that can disorient and trap them during a fire. Periodic inspection must be done because these buildings change rapidly as they rise or are taken down:

- Scaffolding and plywood around the outside of the building can inhibit smoke venting.
- Temporary workers' shanties, inside and outside the structure, grouped together contain hidden combustibles flammables and explosive gases.
- A forest of dried-out wood formwork supporting freshly poured concrete floors is like a maze and slows hoseline advancement.
- Propane gas tanks, and coal and coke fires used for heating concrete (called *salamanders*) laying around open floors can burn firefighters.

- Unprotected electric wiring strung overhead and laying on floors for temporary lighting and powering tools can electrocute firefighters searching in smoke.
- Piles of combustible materials used for construction obstruct exits. During asbestos removal, some exits are sealed or blocked to limit contamination.
- Large amounts of rubbish piles placed near open shafts and stairs create multifloor fires.
- Openings everywhere create fall hazards.

Entrapment, disorientation, falls, falling objects, burns, collapse, and hot, rapidly spreading fires and explosions greet firefighters in these dangerous, half-finished structure.

Special Occupancy Structures

Buildings under construction, or demolition (and vacant) are designated "special occupancy structures" by the National Fire Protection Association (NFPA). Because of the hazards stated above, these unfinished structures kill more firefighters than fires in other type of occupancies, such as residence and commercial occupancies. Statistics show fires in these so called "special occupancy structures" kill eight to 10 firefighters for every 100,000 fires that occur in them. Store and office fires kill four to eight firefighters per 100,000 fires; and residence buildings kill four firefighters per 100,000 fires. Special occupancy structures present the greatest danger to firefighters. The following are detailed collapse and fire dangers firefighters will encounter in a building under construction or demolition.

Crane collapse

Cranes collapse most often when they are overloaded, moved, assembled, or disassembled, not during fires. When cranes are used to construct or demolish a structure, they can fail and collapse on nearby buildings. This is the worst type of incident—the building under construction or demolition and adjacent buildings can collapse under the weight of the falling massive steel tower. If nearby buildings collapse under a crane, broken gas pipes and electric sparks can quickly ignite the damaged buildings, creating a fire. Firefighters will be called to fight fire and search for survivors after a crane collapse. It will require collapse search-and-rescue operations and firefighting operations simultaneously. If a fire in a building under construction is exposing a crane, the incident commander must consider a possible collapse. A collapse zone equal to the height of the crane must be set up. Police should be requested to close off blocks around the fire and evacuate all the nearby buildings when there is a possible crane collapse. During a fire in a building under construction or demolition, where a crane is used, a worker is often trapped in the cab of the crane above the fire. Master streams should be used to protect a trapped crane operator and a secondary search of the crane may have to be conducted if the crane operator is unaccounted for.

Concrete floor collapse

The construction industry uses two different types of framework for high-rise structures: reinforced concrete and structural steel. Nationwide, these *cast-in-place* concrete structures present the most serious fire and collapse hazard and suffer the greatest number of major fires. A serious fire in a structure of this type most often involves the wooden formwork timbers used to support the freshly poured cast-in-place concrete top floor. The lightning fire spread in the formwork supports creates a large fire and a floor collapse danger. It requires a major fire department response to control, is extremely dangerous, and often spreads fire to surrounding exposures by windblown burning embers. The strategy at this type of fire, should be defensive, using outside aerial master streams to protect exposure buildings from fire spread and extinguish the fire from a safe distance.

A cast-in-place concrete building built on the site, floor by floor, has steel reinforcing rods and wires strategically placed, which act as tie rods to make one monolithic structure. The wood formwork shapes the poured concrete into floors, walls, and columns. When completed, the cast-in-place structure is one solid, concrete structure reinforced by steel. The hardened concrete provides the compressive strength to the structure, and the steel reinforcement supplies the tensile strength to the concrete. After each floor is poured and hardened, the combustible formwork and the supporting shoring are disassembled and rebuilt to receive the concrete for the next higher floor. After five or ten floors of concrete are poured and dried by heaters—salamanders—the reused wood formwork becomes extremely dry and ignites very easily and burns rapidly. In some instances, workers apply an oil coating to the formwork to prevent it from sticking to the concrete. In some "fast-track" construction projects, one concrete floor is poured every 48 hours. Although it takes approximately 27 days for concrete to reach its maximum strength, the high-rise building construction process cannot wait so long for each floor to harden. After 48 hours, a concrete floor, depending upon the type of concrete and the temperature, can have sufficient strength to enable the wood formwork below to be removed and reconstructed above. Even though the formwork is removed, bracing remains below the freshly poured concrete floors for support, and portable steel jacks or timber columns will continue to support several floors below. Construction engineers state that, within 24 hours of pouring, the entire concrete floor can collapse on firefighters if the wood formwork below has been destroyed by fire. The fire service believes this is an underestimation. A floor collapse can occur any time during construction and can trigger a multilevel floor collapse of the entire building under construction. Experience has shown the most serious fires in a building under construction and the most serious floor collapses occur during a fire in the formwork.

Spalling concrete collapse

Concrete can also collapse in small sections if heated by a scrap lumber fire or a temporary shed or construction shanty, or by flames spread along the underside of the exposed concrete. Firefighters searching, operating hose streams, or overhauling can be struck by chunks of falling concrete, causing serious head injuries. This *spalling* collapse is the failure of the outer layers of concrete, caused by the rapid heating of moisture inside the concrete. Concrete always contains some small amounts of moisture from the atmosphere. The relative moisture content of concrete is greatest during the first 27 days after pouring. This outer layer of concrete, when heated causes a rapid expansion of entrapped moisture. This moisture expansion can cause a chunk of concrete to collapse. A piece of falling concrete can be a small section or it can be 3 or 4 inches thick, 5 or 6 square feet in area, and weigh over a hundred pounds. Two types of concrete spalling are experienced during fires: an *explosive* spalling, which propels concrete downward with an explosive force and is accompanied by a loud noise, and a dropping spalling, which is similar to a plaster ceiling collapse. Whether the collapse is explosive or not, a firefighter can be killed or seriously injured if he is struck by the spalling concrete (fig. 16–2).

Fig. 16–2. Large chunks of concrete can collapse from a ceiling when heated by a fire.

Worker Shanty Fires

People start fires, buildings do not start fires. Worker sheds, office shanties, and temporary office trailers are the location of many fires started by workers. When you inspect a building under construction, take a good look inside and around the temporary structures called *shanties*. These structures may be trailers or small sheds constructed of plywood or sheet rock. They are used as engineer offices, storage buildings, or workmen's changing areas. Shanties are found around the perimeter and sometimes even inside the building under construction. Shanty fires spread to nearby shanties and to buildings under construction, causing major damage, million-dollar losses, and workmen's death.

Because of their fire history, shanties are required to be noncombustible construction and at least 30 feet from the building under construction. Also shanties must be a safe distance from each other—they can not be grouped together. These shanties are small cramped spaces with large amounts of combustible material inside: clothing, furniture, office supplies, architectural plans, propane storage, and even gasoline for power tools. There are many fire ignition sources also found inside a workmen's shanty and number one is portable heating devices. Electric heaters are often placed too close to combustible material and start fires. Another fire cause is illegal smoking and matches used to light up inside a shanty. Careless cooking with hot plates and open flames are another firestarter. Buildings under construction, shanties, trailers, and sheds should be inspected and firefighters should note the construction methods used, the presence of truss construction, and hazards. Workers' shanties should be considered a target hazard.

Formwork Fires

Many types of fires occur at construction sites: stored lumber fires, wood shanty fires, rubbish fires, electric fires, propane fires, coal and coke heating equipment fires, torch cutting and welding fires, and flammable liquid fires. The most serious of these involve the combustible formwork used to support a cast-in-place concrete floor. In a building with this type of construction, the closely spaced 4×4-inch timbers and 4×8-foot sheets of plywood create a heavy fire load.

The typical formwork system used to support and cast a reinforced concrete floor works like this: The 4×4 inch timbers are used as columns, girders, and beams (called "legs, stringers, and ribs" by construction workers) (fig. 16–3). The timbers support a platform of plywood sheets, which is prepared with steel reinforcing rods and wire. All edges around the perimeter of the building and floor openings are built up to keep the concrete from dripping off the sides. The concrete is poured into this wooden form or mold to create the floor and supporting columns. In and around this lumber yard of wood formwork, there are heaters to dry out the concrete. The entire floor containing the wood and heaters may be covered around the perimeter with plastic sheets to hold the heat. These heaters—propane flames or coke fires—often ignite the nearby wood or plastic sheeting.

Chapter 16 | **Collapse Hazards of Buildings under Construction**

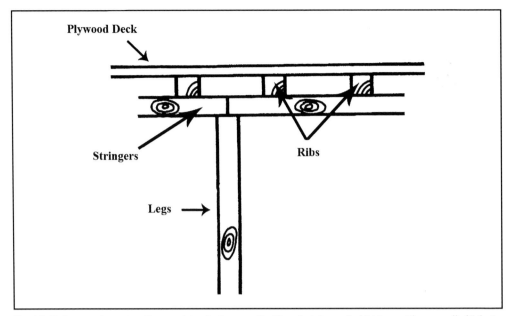

Fig. 16–3. Formwork construction consists of columns called "legs," girders called "stringers," beams called "ribs," and a plywood deck, which holds steel bars and concrete.

When firefighters respond to a fire involving wood formwork after the formwork is in place but before the concrete has been poured, there is still a danger of formwork collapse. Because the 4×4-inch columns, girders, and beams must be disassembled and reassembled 30 or 40 times (when each concrete floor is about to be poured), the construction workers use very few nails or connectors when assembling the formwork structure. These timbers, placed on top of each other and fitted together, will collapse progressively if one or two strategically placed columns or "legs" are destroyed by fire or removed during overhauling. Firefighters trapped beneath a burning mass of wood and wire would be difficult to rescue. For this reason, firefighters should not remove the supporting timbers of formwork during overhaul operations after extinguishing a small fire. If necessary, they should consult with the construction shoring foreman to determine the structural stability of the formwork. Any fire that occurs in wooden formwork will spread rapidly throughout the entire floor, making useless a single hoseline attack from a wooden access ladder through the elevator shaft. Outside aerial master streams are the recommended strategy to extinguish fires in formwork supporting floors. All firefighters should be removed from the building during this operation.

Scaffolding Fires

After all the floors and columns of a cast-in-place reinforced concrete building are in place, the enclosure walls are built. Enclosure walls of glass, stone, brick, or masonry are often of panel wall design, and supported at each level by the outer edge of the concrete floor. The panel wall is constructed by erecting a timber scaffold around the perimeter of the structure. Workers stand on this platform to build the enclosure wall from the outside. The scaffolding presents a serious collapse danger during a fire. It extends beyond the outer edge of the structure and is constructed of heavy 2×8-inch timbers (fig. 16–4).

Fig. 16–4. Scaffolding can extend beyond the edge of a building under construction.

Usually supporting large bundles of brick, wheelbarrows full of sand and mortar, and heavy tools and equipment, scaffolding is combustible; sometimes it is supported by metal frame work, or by cantilevered aluminum beams extending out beyond the structure several floors above. Fires can start or spread to the scaffolding and weaken the scaffolding itself, and then it can collapse around the sides of the building under construction. Hundreds of pounds of scaffolding, timbers, bricks, wheelbarrows, and mortar will rain down around the building, striking apparatus and firefighters and crashing through the roofs and floors of smaller structures nearby.

Burning Embers

After a major fire in a cast-in-place reinforced concrete building, the ground around the construction site is littered with hundreds of pieces of charred timber. (What had appeared to be windblown burning embers or sparks during the height of the fire were actually heavy sections of burning timber dropping to the ground.) Fire companies arriving first at a fully developed construction fire must run an obstacle course through these falling timbers. Only when they are inside the structure, safely beneath the concrete floors, can firefighters advance up the building to the fire. Burning timbers will present the fire chief with another problem: they will be the major cause of fire extension of a framework fire in a cast-in-place building under construction. Swept along by powerful wind currents of flames and smoke, these airborne, red-hot pieces of timber will drop into narrow shafts between buildings, blow into open windows, and ignite grass, brush, and trees. They settle on piles of rubbish, fire escape landings, window ledges, and cornices, igniting window frames and sometimes spreading fire into the building. Most of all, these burning timbers fall on the rooftops of smaller structures around the fire. If the roof has a noncombustible covering there will be no problem, but if the roof covering is combustible, the fire can spread.

When a high-rise building under construction starts to extend above the rooftops of smaller adjoining structures, the construction company sometimes protects the skylights and lightweight roof decking from the impact of falling construction material and equipment by covering them with wooden planks. Often the roof of a smaller building adjacent to the construction project will be completely covered with 2×8-inch, heavy wood timbers. Though this precaution protects the roof from impact damage, it also changes the roof covering of the smaller building from noncombustible to combustible material. The combustible roof surface will be more susceptible to ignition from windblown burning timbers falling from the upper floors of the building under construction. Chiefs in charge of fires where windblown burning embers are a problem must direct firefighters to examine adjacent roofs. Sparks which have fallen on the rooftops or between the cracks of the protective wooden planks can result in exposure fires.

Construction Site Hoist Collapse

When erecting a cast-in-concrete high-rise building, construction workers build "hoists" on the outside of the structure to carry personnel and building materials to the upper stories. Hoists built to lift equipment are designed differently from those built to lift personnel, and, for this reason, construction workers are permitted to use only the hoists made for transporting personnel. Over the years, workers have been killed or seriously injured when improperly riding material hoists. When responding to a fire ten or twenty stories above the street, firefighters must use the hoist designed for the workmen—not the material and equipment hoist, even if a watchman or construction worker suggests that they do so. Firefighting equipment may be sent to the upper floors by the material hoist car, but firefighters should not ride up with it. If two elevators are present at a construction site, the material and equipment car will be the one supported by only one or two steel cables and operated by controls at ground level. A bell system at each upper story signals the ground operator to move the car. The car has no safety brakes, and if flames weaken the supporting cable during a fire, it will drop to the ground and crash into pieces. The bell signal of the material hoist creates another hazard. Because it sounds like the low-air warning on some breathing apparatus units, the operator at street level, hearing the SCBA alarm, might mistake it for the hoist bell and move the car. A firefighter entering or leaving the car at that moment can lose his balance and fall down the side of the building. For these reasons, firefighters should ride the hoist car designed for use by construction workers. Operating controls are inside the car, the car operator rides inside the car, and built-in safety brakes stop the elevator car from falling free.

Asbestos Removal

When a building under demolition is discovered to have asbestos insulation used to fire-retard steel columns girders and beams and ceilings when first constructed, this asbestos (hazardous waste) must be scraped off, packaged and transported out of the building before demolition begins. The Environmental Protection Agency (EPA) must be notified and they have specific requirements to protect the environment during asbestos removal and demolition. The EPA requires an asbestos-contaminated building be completely enclosed in plywood on all four sides to prevent dust caused by the asbestos removal and demolition to spread and contaminate nearby people and buildings. Before the building is demolished, the floors must first have the asbestos removed. One floor at a time will have asbestos scraped, packaged, and taken away in hazardous waste containers. In addition to enclosing the entire building with plywood to present asbestos dust from escaping, the EPA also requires the floor in the process of having the asbestos removed to be sealed tight and placed under a vacuum. To create negative atmospheric pressure, some of the exit stairs are closed off and sealed. When the floor is sealed off, the negative atmospheric pressure also helps to contain asbestos dust during its removal from steel columns, girders, beams, ceilings, and pipes.

These asbestos-removal protective measures, required by the EPA create problems during a fire. First, the plywood encasing the building adds tons of combustible material into the building. Even if the plywood is fire retardant, it still burns. It burns slower and generates large quantities of smoke. Second, the sealing of the building with plywood prevents ventilation to remove smoke during a fire. The plywood can not be quickly vented. The building under demolition becomes a sealed building and when the exit stairs are sealed off, firefighters, disoriented in smoke, become trapped. A building undergoing asbestos removal becomes a windowless building and normal windowless buildings such as computer switching and telephone exchange buildings require automatic extinguishing systems. These were some of the problems that led to the death of FDNY firefighters Robert Beddia and Joseph Graffagino in the Deutsche Bank building that caught fire during demolition on August 18, 2007.

Fire Protection Systems

In most states, laws require a vertical standpipe system to be installed and in operating condition in any building under construction that rises above 75 feet or the seventh floor. It must then be maintained in service during the entire construction process. Likewise, when a building is being demolished, the standpipe system must be in service during the entire demolition process. Responding to a fire in a building under construction or demolition, firefighters must carry four lengths of folded hose up to the floor below the fire and connect them to the vertical standpipe riser outlet. Water is pumped up the standpipe from pumpers connected to a hydrant. During the construction process, the standpipe riser is supposed to keep pace with the height of the structure, usually being one or two stories below the most recently constructed floor. Unfortunately, however, it rarely does keep pace with construction. Standpipe risers in buildings under construction have been found anywhere from five to ten stories below the most recently built floor. During a fire, any dismantled standpipe system or variance between the top of the standpipe and the last floor constructed should be reported to the officer in command as soon as it is discovered. Additional hose will have to be stretched. In some instances, the standpipe riser does keep pace with the floors being constructed, but the threaded top of the standpipe riser is not fitted with a cap or seal to hold the water. In the event of a fire under such circumstances, when hose is connected to the riser and water is pumped into the system, the water overflows out of the top of the riser and down the stairway. If the cap is left off and the top of the standpipe is open, there will never be enough pressure for an effective firefighting hose team. In this situation, a hoseline must be stretched to the fire from the street.

When installing a standpipe in a building under construction, plumbers are not required to make the system watertight. Its water tightness will be tested prior to the building's occupancy. The outlet valves for hose connections need not have control handles (wheels for turning valves) installed, so only the valve stem may be present. These control valves are often installed in the open position. During a fire, water supplied into the standpipe from a pumper may flow out of open valves and into the stairway. Again, no pressure will be developed in the system for an effective firefighting hose stream. The entire first-alarm assignment may have to be diverted from firefighting just to shutting off the open valves on every floor. During such an operation, the one open standpipe valve often overlooked is the one in the cellar or basement, which is frequently used as a condensation drain. If the companies report all control valves closed and still no water is getting pumped to the upper floor of the construction building, someone should check the cellar outlet valve. Engine company members should also remember to bring a Stillson wrench (pipe wrench) up to the fire floor, in addition to hose and fittings. It can be used to open a control valve installed without the wheel.

The laws in some states require that, when high-rise buildings under construction exceed 150 feet or 15 stories, an outside stem and yoke control valve (OSY) must subdivide the vertical pipe riser into sections. This intermediate OSY valve can be installed and left closed during the construction process. When firefighters cannot pump water to an upper floor hoseline and all the outlet valves are confirmed closed, they should examine the pipe riser for the presence of an OSY valve and check its position. It may be closed and preventing water from reaching the upper floors. If the stem of the OSY valve protrudes beyond the wheel, this indicates that the valve is open. If the valve wheel is at the end of the stem, this indicates the valve is closed; it should be opened. Remember, the hose outlets must be closed and the OSY valve open.

A fire department standpipe connection, called a *Siamese* is clearly visible on the front or side of a completed building. It is usually near the entrance and is sometimes marked by a sign or distinctively colored caps. In a building under construction or demolition, however, the fire department connection is often difficult to locate. It is usually concealed by the fence around the construction site, trash dumpsters, or construction materials such as sand, lumber, steel, and bricks. An experienced pump operator knows the location of hydrants around the construction site; he or she should also learn the location of the fire department connection. During a periodic inspection of a building under construction, the standpipe fire system should be closely examined: the top cap, the closed gate valves, the basement valve especially, the open OSY valve, and most importantly, the standpipe supply Siamese inlet.

Lessons Learned

A successful firefighting operation in a building under construction depends greatly upon pre-plan inspections of the site during construction or de molt ion. A fire company should periodically inspect every construction or demolition site in its district to become familiar with hazards. The best time to do a pre-plan inspection in a building under construction or demolition is on a weekend. At this time a slow, through analysis of the structure can be conducted and a fire strategy and tactics preplanned. Firefighters should note dangers presented by construction—like open shafts, stored explosive gases, and collapse hazards—and personnel should examine access to upper floors. If the building is undergoing asbestos removal, firefighters cannot enter the contaminated areas. It is declared a hazardous material site—no entry for inspections. Construction sheds, shanties, and offices will be locked on weekends and must be inspected during working hours.

Many cities require a person to be in charge of fire and worker safety—a "site safety officer." Firefighters conducting an inspection during hectic work hours should contact the site safety officer and have him or her assist you during an inspection. Construction elevators and access ladders to the top floor should only be tested with the building construction site safety officer, who is responsible for ensuring worker safety and fire safety during construction or demolition. Inspections of a building under construction or demolition should check the hazards, fire protection systems, and the hydrants. Firefighters should determine the progress and serviceability of the built-in firefighting protection system during their inspections. They should check the location, accessibility, and serviceability of the standpipe connections; the location of the standpipe riser in relation to the building height; the closed floor control valves; the open OSY valve; and the cap at the top of the riser pipe. The valves must be in proper position and the hydrants must be check for serviceability. Any fire violation discovered during and inspection of a building under construction or demolition should be reported in writing to the site safety officer for compliance.

17
COLLAPSE CAUSED BY MASTER STREAM OPERATIONS

"Say, Chief, how can you justify pulling us out of that burning vacant building last night, having us pour tons of water on the floors with master streams, and ordering us back inside the building to overhaul?" a firefighter asks at an informal critique of a fire, held in the firehouse kitchen. The other firefighters stop talking and turn toward the chief, waiting for his answer.

"I thought about that last night, also," answers the chief. "And here is how I justify it. First of all, the five-story building was vacant. If it was occupied, I would have continued with the interior hoseline attack. But I knew it was vacant because there have been several fires there in the past month. Also, the building stood alone. There were empty lots on all four sides, so there was no exposure problem. It was an ideal situation for using master streams.

"There are advantages and disadvantages to every fireground decision. The disadvantage of using master streams for long periods of time is that the volume of water can cause the building to collapse in the latter stages of the fire. The advantage, of course, is that firefighters will be safe during the initial and most dangerous stages of a vacant building fire. For example, last night, because we used master streams, none of you had to run through the dark interior of the vacant building when flames were spreading rapidly and visibility was zero. And none of you was exposed to the hazards presented by the dilapidated structure; cracked stair treads, holes in the floors, missing stair landings, broken fire escape steps and/or cracked and loose parapet walls. Preoccupied with searching, venting, stretching hose, and avoiding the dangers of the fire and smoke, any one of you could have been seriously injured by the hazards of the dark, vacant structure. So that's how I justified ordering you out of the building and using outside master streams.

"Now, here's how I justified sending you back inside to overhaul the smoldering fire after pouring tons of water into the building. It's a fact that there is always a danger of sudden building collapse after firefighters have used heavy-caliber outside streams on a serious fire for a long time. But there are precautions that a chief can take to reduce that danger and I took those precautions last night. First, I had you shut down the master streams and, for several minutes, I allowed no one to enter the building. This delay gave the structure time to settle after the shock of being hammered by high-pressure water streams. Next, I had the safety chief go inside the structure and survey each floor for signs of collapse. He looked for water accumulations that could overload a floor and structural weakness caused by the powerful, high-pressure master streams, such as columns out of plumb, sagging floors, and bulging walls. He also looked for leftover stock or material that could absorb large quantities of water and cause a floor overload.

"None of these warning signs was present. If any of these hazards had been reported by the safety chief, I wouldn't have allowed anyone to reenter the building. Instead, we would've set up an all-night watch and continued to drown the building with master streams to prevent a rekindle. Eventually, the water would have seeped down and quenched the deep-seated, smoldering fire. After the safety survey, I ordered only one company to reenter and overhaul. I restricted the number of firefighters operating in the danger area to one officer and four firefighters, instead of sending in 25 firefighters from the first-alarm assignment. This time when you entered the building, there was no danger from spreading flame and smoke and there was no great urgency. Portable lights pinpointed the weakened stair sections.

"Personal safety was now the number one priority and the fire—or what was left of it—was downgraded to the number two priority. Even after all these precautions were taken, a tragedy could have occurred—the building could have collapsed—but it did not. And that's why the decisions to go with master streams and then reenter to overhaul proved correct at this fire."

The Aerial Platform Large-Caliber Master Stream

The aerial platform large-caliber master stream is an extremely effective weapon in the arsenal of firefighting equipment. Its large-caliber stream has saved many firefighters. Instead of operating many hand lines inside a dangerous burning building for long periods, incident commanders can withdraw firefighters and quickly set up outside streams. When improperly used, however, the aerial platform stream can result in the death of or injury to firefighters and the destruction of property.

The large-caliber ground and aerial stream is a dangerous fireground machine if firefighters have not been trained to position apparatus correctly, control the supply of water to the apparatus, and properly direct the powerful stream. Three or four tons of water speeding through a nozzle at 100 feet per second is a tremendously destructive force. When improperly directed, streams delivered from ground or aerial appliances have caused ceilings to collapse, overloaded floors, knocked over brick walls and chimneys, lifted roofs off buildings, and made large roof sections of slate shingles explode into the air.

What exactly is a large-caliber stream? A fire department large-caliber stream (master stream) is a ground-based or aerial device with a fog or solid stream, capable of delivering more than 300 gallons per minute (gpm) to a fire. A typical master stream delivers 500 to 1,000 gpm. At this delivery rate, handheld nozzles attached directly to a hoseline are too difficult to control and direct, so mechanical, electrical, or hydraulic assists are required.

Fog streams with a delivery rate of more than 300 gpm and solid-stream nozzles of 1½-inches or more in diameter are considered large-caliber stream nozzles. Ground-based master streams include deck guns mounted permanently on top of apparatus and portable deluge nozzles that can be operated from atop an apparatus or removed from the apparatus and positioned closer to a fire. Aerial-mounted large-caliber streams are ladder pipes affixed to the rungs of ladder, snorkel, and aerial platform nozzles.

Three changes in the design and use of large-caliber streams over the years have increased their effectiveness:

- The hose diameter supplying water to large-caliber streams has increased.
- Radio communications and mutual-aid agreements have improved, enabling fireground commanders to quickly order and put into operation large numbers of large-caliber streams.
- Most importantly, the large-caliber stream no longer is restricted to the ground, but has been elevated 50 to 100 feet above street level. At serious fires 30 years ago, firefighters would operate more deck guns and portable deluge nozzles than aerial streams. Today, firefighters use more aerial streams.

Stream Destruction

While these changes have increased the large-caliber stream's effectiveness, they also have increased the destructive capability of its high-pressure, large-volume stream. The destructive pressure of a large-caliber stream is greatest in the area where the water leaves the nozzle. When the nozzle of a high-pressure, aerial large-caliber stream is maneuvered close to parapet walls, chimney tops, coping stones, cornices, and roof dormers, it can blast sections of these structures away from the building and cause a partial collapse. Firefighters working near a burning building where an aerial stream is operating can be struck by building fragments knocked loose by the powerful stream.

The most serious collapse danger of a large-caliber stream, however, is caused by the large volume of water it discharges into a burning building. One gallon of water weighs a bit more than eight pounds. The average stream delivers 500 gpm into a burning building, which equals two tons of water or 4,000 pounds a minute (fig. 17–1). When an aerial ladder stream has been operating for 10 minutes, it has discharged 20 tons or 40,000 pounds of water weight into the building. Three streams delivering this water into a burning building will introduce 60 tons or 120,000 pounds.

Some of the water will be vaporized by the heat of combustion; most of it will flow through cracks, beneath doors, down the stairs, and back into the street. However, an undetermined quantity of water from the stream will be absorbed into plaster ceilings and walls, dried-out wooden floors, and the porous paper and cloth contents of the building. It is this absorbed water weight that can cause a collapse.

Water from streams also can become trapped inside a watertight, sealed floor area and quickly build up to dangerous proportions—sometimes as high as windowsill level. At one fire, water from stream accumulations was seen spilling over windowsills and running down the front of a fire building seconds before all interior floors suddenly collapsed. Firefighters operating on floors below the fire were buried in the collapse.

Fig. 17–1. Two tons of water delivered by master streams can cause a building collapse.

A firefighter who has used only small, handheld attack streams during his or her career may not be aware of the destructive power of a large-caliber nozzle or the collapse danger presented by improperly using a large-caliber stream. Firefighters should understand the following principles of control and direction of large-caliber streams:

- **Weight of water.** Firefighters directing a large-caliber stream should realize the nozzle is pouring two to four tons of water a minute into the building. When they extinguish the fire in one window, they should move the stream to another window, never directing it at smoke. Firefighters should notify the officer in command when visible fire is darkened down and suggest a possible shutdown or repositioning of the stream.

- **Dangers of the upper portion of building.** The upper portion of old structures, such as chimney tops, parapet walls and cornices, may be structurally unsound and present a collapse danger even before a fire occurs. Exposed to the elements of wind and rain on more surface areas than other parts of the building, the upper portion deteriorates more rapidly and its maintenance is expensive and often neglected. When a deck gun or an aerial stream sweeps the upper portions of a burning structure at close range, it can knock over a section of a chimney or collapse a side or rear parapet wall or loose coping stones. Firefighters operating in the vicinity of the master stream can be struck by building parts blown off the building by the force of the stream.

A large-caliber stream delivers 500 gpm into or onto a burning building. This equals two tons of water or 4,000 pounds a minute. This destructive force is shooting from the nozzle at 100 psi (pounds per square inch) around 100 feet per minute.

Stream Direction

Today, inexperienced firefighters must be trained to operate master stream nozzles. An inexperienced firefighter who has used only small, handheld streams during his career may not be aware of the difficulty of directing the powerful master stream. This problem was highlighted when a firefighter was ordered to operate a deck gun into a window of a tenement to protect people trapped by flame on a fire escape. The first attack hoseline inside had suffered a burst length. Flame coming out a window prevented the people from descending the fire escape. The chief urgently ordered a quick knockdown of the fire in the window as the burst hose was being changed. He ordered the pumper repositioned in front of the fire building and the booster tank water to supply the deck gun atop the truck. The booster tank had 500 gallons of water. It provided only about two minutes of deck gun use. The order from the chief was a quick knockdown, to drive the flames back into the window below the people on the fire escape.

When water came out of the nozzle, the young firefighter, unfamiliar with the difficulty of maneuvering a large master stream, directed the water stream up the face of the wall along side the flaming window. Then he moved it across the top of the wall above the flaming window and finally down the side of the wall next to the flaming window and ran out of water. Then the water supply ended. He completely missed the flaming window. Needless to say, the chief was upset. Fortunately, the interior hose stream was restarted and extinguished the fire coming out the window and the people were safely removed from the fire escape. All firefighters should be trained in the control and direction of large-caliber aerial and ground stream nozzles.

Water Accumulations

With a greater overall view of the fire scene than personnel working on the ground, firefighters operating elevated streams high above street level are usually the first to detect hazards, such as water buildup on a floor or roof. They should notify the incident commander immediately of any water accumulations.

Roofs surrounded by parapet walls on the four sides of the building are especially prone to water buildup. If drains are clogged and several ground-level, large-caliber streams are operating, a roof area may fill up with water quickly and collapse on firefighters inside the building.

The hollow area inside a marquee or canopy attached to a building is another place where water accumulates. The weight of the water built up inside the void of a marquee where drains are clogged can make the marquee collapse and also bring down the front facade wall to which it is fastened.

Stone or Brick Veneer Wall Collapse

When a large-caliber stream is redirected from window to window on a burning masonry building, it strikes the brick wall between openings at close range. If the cement bonding between the finished stone or brick veneer and the back wall to which it is attached has lost its adhesive qualities over the years, the impact of the stream can cause large sections of stone or brick veneer to collapse into the street below. A large-caliber stream continuously directed at a brick or wood shingle wall at close range can blast away the wall and throw fragments of the wall into the air.

Sounds of a Large-Caliber Stream

At a fire where smoke reduces visibility to zero in the street, firefighters must rely on the sound of the large-caliber stream striking objects to determine its effectiveness. For example, when a large-caliber stream strikes a brick wall in smoke, a "splattering" sound is made by the stream striking the wall. When it strikes the side wall of a wooden building, a "drum" sound is made by the stream striking the wall. When it enters a window, the sound of the large-caliber stream is reduced and only a distant rumble is heard.

Collapse Zone for Aerial Streams

In the past, some chief and company officers used a dangerous firefighting strategy when operating large-caliber streams at major fires. When the wall of a building was in danger of collapse, all firefighters working at ground level were ordered to withdraw from the perimeter of the building and a collapse danger zone was established. Then, firefighters operating tower ladder streams were ordered to begin operations, with the elevated stream operating close to the perimeter of the building and within the collapse danger zone. This no longer is the strategy (fig. 17–2).

When chiefs establish a collapse zone for firefighters operating in the street, they now take a similar safety precaution for firefighters operating aerial large-caliber streams. This is imperative because, in recent years, an increasing number of building collapses have seriously injured firefighters operating aerial master streams. The collapse zone for an aerial stream will vary slightly because of the height of the nozzle above ground level. The tip of the aerial ladder or platform basket should be kept away from a weakened wall a distance greater than the height of the wall above the bucket floor.

Chapter 17 | **Collapse Caused by Master Stream Operations**

Fig. 17–2. This aerial master stream was operating in the collapse danger zone.

Master Stream Close Approach in Windows

When an aerial master stream is to be used for a quick knockdown, the master stream nozzle is most effective when placed close to the window of the building. This close-up position can give the stream deep penetration and its widest horizontal range inside the floor area. A large, burning open-floor area of a supermarket or factory floor sometimes can be quickly extinguished when the master stream is positioned near the window opening and the stream sweeps the floor. However, this close approach should not be ordered if the wall over the window is in danger of collapse. When there is a danger of structural collapse, no part of the aerial stream bucket and/or nozzle should be positioned where it could be struck by a falling wall. There have been instances when this fireground precaution was not heeded, resulting in firefighters in the bucket of an aerial platform being buried with falling bricks, the tower ladder bucket being torn from the apparatus boom, or the ladder being tipped over on its side by the weight of a collapsing wall.

At one recent fire in a row of stores in the Bronx, New York, a parapet wall collapsed on firefighters in a tower ladder bucket. Everyone at the scene was notified by the safety chief of a dangerous parapet wall leaning out over the burning stores. It was a five-foot-high, 75-year-old brick parapet. A collapse zone was ordered. Some time later, the ladder was ordered to reposition for use of the aerial master stream in front of the building and so the apparatus was moved in front of the fire building. The chauffeur safely positioned the ladder truck outside the collapse zone, away from the dangerously leaning parapet wall. Supply lines were being connected.

However, to enable two firefighters to climb into the bucket for operations before raising, the chauffeur raised the boom and bucket from the bed and lowered it onto the sidewalk inside the collapse zone. As the firefighters walked in the collapse zone and opened the gate to step into the bucket, the parapet wall came crashing down on top of them. One firefighter suffered a concussion; the other a broken collar bone and a dislocated shoulder. The lesson learned here is when a wall is in danger of collapse, the fire apparatus and the tip of the tower ladder must remain out of the collapse zone. The priorities of positioning apparatus are life safety first, which includes the lives of firefighters; fire containment is the second priority. That rule never changes.

- **Renovated Buildings**. Renovated brick buildings being gutted have all interior partitions removed, with only the four brick walls and wooden floors remaining in place. An aerial large-caliber stream is particularly effective when there are no interior partition walls to obstruct the stream reach, as it can penetrate through the entire depth of the burning floor. However, when a fire occurs in a renovated building that does not have great floor depth from front to rear wall, the powerful large-caliber stream directed through a front window can travel through the burning floor area and strike the inside of the rear wall with sufficient impact to collapse it into the rear yard. Firefighters operating a hoseline in the rear yard can be buried under the tons of falling brick that are collapsed by the stream.

- **Flanking a Building**. When there is a danger that a wall will collapse outward with explosive force and be driven beyond the normal collapse zone, firefighters should operate streams from a flanking position. If a portable deluge nozzle or an aerial platform is placed on one side of a weakened wall of a building in front of an adjoining building, the stream range operated on an angle into a window or doorway of the burning building may be limited. However, the firefighter will be protected from the collapse (fig. 17–3).

Chapter 17 | **Collapse Caused by Master Stream Operations**

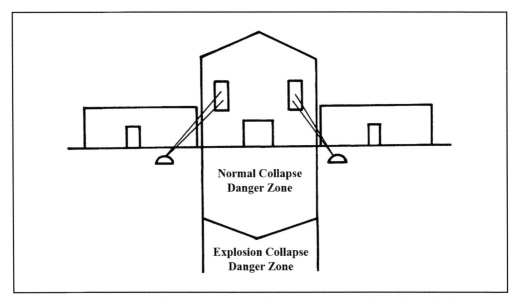

Fig. 17–3. Master streams must often be positioned in a flanking position to avoid a possible wall collapse.

Aerial streams operated by firefighters above the roof of a burning building will also be less effective, but they will be safely above the collapse zone of a weakened wall. Again, when there is no danger of wall collapse and aerial master streams are needed because of the size of the fire, firefighters may operate the aerial nozzle close to the flaming window of the building for effective penetration of the master stream. But before this tactic is used, the incident commander must size up the wall above the windows. If the wall appears unstable, this tactic should not be used. Instead, withdraw the tower ladder bucket back out of the collapse zone of the unstable wall and use the reach of the stream from a safe distance to penetrate the flaming window, or operate the bucket of the aerial master stream above and attack the fire at roof level.

Ceiling Collapse

Firefighters often use large-caliber streams when fire spreads through concealed spaces above suspended ceilings in supermarkets and rows of stores. These streams are effective when they are directed from below the ceiling. This works well in large, open areas. If part of the suspended ceiling has fallen or burned away the concealed space above, the area can be swept with a large-caliber stream. In other instances, the force of a large-caliber stream directed from below can break through a ceiling and expose flame above. This can be accomplished effectively when firefighters operate large-caliber streams near the front of a store or supermarket where they previously have removed a large open display window to vent the fire. When an interior partition or the absence of large windows prevents this strategy, firefighters must operate aerial large-caliber streams from above, through a burning roof, where the fire is beyond the control of an inside hoseline attack.

When either of these strategies is used, water from large-caliber streams may become trapped in pools above a watertight suspended ceiling or absorbed into sound- or heat-insulating material in or above the suspended ceiling. If the water weight becomes excessive, a large section of the ceiling can collapse suddenly. This is one reason why it is unsafe to direct outside large-caliber streams into burning buildings while firefighters are inside the structure. Even after they have shut down and reentered a building with a suspended ceiling, firefighters should examine the space above the ceiling for water accumulation or absorption.

Strategies for Master Streams

There are two strategies for master stream use. One is using a master stream for a "temporary knockdown" of a large body of flame. In this strategy, firefighters inside the burning building are withdrawn and then reenter to conduct interior firefighting when the master stream is shut down. During this strategy, a temporary knockdown, the incident commander must confirm that all firefighters have withdrawn to a safe position before the master stream is used. Interior forces must retreat to the floor below or out of the building before the large-caliber stream is directed into the burning building.

The second strategy is an exterior master stream attack used for final extinguishment. When this defensive strategy is used, no firefighter should remain inside the burning building during the master stream attack.

When master streams are put into operation for total extinguishment of a fire, the force of the high-pressure streams, the weight of the water absorbed into the building and the fire destruction over a long period of time can weaken the structure. The incident commander must prepare for collapse of the building. The chief must withdraw all firefighters operating inside the building, as well as order those working around the perimeter of the building to move beyond a collapse zone. A collapse zone is the distance away from the wall equal to one, one and a half, or two times the height of the unstable wall.

Overhauling Master Stream Strategy

After master streams have been used to extinguish a fire, overhauling becomes a dangerous strategy because of collapse danger. Several deadly building collapses have occurred after master streams were used and firefighters returned into the building to extinguish spot fires and prevent rekindle. Collapse happens during overhaul because the building has been destroyed by flame, the impact of the master streams has pounded the building, and tons of water have been absorbed by the plaster, wood, and concrete of the structure.

After a long-duration master stream operation is completed and the main body of fire has been extinguished, the incident commander should order master streams shut down. The large-diameter streams used for total extinguishment have accomplished their task. Now, before any firefighter is ordered to reenter the burned-out, smoldering, water-soaked structure to overhaul, the following safety actions should be taken. The building should be allowed to drain. Next, the incident commander must order a safety officer to conduct a safety survey.

The safety chief should determine if it is safe to allow firefighters to enter and overhaul. During this inspection, the safety chief looks for the following warning signs: broken stair treads and cracks in marble intermediate landings; floor and ceiling sagging due to the weight of water; fire damage to steel columns and girders that are twisted, warped, bent, or elongated. The presence of trusses and/or lightweight steel bar joist or wood I-beam construction must be reported to the incident commander. Water accumulations and heavy machinery are warning signs that could trigger floor failure. Water-absorbing content, such as paper bales, baled rags, or plumbing supplies, are heavy loads that could cause collapse during overhauling.

If the safety officer decides that the structure is safe, the incident commander may start overhauling, but limit the number of firefighters who reenter to extinguish spot fires. However, if the safety officer decides the structure is unsafe, no firefighter should reenter the structure. Instead, the incident commander should order *defensive overhauling*. Defensive overhauling is a strategy of using a master stream or several hoselines directed from outside the burning building and outside the collapse zone, into the burning structure for several hours or days, if necessary. Firefighters may be rotated each tour. Defensive overhauling—sometimes called a "watch line"—is continued for however long it takes until the smoldering fire is quenched.

Lessons Learned

Chief and company officers have a personal responsibility to safeguard firefighters when large-caliber streams are put into operation at major fires. Even though firefighters direct and control the large-caliber streams, chiefs and company officers constantly must monitor and evaluate their effectiveness and safe use. The incident commandeer also must confirm that all firefighters have withdrawn to a safe position before allowing them to use a large-caliber master steam. It is not enough to simply order all companies to withdraw. The chief must wait for the officer directing operations inside the building to confirm the safe withdrawal. Over the years, firefighters retreating from a fire building have been burned and scalded with steam created by the outside master streams directed into the building. They were not given sufficient time to leave.

The safe transition from interior to exterior attack of a structural fire requires four elements:

- Effective communication between interior and exterior sector commanders
- An interior sector who has effective command and control over the firefighters
- A pump operator who waits for the order to start water from the incident commander and does not prematurely start water to the master stream

- An incident commander who understands the priorities of fireground safety—protection of life first, including firefighters; fire containment second; and property protection last

When master streams are put into operation, not for a quick knockdown, but for total extinguishment, the force of the high-pressure streams and the weight of the tons of water poured into the building will weaken the structure. The officer in command must prepare for eventual collapse of the building. The incident commander must withdraw firefighters operating inside the building, as well as order those working around the perimeter of the building to move beyond a collapse zone.

Finally, the officer in command should order the master stream shut down as soon as they have accomplished their task. Then, before any firefighter is ordered to reenter the burned-out, smoking, water-soaked structure to overhaul, a safety chief first should inspect the structure for stability. If this officer decides that the structure is unsafe, no firefighter should reenter the structure. Instead, the incident commander should implement a defensive overhauling strategy using master streams. A watch line should be established.

18

SEARCH-AND-RESCUE AT A BUILDING COLLAPSE

On a hot summer night, an alarm is received in an older, downtown section of town. As two engine companies arrive simultaneously at the fire, they are greeted by a blazing vacant three-story wood-frame residence building 20 feet wide and 40 feet deep. The structure has one-story commercial buildings on either side of it. The dark street is aglow with the flames coming out of every window of the building's front. A thermal column of heat showers sparks into the night sky. The firefighters of the first engine connect to a nearby hydrant and a 1¾-inch hoseline is stretched to the front of the building. The captain of the first-responding engine calls for water, and the stream is directed into the flaming open doorway. The flames absorb the stream of water; there is no extinguishing effect. The captain tells the officer of the second-arriving engine company to back them up with a portable deluge nozzle. The captain surveys the area. Looking up, he sees a rusted metal fire escape attached to the front of the building. Flames shoot out every top floor window and are about to spread to the decorative wood cornice at the roof level.

The captain and three firefighters are standing inside an old cast-iron fence that once enclosed a garden in front of the abandoned burning structure. Realizing his men are too close to the front of the fire building, the captain taps the nozzle firefighter on the shoulder and says, "Move back. If that wall falls, we will be buried." Just then there is a loud cracking noise. He shouts, "Watch out!" The front wall of the burning wood-frame building cracks horizontally across between the first- and second-floor windows and collapses outward. The lower one-story portion falls outward on top of the firefighters operating the hoseline. The upper two-story portion of the front wall falls back into the building and pancakes down on top of the collapsing floors. The chief officer, who is walking up to the front of the building to assume command of the fire at the moment of collapse, sees the falling building engulf the firefighter.

When the rumbling noise stops and the dust, smoke, and sparks covering the building clear, shouts of firefighters trapped and firefighters running to their rescue can be heard.

"Help! Help! Get me out of here. I'm burning!"

"Help! We are trapped."

"Get this building off me."

"Hey, there are firefighters trapped there."

"Give us a hand."

"Come on! Let's go!"

The scene is chaotic. Firefighters are caught in a small crawl space behind the bars of the cast-iron fence. Flames and sparks are beginning to fall down into the small space where the firefighters are caged. The heavy, burning front wall, which has collapsed outward, is prevented from crushing the trapped captain and the firefighters only by the cast-iron fence. A 2-foot lean-to void has been created. The trapped firefighters are prevented from crawling out from under the collapsed building by the metal fence, and the rescuing firefighters are prevented from reaching the trapped men by the metal fence. Part of the fallen wall on top of the firefighters has filled the space where the fence gate had been, blocking this avenue of escape. Immediately there is a rush of all firefighters to the pile of burning rubble. They attempt to rescue the trapped men. There is no organization to their efforts, however. Some firefighters jump on top of the fallen wall to search for surface victims; others attempt to lift the wall up by hand. Some firefighters call for the hose stream to cool off the trapped members. Other firefighters shout to stop the hose stream because it is getting them wet. Some firefighters are able to lift one end of the fallen wall upward off the fence, but there is no one to apply shoring beneath the heavy wall. The wall falls back down, collapsing the fence slightly and causing further injury to the trapped firefighters.

Search-and-Rescue at a Building Collapse

Control and organization is one of the most important objectives of collapse rescue operation; without it, rescue and removal can become a mob scene. Admittedly, it is extremely difficult to control and organize a rescue effort at a collapse that suddenly occurs during a fire, where a large number of firefighter on the scene rush to the collapse area to dig out buried firefighters. This first reaction to search for victims is ineffective when it is done without knowledge of where the victims are buried. In fact, without coordination or a plan, the rescue efforts could cause further injury or death. Even when the exact location of the victims are known, the fewer men engaged in actual rescue work the better. It is the quality of effort not the quantity that is needed in such situations. The best way to gain control of a collapse rescue effort is for the incident commander to issue specific assignments to officers who should assemble their company members and carry out assignments; expand the incident command system assigning division officers and group officers areas and functions of responsibility; designate a victim tracking officer; and most important, use a collapse rescue plan.

There have been changes to the collapse rescue plan used by firefighters in this country. The collapse search-and-rescue operations in Oklahoma City and New York City highlighted needed improvements in the plan of action. Terror attacks at the Alfred P. Murrah Federal building on April 19, 1995, and the World Trade Center on September 11, 2001, required the fire service to reconsider and reevaluate its collapse search-and-rescue procedures. Lessons learned at these disasters must be included in our everyday collapse search-and-rescue operations. New lessons learned at Oklahoma and New York City are examined in the following sections.

Secure the collapse site

An important lesson learned is to secure the area around the collapsed building. After establishing command, one of the first actions the incident commander must direct is to restrict access to the scene (fig. 18–1). The police department should secure the collapse area as soon as possible. The area should be sealed off by police officers, street barricades, even barbed wire. The access points, leading in and out of the collapse site, should be guarded. Only emergency vehicles and personnel should be allowed entry. The people within the secured collapse area should be escorted out of the collapse area. If the collapse is a suspected terror attack, people in the area will have to be screened and questioned. No one except fire, police, medical, and authorized governmental officials and construction workers should be allowed to enter the operation zone. This action is critical for safety of the community and for maintaining control inside the collapse site. Police must keep the streets around the collapse area open for emergency and government officials.

Fig. 18–1. Securing the scene

At the Oklahoma City collapse, the fire department was able to secure the area around the collapse site. In New York City, the fire department lost control of the area and streets. The Oklahoma collapse site was initially out of control as townspeople ran into the bombed building site to rescue people. Everyone in nearby stores and offices wanted to help. Very shortly after this initial rush to rescue, there was an erroneous report of a second bomb about to explode. During this period everyone retreated. This provided time for fire and police to get control of the area around the collapse. The area around the smoldering Murrah building was quickly surrounded with barricades and fencing and locked down. No one except authorized personnel was allowed back into the secured area for the 16-day duration of the collapse search-and-rescue. Only the FBI, police, fire, and construction personnel were inside the rescue zone. Unlike Oklahoma City, the 16-acre disaster area of the World Trade Center was not secured. Due to the enormity of the devastation and loss of life, it was impossible to secure the large area for several days after the collapse.

Incident command system

Another lesson learned at the two terrorist bombings was that when there is a collapse and fire spread, the incident commander should consider establishing an incident command system dividing "operations" into two branches, one for firefighting and one for collapse search-and-rescue (fig. 18–2). An "operations" officer is placed in charge of the two branches. The firefighting branch supervises a firefighting plan, while another branch supervises a collapse rescue plan. This division of operations into two branches is necessary at a large operation when there is a collapse and a spreading fire at the same time. Overseeing two urgent complex tasks, firefighters battling a blaze, side by side with firefighter performing collapse search-and-rescue will overwhelm a single fire ground commander. When this happens, responsibility of fire strategy and collapse strategy should be divided and delegated to commanders.

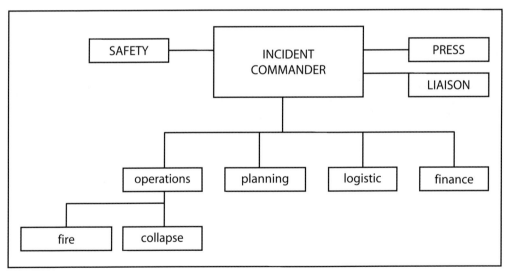

Fig. 18–2. Incident command structure with fire and collapse branches reporting to Operations

Fire extinguishment

At every structural collapse the potential for fire is great. Broken gas pipes and electric wiring will be damaged during a building failure. Rapid fire spread in the high surface to mass rubble of a broken structure, fallen partitions and content strewn across a large area must be stopped. To do this, large diameter hoselines must be stretched and an aerial platform master stream must be quickly positioned near the collapse site. There could be an explosion or rapid fire throughout the collapse rubble. For example, when a collapse occurs in a downtown area in a row house, hoselines should be positioned at the front and rear of the collapsed building to protect trapped victims, protect adjoining exposures, and to prevent fire spread in the pile. In addition, an aerial master stream with greater reach should also be positioned near the collapse site. If the aerial platform is not needed for the fire extinguishment, it can be effectively used for aerial reconnaissance and viewing the overall collapse site fire and rescue operations (fig. 18–3).

Fig. 18–3. Aerial devices cover the front and are positioned at the corner outside of the collapse zone.

Collapse search-and-rescue

A collapse search-and-rescue plan should be put into action at a collapse operation. The collapse search-and-rescue plan used in Oklahoma and New York City was:

1. Secure the area.
2. Survey the collapse site.
3. Shut off utilities.
4. Remove surface victims.
5. Search voids and remove trapped victims.
6. Take a time-out to reexamine safety and determine locations of buried victims.
7. Tunnel and trench to buried and trapped victims.
8. Begin general rubble removal, which includes tearing down the rest of the structure and hauling off the rubble.

This is an expansion of the collapse search-and-rescue plan that originated in London during the World War II bombing. When the bombs fell, the buildings collapsed and buried people in the rubble. The London Fire Brigade organized a basic five-step procedure to search and dig for buried victims in the rubble after a bombing:

1. Survey the area.
2. Remove surface victims.
3. Search voids and remove victims.
4. Tunnel and trench to reach buried victims.
5. General rubble removal.

This plan has been expanded by firefighters in the United States into an eight-step collapse search-and-rescue procedure. The changes are based on collapse search-and-rescue experience gained by the American fire service. Securing the area around the collapse has been added; shutting off utilities has been added; and taking the time out to reassess the risks and plan the tunneling and trenching has been added. In London during the bombing, securing the area was not a problem, there were not many people leaving the bomb shelters to sightsee a collapse search-and-rescue operation. In World War II London, there was not as much piped gas and electric appliances in the buildings. In this country, shutting off the utilities in a modern building to prevent a gas explosion or electrocution or drowning is a major part of a collapse rescue. Shutting off utilities saves many lives from fire and explosion after a collapse has happened.

Finally, taking a time-out has also been included in the collapse rescue plan. Risk management and the safety of rescuers is a major concern at a collapse rescue. Experience has shown that after removing surface victims and a void search, the risk benefit between, finding buried victims versus danger to rescuer changes dramatically. After surface victim removal and a void rescue, 75% of the survivors have been found. Now, danger to rescuers increases and the chance of finding a survivor decreases. Today, the incident commander orders a time-out to reassess the rescue operation and improve site safety.

Site safety size-up

After ordering police to secure the area, the incident commander must order a site safety survey of the six sides of the collapse site—the four sides and above and below—to size up the scene. For example firefighters must be ordered to examine the entire collapse site, not just the front of the collapsed building. During this survey, firefighters look for four important items: secondary collapse dangers, fire spreading in the structure, access areas to the interior of the collapse structure that may be used by rescuers, and trapped victims out of sight in the rear or side of the collapse area.

A safety survey may reveal many more trapped victims in need of rescue at the sides and rear of the building. Any potential fire spread problem must be determined at the sides and rear of the collapse site—this could spread quickly and trap firefighters and victims (fig. 18–4). The site survey may uncover valuable avenues of access to the interior of the collapsed structure that could save time reaching buried victims in a cellar. The results of the six-side site survey must be reported to the incident commander.

Fig. 18–4. Step 2 of a collapse rescue plan is to make a size-up of the scene.

Utility shut-off

After securing the area around the collapse area and conducting the site safety survey, the incident commander must order building services—gas, electricity, and water—shut off. After a building collapse, a broken gas pipe may cause an explosion and fire. If gas pipes are ruptured by the collapse, gas and flame may spread rapidly and must be controlled before work on the collapse pile can begin. The local utility companies must be called to the scene, but before they arrive, firefighters should be assigned to shut down utilities. Utility companies are often delayed at night or on weekends. Leaking gas could also accumulate in a concealed space where a victim is trapped and displace life-giving oxygen.

At any collapse site, even if there is not fire, hoselines must be stretched to the collapse and an aerial master stream must be positioned near the building in case a gas explosion or fire erupts. These lines protect firefighters and exposures. Just as important as gas shut off is the electric utility shut off. If electric power is not cut, "live" wires threaded throughout the collapse rubble may ignite leaking combustible gas or electrocute victims and rescuers. Modern buildings are a maze of electrical wiring. If a rescuer gets a metal tool entangled with the live wire, it could mean a slow and quiet electrocution. Water is another utility that must be controlled during a collapse rescue. Victims trapped below grade will often be the last to be rescued. Broken water pipes or long use of master streams may cause water accumulation in below-grade areas. Drowning is a possibility as people take refuge in below-grade areas. Water from hose streams may be needed to keep flames from trapped victims; however, long-term use of water must be considered a danger. Water removal operations should be started at a major collapse.

Rescue Surface and Victim Trapped In Voids

After a collapse, some survivors will have escaped from the rubble by their own efforts. They may be seen staggering around injured, lying on the ground or on the collapse pile calling for help. First responders should remove these victims to safety and administer first aid. At the same time as surface victim rescue, if sufficient resources are on scene, other firefighters may search voids that could contain trapped victims (fig. 18–5). Several steps of a collapse rescue plan may be carried out simultaneous if sufficient resources are available—it is not necessary to complete one step before starting the next.

Large sections of collapsed floors or walls sometimes form large spaces, crevices, and voids in the collapse rubble where victims are trapped, but alive. If voids or spaces are stable, and accessible through openings, they may be quickly entered and searched. In smaller voids, firefighters may shine lights and call out to any victims pinned inside these small voids. If a call for help is heard, a tunnel or trench search is started. Some voids have no avenues of access and are completely sealed by the falling structure. People trapped inside can only be heard calling for help. These concealed spaces should be cut open and examined during the void search of a collapse rescue plan. The results of all void searches must be reported to the incident commander.

Chapter 18 | **Search-and-Rescue at a Building Collapse**

Fig. 18–5. As victims are removed from the surface, a search should also be done of void spaces.

Time-out

After all surface victims have been removed, and all voids searched there should be a *time-out*. During this time-out stage of collapse rescue all firefighters are withdrawn from the collapse rubble pile and reassessment of all safety conditions and a planning session are conducted. Experience has shown us, that at this time, 75% of the survivors have been rescued from the rubble. Now the risk/benefit ratio turns against the rescuers. The risk of death or injury increases to the firefighters and the reward of possibly finding victims alive decreases. During this time-out break, another more thorough safety survey is conducted of the entire site. A sounding and listening period for buried victims should be conducted. A victim could be buried in rubble calling out for help. To do this, there should be a period of silence, all noises should be stopped so rescuers can call out to buried victims and listen for any responses. More effectively, sound-sensing electronic devices are available today for use by firefighters to monitor sound from below tons of rubble. Search cameras and search dogs can be safely used during this time-out period. The entire collapse site is analyzed. A conference is held between the incident commander, operations officer, planning officer, and the victim-tracking officer to determine who is missing, where in the collapse rubble the suspected victim may be buried and the best method to tunnel or trench to the pinpointed area.

During the initial stages of a collapse operation, a victim-tracking officer should be assigned to the planning team. The victim-tracking officer's exclusive duty is to attempt to determine the exact number of missing victims, those at first aid stations and ambulances, in hospitals and home, and those possibly trapped in the rubble. The victim-tracking officer gathers information, analyzes the building collapse and makes a determination where individuals are buried. This information is necessary before tunneling and trenching to a specific location begins. Before rescuers start tunneling and trenching, a victim-tracking officer must be able to state the following:

1. The missing person is confirmed by a coworker to have been inside the building during the collapse.
2. The person is not safe at a nearby hospital or in an ambulance.
3. The missing person is not being treated at a first-aid station or has not left the scene and gone home.
4. The exact floor location where the victim was last seen before the collapse.
5. The victim-tracking officer must pinpoint the area where collapse could shift the victim in the rubble.

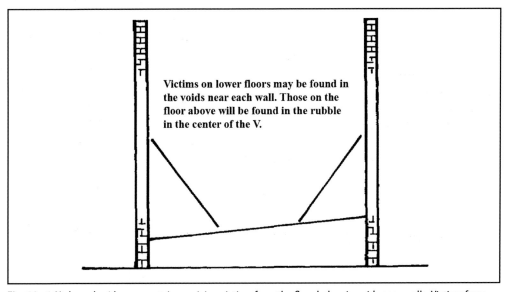

Fig. 18–6. V-shaped voids may contain surviving victims from the floor below in voids near walls. Victims from the collapsed floor may be found in the rubble at the center of the "V."

For example, if a person is determined to have been on a floor that collapsed, a V-shape collapse of the floor shifts the victim to the bottom of the V (fig. 18–6); a lean-to collapse of the floor shifts the victim to the lowest end of the lean-to collapse (fig. 18–7 and fig. 18–8); a tent or A-frame-type collapse of a floor shifts the victim to the lower, outer ends of the collapse (fig. 18–9); a pancake collapse of several floors may not shift the victim. Instead, the victim falls straight down with the collapsing floors (fig. 18–10).

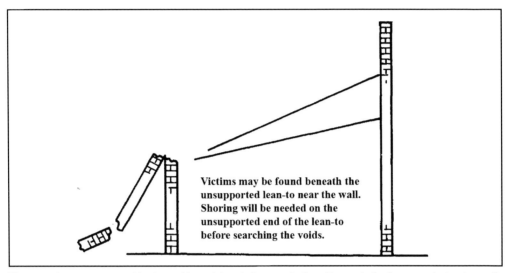

Fig. 18–7. In an unsupported lead-to void, surviving victims may be found beneath the floor near the bearing wall.

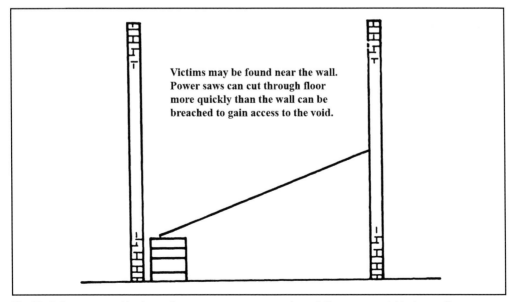

Fig. 18–8. In a supported lead-to collapse, power saws can cut through floors more quickly than walls can be breached.

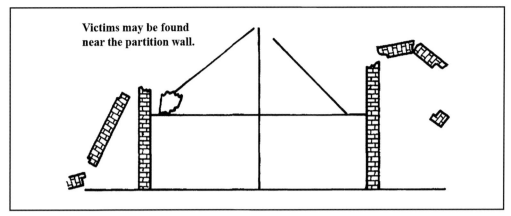

Fig. 18–9. When a tent void is created by a collapse, surviving victims may be found in the voids near the partition wall.

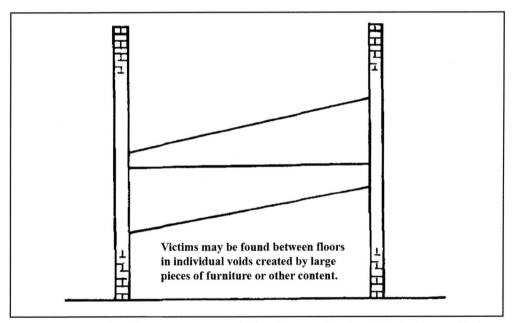

Fig. 18–10. In a pancake void, surviving victims may be found between floors in individual voids created by large pieces of furniture or content.

The victim-tracking officer has one of the most complex duties at a collapse operation. It is also one of the most important duties. If a firefighters is killed during tunneling or trenching operations, the incident commander can not state: "We had a report a person was trapped." There must be solid investigation to determine victims are trapped. Police, fire marshals, and firefighters must be assigned to assist a victim-tracking officer. These investigators must question survivors and check hospitals, ambulances, and first-aid stations. They may have to visit residences of reported missing victims to confirm that those people did not leave the scene and go home.

During the time-out, the incident commander orders another safety survey and ensures utilities are shut off. Any unstable portions of the collapsed building must be shored up. Lighting is positioned around the site in preparation for the onset of darkness. A transit to monitor unstable walls is set up. A *transit* is a surveyor's tool that has a telescope that can be sighted on a crack in an unstable wall. The transit can see a crack in a wall or a wall movement not visible to the human eye from a safe distance. If the crack widens or enlarges, it can be detected through the transit telescope. The firefighter manning the transit should continually monitor an unstable part of the structure and report back to the incident commander if there's the need for additional shoring, or removal of the unstable structure (fig. 18–11).

Fig. 18–11. A transit can monitor the slight movement of an unstable secondary collapse danger.

Secondary Collapse

After every structural collapse where portions of the building remain broken but upright, there is a danger of secondary collapse. A masonry wall may be cracked and leaning over the area where people are trapped, or a partially collapsed roof, unsupported at one end, may be hanging above rescuers. A second collapse can kill more rescuers than the number of firefighters originally trapped or buried. The danger of a secondary collapse increases as firefighters work and move about the fallen building. They will cause vibration of the remaining structure. Moreover, when large structural elements are mechanically removed from the wreckage, a second collapse may occur.

At an extensive collapse where prolonged rescue operations are to be conducted for a considerable period of time, steps must be taken to shore up and secure unstable portions of the collapsed structure. A company should be ordered to survey the structure and report back the need for shoring. Emergency shoring may consist of portable jacks or cut timbers placed beneath an unsupported lean-to roof or bracing an unstable wall. Rope may be used to secure the end of an unsupported roof beam to a higher stable structure. The important point to remember when shoring is not to move or attempt to restore the unstable structure to its original position. This attempt could cause a collapse. If the supporting object or shoring is gently placed beneath the unstable structure, barely in contact with it, as the rescue work proceeds, movement of the building will tighten the point of connection at the shoring. When shoring is placed at an unstable structure, an observer should be positioned at that point to monitor the structural member continuously for further movement that might endanger workers below. If the wall leans out farther, or the unsupported floor appears to move, rescue work must be stopped in the area and additional shoring or removal undertaken.

Firefighting

When a collapse occurs during a fire, the first duty of the officer in command is to determine who is trapped or missing. The company officer must have a record of all the firefighters riding on the apparatus. A list of firefighters' names on a "riding list" or a tag with the name of each firefighter should be kept on the apparatus. After a collapse, this riding list or tags containing names of responding firefighters must be checked during a roll call. The officer of every fire company must know how many firefighters are responding on the apparatus and must be able to identify by full name any member of his company who is reported missing. The roll call results are reported to the incident commander.

Support Personnel

A rescue team engaged in victim removal requires supporting firefighters to supply rescue material and rescue equipment to them. Just as a doctor requires a team of assistants during an operation, so a rescue team in the process of removing buried victims requires cooperation and help from other firefighters. Instead of rushing in an overcrowded mob scene to dig out surface victims, some firefighters should act as support personnel, bringing needed tools, lights, and first-aid equipment from fire apparatus to the collapse site. All available rescue tools must be brought to the vicinity of the rescue. At night, portable lights must be set up and supplied with a power source. Someone must be ready to cut timbers to any size requested when these are needed for shoring. Rescue tools such as air inflation bags, portable jacks, and spreading bars must be made available at the scene. First-aid equipment should be ready by medical personnel before the victim is freed from the rubble. Stretchers, resuscitators, and bandages will be required. The support team will have to locate all of these essential items and bring them to the rescue site. Behind every successful rescue operation are many unheralded firefighters and medical personnel who supply the rescue team with everything they need. Immediately after a collapse during a fire at which firefighters are trapped and rescue is required, the officer in command must summon additional help. A rescue company and medical assistance will be needed immediately. Additional fire companies must also be called to continue the fire extinguishment effort. After a large-scale structural failure, the members at the scene will need replacements. Some will be injured or in shock, others will be engaged in rescue work or searching for victims. Other companies may be directed to cease firefighting and begin collapse rescue operations such as site survey, utility shutoff, searching voids, and shoring unstable structural portions of the collapse.

Tunneling and Trenching: Digging to Buried Victims

During World War II, the London firefighters called this phase "selected debris-removal." We call it tunneling and trenching. When the collapse search-and-rescue resumes for tunneling or trenching, the incident commander should limit the number of rescuers for safety reasons.

This phase of the collapse plan is accomplished by hand-digging tunnels or open trenches to predetermined areas. Trenching is carried out most often. It is slower but safer. There is less chance for secondary collapse. Open trenches are dug to specific locations in the rubble where victims were last seen, or where they could have been moved by the shifting structure. During the tunneling and trenching, digging firefighters use a pass-along method to remove the material. Firefighters in the pass-along method form a single line from the digging area out to the street where rubble is being piled. The collapse rubble is passed from one person to another. This pass-along method is part of the tunneling and trenching operations and reduces the potential for fall injury, because firefighters remain in place while they pass along rubble. They do not have to walk on the unstable uneven surface of the collapse rubble. Also more material can be effectively removed, more quickly, with this pass-along method of tunneling and trenching.

Remove all collapse site rubble

The final phase of the collapse search-and-rescue operation is general debris removal. The remaining structure is demolished and carted off to a specific dumping area. This may take days or months. After all the victims have been discovered or accounted for, and all the pinpointed locations of tunneling and trenching have been reached, the incident commander orders the start of the final stage of a collapse plan—general debris removal. The fire service remains in charge of the scene during this entire phase. Even if all victims are accounted for, the general debris removal should be carried out. There may be a person buried in the collapse whom no one reported missing. At this stage of rescue, there is little chance of survivors, so mechanical cranes and pay loaders may be used for demolition and to lift large structural pieces from the collapse to waiting dump trucks. However, each load of wreckage lifted out by a crane's bucket should be first spread out on the ground and examined by firefighters (fig. 18–12).

Fig. 18–12. General debris removal in progress

Lessons Learned

The first order an incident commander issues at a collapse is to secure the area. When directing a collapse search-and-rescue operation caused by an earthquake, tornado, hurricane, gas explosion, renovation, demolition, snow load, or fire, the incident commander must first have the police secure the area. Request barricades around the entire collapse area. Police should be stationed at entrances and exits from the collapse site. Remove all people from the collapse site, checking for possible terrorists, and prevent sightseers from entering the secured area. Allow only fire, police, and medical and construction workers inside the fire lines.

Hoselines and aerial platforms must be positioned and charged with water at a collapse rescue operation. If a gas pipe breaks or a container of flammable gas or liquid is ruptured during the collapse, there may be a fire. To protect rescuers searching or digging at a collapse site a hoseline or, better, a tower ladder master stream must be ready to extinguish a sudden fire.

There must be a collapse rescue plan of action: 1. secure the area, 2. survey the entire site, 3. shut off utilities, 4. rescue surface victims, 5. search voids, 6. take a safety time-out, 7. tunnel and trench to buried victims, and 8. start general rubble removal with cranes.

The incident commander must realize that the first reaction at an explosion or collapse is to rush onto the collapse pile and assist victims. Important parts of a collapse rescue plan may not be accomplished. The area may not be secured. There may not have been a six-sided survey conducted. Also, the utilities may not have been shut off. Firefighters and citizens will rush to the pile to rescue surface victims. This is the fourth step of the rescue plan. So the incident commander must ensure the first three steps of a collapse rescue plan are accomplished: 1. Have police secure the area. 2. Size up the entire area. 3. Order the utilities shut off. First responders often overlook these important duties and instead rush to the collapse rubble pile to rescue victims.

Take a time-out. During this time-out break, another more thorough safety survey should be conducted of the entire site. A sounding and listening period for buried victims should be conducted during this period. To do this there should be a period of silence, all noises should be stopped, so rescuers can call out to buried victims and listen for any responses. More effectively, sound-sensing electronic devices can be used by firefighters to monitor sound from below tons of rubble. Search cameras and search dogs can be safely used during this period. During this time-out, the entire collapse site is analyzed. A conference is held between the incident commander, the operations officer, the planning officer, and the victim tracking officer to determine exactly who is missing, and where in the collapse rubble the suspected victim may be buried.

Use a surveyor's transit telescope to measure any movement of walls, columns, or floors of the collapse structure. If movement is discovered, remove rescuers, shore up the structure, or remove the unstable structure.

During a major collapse where there is spreading fire, divide operations command into two branches, a fire branch and a collapse branch.

As soon as possible, assign a victim-tracking officer to determine locations of missing victims. The victim-tracking officer can work with the incident command planning officer, gathering information, analyzing the building collapse configurations, and making decisions about where individuals are buried in the collapse rubble.

A building collapse at which firefighters are reported trapped or missing is a terrible occurrence; however a fire department is an emergency service that must plan for all types of disasters, even ones we all hope will never occur.

19

SAFETY PRECAUTIONS PRIOR TO COLLAPSE

Around midnight, an engine and a ladder company respond to a fire in a shopping center. Street lights reveal large volumes of smoke drifting from several stores near the center of a one-story building. The engine officer radios communications headquarters: "Engine 8 to Communications Center. We have a working fire in a shopping center at King and 3rd Streets. Request a full assignment of companies to respond. Smoke is showing from a one-story, 200×50-foot, brick-and-joist structure containing nine stores. Ladder 6 and Engine 8 are arriving on the scene."

"Communications Center to Engine Company 8, ten-four."

Driving into the deserted parking area, the pumper stops at a hydrant, leaves one firefighter, and lays a supply line to the front of the stores. In front of the fire building, two firefighters pull 200 feet of 1¾ inch hose off the transverse hose bed. They remove the kinks and lay the hoseline out in folds to allow a quick advance on the fire. The pump operator disconnects the 5-inch supply line, connects it to a gated inlet, opens the intake gate, and signals the hydrant man to start the supply water flowing. Then he looks around and sees the hoseline stretched and unkinked. The officer orders the hoseline charged; the discharge gate is opened, and the pump pressure increased.

Meanwhile, the ladder chauffeur parks the aerial ladder away from the one-story structure. Two firefighters unlock a 20-foot portable ladder from the truck, carry it to the front of the stores, raise it, and place it against the building. Carrying a power saw, an axe, and a pike pole between them, the two firefighters climb the ladder to the roof to begin ventilation. A ladder company officer and two other firefighters with forcible entry tools run to the front of the middle three stores, spread out, and quickly examine the interior of the stores through the glass display windows. They pinpoint the fire inside the center store, a bank and check-cashing establishment.

First, after feeling the blackened, stained, glass panel door for heat, the forcible entry team verifies that the door is locked. They quickly force the door with a rabbit tool. As the door is chocked open, a burglar alarm sounds. The officer and forcible entry team crawl inside the bank to search for maintenance workers, but heat forces them back to the entrance. The firefighters with the charged hoseline enter the store doorway and disappear into the smoke. For deeper penetration, the mask-equipped firefighter on the hoseline opens the fog nozzle to the solid stream position.

The captain of the hose team coaxes the firefighter to crawl several feet into the heat and smoke. "Come on, that's it," he says. "You're getting it." In the pitch-black smoke there is no visible fire, but the crackle and roar of flame can be heard from the back of the store. The firefighters temporarily redirect the hose stream at the ceiling to cool down the heated convection currents over their heads, and scalding hot water cascades back down over their helmets and shoulders.

The radio blares: "Ladder 6 to roof team, go ahead."

"Lieutenant, we're having trouble cutting a vent opening in the roof. The roof deck is covered with steel, burglar-proof plating sheets. We can't vent the rear of the building from the roof, either. There are no windows or doors."

"Ten-four, Ladder 6 roof team," replies the truck officer, who then shouts to his men over the din, "Take out those glass display windows!"

Glass shatters as a firefighter smashes the top part of the smoke-blackened glass with a pike pole. He uses just enough force to break the window top without showering the firefighters inside with pieces of broken glass. He breaks the middle and lower portion of the window next, but smoke does not flow out of the opening. He thinks, "There must be a partition behind the window," and attempts to pull a display sign and curtains out of the window opening with a pike pole.

Sirens of responding apparatus are wailing in the distance, when suddenly the interior of the bank erupts into a huge red ball of fire. The smoke flowing out of the door and window openings instantly turns to flame. Two hose team members, followed by the captain, stumble, fall, and crawl out of the blazing store. The chief arriving on the scene runs over to the firefighters on the sidewalk. "Are all of the men out of the store, Captain?" he asks and the captain nods his head.

"Transmit a second alarm!" shouts the chief to his aide.

Firefighters set up a defensive exterior hoseline operation. The fire has spread throughout the entire nine-store structure via the concealed roof space that extends over the ceilings of the stores. Several fire companies are operating hoselines from the sidewalk around the perimeter of the burning structure. The captain of the first-arriving engine and his men are still operating a hoseline—this time from a safe distance, back from a cracked exterior masonry wall of the first building. The young firefighter directing the hose remarks to his officer: "Captain, if we move the hoseline closer to the building, we can hit more fire and get back to the firehouse."

"I don't like the look of that wall," replies the officer. "We already had one close call—let's not press our luck."

"Captain, I have never seen one of these walls collapse in all my time on the job," says the young man. "I don't think that wall will collapse on us if we move closer and get a better shot at the fire with this hoseline. Besides, the chief would know if that wall looked like it would collapse. He did not warn us to stay back."

"Let me tell you something," replies the captain. "You can't predict if that wall will collapse, I can't predict if that wall will collapse, and the chief can't predict if that wall will collapse. That wall will probably stay up for the entire fire operation, but it could also collapse the second we step under it. Relax and stay where you are!"

Collapse Precautions

There is a saying in the fire service. It goes like this, "If you see something, say something." Lives have been lost because firefighters did not report firefighting dangers to the chief in command. However, there is a second half to this saying, "If they say something, you must do something!" When a danger is reported to the command post, the incident commander must do something. The chief must take action when receiving reports of danger from firefighters: order increase supervision, establish safety equipment to the danger area, or withdraw firefighters from the danger area. If we tell our firefighters to report dangers, we must tell incident commander what to do.

What actions should a commanding officer take when a danger is reported? The response must be balanced. The response must match the severity of the danger reported. A response by an incident commander can not be too extreme, or too restrained. The response must equal the danger reported. The degree of danger reported at a fire will vary, and so must the commander's response. The response to a reported danger may depend upon several factors: who transmits the danger; the seriousness of the warning; the firefighters available to respond to the crises, and the urgency of rescue operations being carried out at the time of the report. A response to a reported danger transmitted to the command post may range from a simple verbal acknowledgment, to an emergency evacuation of all firefighters from the building. There may be several responses by the incident commander to a single reported danger. As the danger becomes more severe or less severe, as resources become available, and the rescue effort is resolved, the responses of the incident commander may increase and/or decrease to reflect the changes.

We train firefighters to report all dangers observed, so we must train incident commanders in the various safety precautions that may be taken in response to a reported danger. The following is a list of safety actions an incident commander may consider when receiving a reported danger. These safety actions progress from a simple acknowledgment to extreme emergency evacuation of all firefighters from the burning building.

1. Acknowledge the report and take no immediate action.
2. Light up the danger area.
3. Assign an experienced officer or chief to investigate (fig. 19–1).
4. Increase supervision.
5. Alert a rapid intervention team to respond.
6. Evaluate an unstable structure with a telescopic lens.
7. Rope or tape off a danger area.
8. Establish a collapse safety zone around a danger area.
9. Evacuate people and firefighters near a reported danger area.
10. Order a partial withdrawal of firefighters from a section of a building.
11. Change strategy from offensive to defensive by withdrawing firefighters from a fire building.
12. Order an immediate emergency evacuation of all firefighters.

Fig. 19–1. The incident commander can direct a supervisor to investigate a reported collapse danger.

Acknowledge receipt of the reported danger by radio

A chief in command of a fire can simply acknowledge a report received of a danger and take no action. This situation may occur when receiving what is considered a known minor danger report, such as smoke seeping through an exterior brick wall during the early stages of a firefighting operation. If it is known there is only a content fire and not a structure burning and the crack in the mortar between bricks is not due to fire-weakening of the interior structure, but to improper foundation settling prior to the fire this action may be taken. If the fire is not extinguished by the initial hoseline and, instead, grows and involves the structure, the incident commander may reconsider the reported structural defect and take additional precautions. All reports of danger must be acknowledged by the incident commander.

Light up the reported danger area

During a night fire, if a danger is reported, the incident commander may order increased lighting to be placed near the area (fig. 19-2). A spotlight could be directed on a cracked wall or partly collapsed cornice; a floodlight could be ordered to illuminate the entire area in front of a burning building where a structural defect has been discovered. This action is often taken during an initial stage of a nighttime fire when a chimney or advertising sign has become unstable or partially collapsed and is hanging down on one side.

Fig. 19–2. Lighting is an important safety precaution at a fire operation.

Because of the need for a search-and-rescue effort, and initial hoseline placement, this initial action may be taken when there are no available firefighters at the scene to secure or remove the dangerous structure. When all available firefighters are preoccupied with other duties, having a rapid intervention team illuminate the danger can bring the hazard to most everyone's attention. As soon as the life hazard is controlled and the rescue effort completed, the incident commander may order increased safety measures when additional firefighters are available. The chief may order the dangerously hanging cornice or sign to be secured with ropes or cut and removed.

Investigate the danger area

An incident commander may order a chief or experienced company officer to investigate a reported danger. This officer would investigate the condition and advise whether to continue or discontinue firefighting in the vicinity of the danger. This action might be taken during overhauling when the stability of a floor is in question.

For example, if the floor is reported to be "springy" or bounce when walked upon, this action can be taken. In some older buildings, the floors vibrate when walked upon. This effect may be caused by the absence of cross-bracing or bridging between the joists, or be due to overloading with firefighters or stock. Also, after a serious fire, the charred floor joists may be reduced in thickness.

A frequent type of structural investigation directed at fires in badly damaged buildings is to have a safety chief inspect the building's interior after a fire where master streams are used. Before overhauling is started, a safety chief can be directed to inspect the building's structural interior stability to determine the destructive effects of the fire damage, the powerful master streams, and the water accumulations. No one would be inside the structure but the safety officer and a team member. This safety officer would note floors containing storage or heavy machinery, look for excessive vibration of floors, rope off charred and weakened floor decks that could disintegrate when stepped upon, observe the condition of stair treads and landings, and estimate the weight of water accumulations absorbed into broken and burned plaster that had fallen to the floor, in addition to water that had been absorbed into storage materials. Based upon these factors and other safety conditions, the safety officer would recommend to the incident commander whether to have firefighters reenter for overhauling. If the investigating officer finds the structure unstable, the incident commander could increase the safety precautions taken and order firefighters not to conduct overhauling. Command could order them to remain outside the smoldering building and outside the collapse zone, and direct master streams into the damaged building for several hours or days.

Increase supervision in the reported danger area

At serious fires, an incident commander may assign a battalion chief or veteran officer to operate in a danger area to increase the supervision of firefighters from a position close to the operations (fig. 19–3). This supervising officer would bring a safety-conscious attitude of a commander, to balance the aggressive attack attitude of a company officer. A supervising officer's priorities at this situation would be life safety of firefighters over fire containment and property protection.

The chief, or veteran officer, assigned to supervise a dangerous firefighting operation, has the advantage of being close to the danger. This officer would keep the incident commander informed firsthand of conditions by frequent radio communications. This assigned chief or company officer has the authority to withdraw the firefighters from the danger area, whenever the danger becomes too great. The assignment of a chief to supervise a danger area is often directed when the continuation of a rescue or firefighting effort is critically important to the overall operation. There are several operations in burning buildings where dangers are always present and where the assignment of a chief is often directed. These dangers include: cellar fires, operating on the floor above a fire, roof venting, top floor fires, and hoseline advancement at a rapidly spreading fire.

Chapter 19 | **Safety Precautions Prior to Collapse**

Fig. 19–3. The incident commander may assign another chief officer to direct operations in a danger area.

Alert the rapid intervention team (RIT) of the reported danger

Whenever there is a danger reported to the command post, this information should be relayed to the rapid intervention team (RIT). Based on the reported danger, the incident commander might alert the RIT officer of the need to obtain additional equipment preparing for a potential trapped firefighter. There should be a RIT standing by at every working fire. They should be put on alert when a danger is reported but only respond to rescue trapped firefighters. They may raise ladders in front of the building, reposition an aerial ladder for rescue, and set up lights in front of the building, and remove hoseline kinks in front of the building. However the RIT team should not enter the building or wander from the command post. They must be ready for quick response to a trapped firefighter. If, for some extreme reason other than firefighter rescue, a RIT team is used, the dispatcher should be notified and another company sent to replace the team.

Assign a monitor

At most building explosions and collapses, there is a danger of secondary collapse due to a cracked or leaning wall, unsupported lean-to floor, or unsupported content sliding out of the broken building. When it seems that a secondary collapse could occur, yet the firefighters must continue to rescue surface victims and operate in the secondary collapse danger area, the incident commander should order an officer or firefighter to observe the structural defect continually. This monitor must warn the incident commander of any change in the unstable wall, floor, or content. This safety precaution is often directed during an explosion or collapse where firefighters are searching the rubble for live, buried, or partially buried victims.

A transit—a telescopic surveyor's tool that can see small movements of a building, undetectable by the human eye—is now carried on many rescue companies. A transit is more effective than a firefighter when a monitor is needed. A firefighter must continually monitor the transit by looking through the sight glass anytime firefighters are operating in a secondary collapse danger area. The firefighter must have a radio to notify the incident commander of any movement of a monitored wall, floor or roof.

Cordon off the reported danger area

Another action often directed when a danger is reported to the command post is to cordon off the reported danger area. A utility rope or police tape is placed around the danger to restrict entry into the area. The rope or tape could be used to restrict entry into an area near a leaning masonry chimney or a section of loose bricks on a parapet wall. This action can be taken when the danger area beneath the unstable structure is small. Firefighters would estimate the ground area that would be covered by the falling structure and then cordon off the area with a utility rope or orange marking tape. The rope or tape is tied to parts of apparatus, building parts, or utility poles and should be waist to chest high above the ground. Incident commanders should be aware, even after an area is cordoned off, that continuous monitoring is required to ensure the roped-off area is not entered by pedestrians and fire personnel.

Establish a collapse zone around the danger area

When a large part of a building is in danger of collapse and requires more than a small area to be roped off, a collapse danger zone should be established by the incident commander (fig. 19-4). This collapse zone is directed to keep all firefighters out of the area where the building could fall and cover a large area of ground with bricks and steel. The collapse zone area should be an area equal to one, one and one half, or two times the height of the wall, depending on the incident commander's orders. A wall collapse danger from a truss roof failure with hip rafters or a mansard roof with hip rafters can be greater than just the height of the falling wall. The force of the collapsing hip rafters of the truss or mansard roof could kick out a wall farther than normal. A collapse area distance should be specified in every fire department's standard operating procedure (SOP) and firefighters trained to comply with the order to establish a collapse zone.

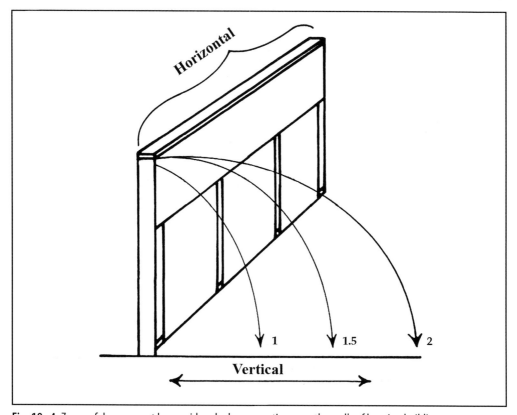

Fig. 19–4. Zones of danger must be considered when operating near the walls of burning buildings.

Some times after a collapse zone has been ordered, freelancing firefighters try to enter a danger zone to obtain better hose stream penetration. Company officer supervision is required in addition a collapse danger zone. When a collapse zone is ordered by the incident commander, responsibility to ensure firefighters comply with the order rests with the company officers and the sector, or division officer, in command of the area. An incident commander must also considered the horizontal frontage of the wall which could collapse. In addition to the outward distance that the wall may fall, an incident commander must estimate the horizontal danger.

A parapet wall is a section of wall that extends above a roofline. The incident commander must estimate whether just the unstable portion will collapse, or will the unstable portion drag down the 50 or 100 feet of wall on either side with it? The most dangerous type parapet wall is an ornamental cast stone wall. This is because a hollow ornamental cast stone wall is tied together by reinforcement bars and angle irons. When an ornamental cast stone parapet appears unstable, bulging or leaning outward, assume the horizontal collapse zone to be the entire parapet wall length. Miscalculation of the potential horizontal length of wall collapse could be just as deadly as miscalculation of the outward area of a wall collapse.

Evacuate occupants and firefighters from buildings adjacent to the danger building

An incident commander should consider many factors when analyzing a potential building collapse. One important consideration is the possibility that the falling structure could cause the failure of a smaller structure nearby. If a three-story structure falls on an adjoining one-story building, it will collapse the one-story building. Even the failure of a rooftop parapet wall can collapse the roof of a small structure nearby if tons of bricks crash down on top of it. At one incident where FDNY Lieutenant Robert Dolney died, an exterior wall of a three-story building collapsed, fell on the roof of a one-story building, and collapsed the roof. When the roof collapsed, it pushed out the front wall that killed the officer and injured a firefighter. The post-fire analysis revealed the original building collapse caused a second roof collapse, and the falling roof caused a third wall collapse. When analyzing a potential collapse, look beyond the initial danger. Try to determine if the falling building could trigger a series of subsequent structure failures. You might have to consider the collapse danger zone for the structures surrounding the original weakened building.

Partial Withdrawal of Firefighters

When firefighters inside a burning building report a collapse danger, the chief may order withdrawal of firefighters from a portion of the building. A partial withdrawal of firefighters to a safe location inside the building may allow interior firefighting to be continued. This practice can be followed only when the construction of the building is known in detail, and partial collapse will not cause the entire structure to fail. For example, when a ceiling collapse is possible, the building has a stairway enclosed by masonry walls, and the stair landings are independent of the floor landings, firefighters can be withdrawn to the safety of the stair enclosure before the ceiling collapse occurs. Also, some structures are divided into building sections, which are separated by fire walls and have separate roofs and floors. A party wall where roof beams from both sections are supported by the wall is not a true fire division. Collapse of the roof on one side could cause the party wall to collapse on the other side.

According to a fire protection design standards, a true fire wall, separating two sections of a building, has an independent foundation and is designed to allow collapse of the roof and floors of one section without affecting the integrity or stability of the fire wall and other section of the building. A fire wall where roof and floor beams run parallel with the wall can be considered a fire division. In this situation, a fire company could be safely withdrawn from one section of the burning structure and continue interior firefighting from a horizontal opening in a fire wall from a safe section of the structure. Attached two-family houses with a plaster board or masonry fire wall between the houses must be considered one structure. When one half of the attached structure becomes unstable and in danger of collapse during a fire, the entire structure is in danger. Occupants and firefighters must be withdrawn from both sections.

Complete Withdrawal of Firefighters

An incident commander may direct complete withdrawal of firefighters from a fire building and its surroundings any time a serious structural defect is reported by a firefighter. When this happens, an offensive strategy is changed to a defensive strategy. Firefighters are withdrawn and outside master streams are put into operation—this is a common strategy change. An incident commander may order an orderly withdrawal and an exterior defensive attack on a burning building whenever a fire is too large for the resources at the scene or when a structural defect is reported to the command post. Some structural hazards or warning signs that would justify this action are:

- An expanding crack in a masonry wall
- Bricks falling out of a wall
- Walls separating at the corners where exterior walls meet
- A structure leaning to one side
- Large volumes of runoff water from exterior hose streams through mortar joints or over tops of windowsills
- Roofs from which joists have fallen or which are extremely bouncy when walked upon
- Floors and roofs that give the sensation of swaying or moving with supporting walls
- Floor or roof joists that appear to be pulling away from masonry wall supporting enclosures
- Vibrating floors that are lower at the center because of deflection or overload
- Steel columns or girders that are out of plumb, warped, sagging, twisted, or fallen
- A building that has sustained any interior collapse
- Severe cracking and slanting of plaster interior walls or ceilings, which may indicate a shifting of structural framework behind the sheathing
- Creaking, rumbling, or cracking sounds coming from a structure
- Any fire where steel bar joist, timber or lightweight truss construction is exposed
- A large body of fire on several floors in an old building

This last warning sign is the most frequent event that causes incident commanders to withdraw firefighters from a burning building. Any large rapidly spreading fire throughout a structure is a danger to firefighters because of the possibility of structural collapse. A large, uncontrolled body of fire in an old building of combustible structure must be considered a warning sign. In many older cities and towns, fire chiefs have developed informal guidelines as to when to withdraw firefighters from certain collapse-prone old structures. These guidelines, usually passed down from veterans to newer chiefs and company officers, have been developed after years of experience—sometimes years of tragic experiences. are: "Firefighter may be withdrawn from a burning building when there is a prolonged burning of a serious fire on one or more floors." The age of a structure, large combustible content, large structural fuel load, undivided open floor spaces, use of unprotected cast-iron columns, lightweight steel and wood trusses, and long span roof and floor systems are additional factors that cause an incident commander to withdraw firefighters and order a defensive outside firefighting strategy.

Emergency Exit Evacuation

There is a difference between changing from offensive to defensive strategy by withdrawing firefighters from a burning, and ordering an immediate emergency exit evacuation of all firefighters. The less urgent withdrawal strategy change is ordered when fire is beyond extinguishment or when a possible collapse is anticipated. The more urgent emergency evacuation is ordered when there is imminent danger discovered or reported, such as an explosion, collapse, terrorist event, or hazardous material incident threatening to overtake firefighters and trap them. In an emergency exit evacuation, unlike a withdrawal, a special evacuation signal is sounded, fire department tools and hoselines are left behind, all firefighters report to a predesignated area, a roll call or a head count is conducted, and the results reported to the incident commander. An emergency exit evacuation is a rare occurrence in the fire service, and, because of this, there is usually confusion and delay when it is ordered. Fire departments should train members for such an emergency exit evacuation. There should be a specific signal used only for an emergency exit evacuation, firefighters should exit the building upon the receipt of this prearranged signal, leaving behind tools and equipment. The should report to a prearranged area for roll call by their immediate supervisor, and the results of the roll call are reported to the incident commander.

Exit strategy training

Over the years, I often told firefighters how to enter burning buildings. I never once told them how to get out of a burning building. I never explained to firefighters how they may have to quickly exit a burning building in case of a possible sudden collapse, explosion, terrorist "dirty" bomb residue, or a hazmat release. Neither did any other fire chiefs I know. The words *retreat*, *withdraw*, *fall back*, *evacuate*, *escape*, and *draw back* are rarely mentioned in the fire service. There's no exit strategy training in the fire service. There is a firefighting strategy, a search-and-rescue strategy that explains how firefighters should enter into a burning building, but there is no exit strategy, no plan describing how firefighters should quickly withdraw from a burning building. There is no standard operating procedure that explains exactly how firefighters should evacuate a burning building.

This omission was evident on 9/11 at the World Trade Center fire and collapse. When Fire Chief Joseph Callan, at the lobby command post, Tower 1, ordered everyone to evacuate the building, even before the first collapse of Tower 2, some firefighters did leave, and some did not. No one knows why. One theory is some firefighters did not hear the message to withdraw because the radio signal was blocked by the steel and concrete in the high-rise building. Another theory is some firefighters heard the message, but chose not to leave, and instead continued to search for victims. If the latter theory is true or not, the fire service must do a better job of training firefighters, how and when to leave a burning building.

The fire service must develop an emergency exit strategy, and train firefighters how to withdraw from a building when there is an imminent danger of collapse, explosion, gas release, or fire spread (fig. 19–5). The fire service must develop a standard operating procedure for emergency evacuation, and enforce its compliance, just as vigorously as it does an aggressive firefighting attack and rescue strategy.

Fig. 19–5. The fire service must have an exit strategy and train firefighters how to evacuate a burning building.

This emergency exit plan must instruct firefighters how and when to leave a burning building. There must be a universal, specific signal to alert firefighters to evacuate a building immediately, no questions asked. The fire service exit strategy must specify what tools and equipment are left behind. After leaving a building, the exit strategy must designate an assembly point where firefighters go to conduct a head count. The results of the head count are relayed to the incident commander at the command post by radio message.

Evacuation signal

A fire department must have a specific signal to alert firefighters to start an immediate evacuation of the building. Upon transmission of this exit signal all firefighters should withdraw from the building in a disciplined manner and report to the assembly area where a roll call is conducted. This signal must effectively reach and be heard by all firefighters. For example, in a low rise building a continuous air horn sounding could be heard by all firefighters at a scene. In a high rise building this would not be effective. In a high rise building a piercing tone signal transmitted over the portable radio by the officer in command should be considered, or any other type signal that would reach all officers, and most importantly, all firefighters operating out of sight or sound of a supervisor. A tone signal rather than a radio voice message would lessen the chance of misunderstanding. The Mayday announcement does not effectively cause firefighter to leave a building. There are too many reasons to use the Mayday signal today. The Mayday signal is used for firefighters trapped, when victims are found, and many other emergencies. Because the Mayday signal is frequently used and has other meanings, it should not be used as an emergency evacuation signal. If a radio tone signal is used to order a quick evacuation of a burning building it should be used for no other purpose. If the radio signal is used, it is imperative all firefighters who operate out of sight or sound of a supervisors be equipped with a radio to hear this evacuation tone signal. Firefighters working in proximity with the officer will be aware of the signal transmission, but firefighters assigned to operate alone, such as roof operation or outside venting, must have a radio to ensure they receive an emergency exit signal. The evacuation signal should never be used for any other reason except ordering a firefighter evacuation.

Tools and equipment

When the evacuation signal is given, all tools that could slow down a firefighter's exit should be left behind. The location a piece of equipment is left should be noted for future retrieval. Hose should be left. Power saws and heavy forcible entry tools should be left. Self-contained breathing equipment also left behind if it would slow a 10- or 20-story descent. This is an option as long as the exit path is smoke free. Anything that would slow down a quick, orderly, safe exit from a building should be left behind.

Assembly point

During fire prevention week, the fire service advises citizens when conducting a family fire drill in the home to identify an assembly area where the family meets after leaving the house during a fire or a family escape plan drill. The fire service should require the same. After an emergency evacuation signal is transmitted and firefighters exit a building, they too should have an assembly point. This assembly point should not be in the front of the building and it should not be at the command post. These assembly areas could be near the fire apparatus parked in the staging area, or in the street away from the building. At this assembly area the company officer should conduct a roll call and report the result to the officer in command.

Time of evacuation

The fire service does not have a clue how long it would take to get all the firefighters of a second- or third-alarm assignment out of a building during an emergency. The one or two times I have ordered an evacuation of a burning building, it took much longer to get everyone out than I expected. The time of evacuation must be recorded. The exit time is the time from when the signal is transmitted until the time when all fire companies at the scene report the results of the roll call. The fire service does not know how long it takes for firefighters to exit a burning building. We know our response time, we know our "set up" time, but we do not know an exit time. Knowing how long it would take for all firefighters on a working all-hands fire, a second-alarm assignment, a third-alarm assignment, a fourth-alarm assignment, and a fifth-alarm assignment to quickly leave a burning building would assist the incident commander's strategy. This knowledge of exit time might reduce the possibility of an incident commander's overcommitment of resources into a fire area. Overcrowding on a stairway or behind a nozzle firefighter is a problem at some fires. The more firefighters I send into a burning building, the longer it will take to get them out of the building. In this age of terrorism, IEDs, secondary devices, and "dirty" bombs, this exit strategy information is invaluable. Knowing firefighting emergency exit time might help an incident commander assess the time necessary to change strategies from offensive to defensive.

Training

A recent law change in New York City being considered as a result of 9/11 is the requirement of a full-scale evacuation drill of all high-rise office buildings once a year. The fire department must know how long it takes to evacuate all persons from a high rise office building in order to effectively decide the risk or rewards of whether to fight a fire or to evacuate the building. This information is critical when determining strategy during a fire, terrorist bomb threat, hazmat threat, or collapse danger. This law will require all occupants of a high-rise building to descend the stairs and leave the building entirely, and not just assemble near a stairway as was required for a fire drill before 9/11. The fire service must do the same. We must train firefighters for full-scale emergency evacuation of a building.

During rookie school, or during annual evaluation training exercises, fire companies should be trained in procedures for conducting an emergency evacuation. When conducting a firefighting exercise for new recruits during rookie training school or veterans during company evaluation training, an evacuation signal should be part of several training events. For example, during an exercise, of hose stretching or search-and-rescue in a training tower, the evacuation signal should be sounded. Firefighters should leave tools, and quickly, in an orderly manner, and safely leave the training tower. Then they should assemble back at the apparatus or designated area. There a roll call should be conducted and the results transmitted by radio to the training incident commander. The exit should be timed and recorded. This timing is not to increase speed of evacuation. The time recordings should be to show the fire officers and fire chiefs how long it will take for an emergency evacuation to take place.

A strategy for exiting a building should not be left to chance. Firefighters must be trained how and when to leave a burning building during an impending sudden danger. If we do not train, it will not be carried out during a fire or emergency.

Lessons Learned

We ask firefighters to "say something when they see something," so we better know how to "do something when they say something."

20

WHY THE WORLD TRADE CENTER TOWERS COLLAPSED

Terrorists crashed two 767 jetliners into the World Trade Center buildings, killing 2,752 people, including 343 FDNY members, and creating the worst high-rise fires and burning building collapse in the history of this country (fig. 20–1). A resulting fire burned for 56 minutes inside the World Trade Center Building 2. Fire caused this 110-story building to completely collapse in eight seconds. A fire raged in WTC Tower 1 for 102 minutes and this 110-story building pancaked down in 10 seconds. You can

Fig. 20–1. The World Trade Center following its collapse

say they collapsed because of the impact of the 185-ton jet airliners and the heat of the jet-fueled fire, which weakened the steel and caused the collapse. Or, you can take a closer look. The reason these structures collapsed so quickly and completely can be found by examining high-rise construction in America during the past 50 years.

World Trade Center Tower Construction

In terms of structural system, the twin towers departed completely from other high-rise buildings. Conventional skyscrapers since the 19th century have been built with a skeleton of interior supporting columns that support the structure. Exterior walls of stone, glass, steel, or synthetic material do not carry any load. The twin towers were radically different in structural design with the exterior wall used as the load-bearing wall. (A load-bearing wall supports the weigh of the floors.) The only interior columns were located in the core area, which contained the elevators. The outer wall carried the building vertical loads and provided the entire resistance to wind. The exterior bearing walls consisted of closely spaced vertical columns (21 columns, 10 feet apart), tied together by horizontal spandrel beams that girdled the tower at every floor. On the inside of the structure, the floor sections consisted of 60-foot steel bar trusses spanning from the core to the outer wall (fig. 20–2).

Fig. 20–2. 60-foot open-web bar joists

Bearing walls and open-floor design

When the terrorists crashed the jets into the towers, based upon knowledge of the tower construction and high-rise firefighting experience, the following happened. First, the plane broke through the tubular-steel bearing wall. This started the building failure. Next, the exploding, disintegrating jet plane slid across an open, office floor area and severed many of the steel interior columns in the center core area. Plane parts also crashed through the plasterboard-enclosed stairways, cutting off the exits from the upper floors. The jet collapsed the ceilings and scraped most of the spray-on, fire-retarding asbestos from the steel trusses and thousands of gallons of jet fuel exploded into flame. The steel truss floor supports probably started to fail quickly from the heat of the fire and the center steel supporting columns severed by plane parts and heated by the flames, began to sag, warp, buckle, and fail. Then, the top part of the flaming tower crashed down on the lower portion of the structure. This collapse triggered the entire cascading collapse of the 110-story structure.

Steel framing

The most noticeable change in modern high-rise construction in this country is a trend to using more steel and shaping lightweight steel into tubes, curves, and angles to increase its load-bearing capability. The WTC had tubular-steel bearing walls, fluted, corrugated steel flooring, and bent-bar steel truss floor supports. To a modern high-rise building designer, steel framing is economical and concrete is a costly material. For a high-rise structural frame—columns, girders, floors, and walls—steel provides greater strength per pound than concrete. Concrete is heavy and creates excessive weight in the structure of a building. Architects, designers, and builders all know that if you remove concrete from a structure, you have a building that weighs less. So, if you create a lighter building, you can use columns, girders, and beams of smaller dimension or, better yet, you can use the same size steel framing and build a taller structure. In most cities where space is limited, you must build high (up, rather than out).

The trend during the past half-century is to create lightweight high buildings. To do this, you use thin steel bent-bar truss construction instead of solid-steel beams. To do this, you use hollow-tube steel bearing walls and curved sheet steel (corrugated) under floors. To do this, you eliminate as much concrete from the structure as you can and replace it with steel. Lightweight construction means economy. It means building more with less. If you reduce the structure's mass, you can build cheaper and build higher. Unfortunately, unprotected steel warps, sags, bends, and collapses when heated to normal fire temperatures of about 1,100 to 1,200 degrees Fahrenheit.

The fire service believes there is a direct relation between fire resistance and mass of structure—the more mass, the more fire resistance. The best fire-resistive building in America is a concrete structure. The structures that limit and confine fires best and suffer fewer collapses are reinforced concrete; examples are pre-WW II buildings—housing projects and older high-rise buildings, such as the Empire State Building. The more concrete, the more fire resistance; and the more concrete, the less probability of total collapse. The evolution of high-rise construction can be seen by comparing the Empire State Building to the World Trade Center. My estimate is the ratio of concrete to steel in the Empire State Building is 60/40. The ratio of concrete to steel in the

WTC is 40/60. The Petronas Towers in Kula Lumpur, Malaysia, one of the tallest buildings in the world, is similar to the concrete-to-steel ratio of the Empire State Building rather than the concrete-to-steel ratio of the WTC.

The Computer-Designed High-Rise Building

The computer has allowed engineers to reduce the mass of a structure with its ability to more accurately determine the load-bearing capability of structural framework. Years ago, the pre-computer builders were not sure of a structural element's load-bearing capability, so they went beyond the so-called "safety factor." Because of computer calculations, this no longer occurs. The safety factor of two-to-one is exactly what you get, no extras, so if the building code requires a load-bearing factor of 40 pounds per square foot, that is exactly what you get.

Effects of Jet Crash and Fire on a Skeleton Steel High-Rise

In 1945, an Air Force B-25 bomber that only weighed 10 tons struck the Empire State Building and the high-octane gasoline fire quickly flamed out after 35 minutes. When the firefighters walked up to the 79th floor, most of the fire had dissipated. The Empire State Building, in my opinion and that of most fire chiefs in New York City, is the most fire-safe building in America. I believe it would have not collapsed as the WTC towers did. I believe the Empire State Building and, for that matter, any other skeleton steel building in America, would have withstood the impact and fire of the terrorists' jet planes better than the WTC towers. If the jetliners struck any other skeleton steel high-rise, the people on the upper floors and where the jet crashed may not have survived; there might have been local floor and exterior wall collapse. However, I believe a skeleton steel frame high-rise would not suffer a cascading, total pancake collapse of the lower floors in eight and ten seconds.

The Empire State Building

Perhaps builders should take a second look at the Empire State Building's construction. There might be something to learn when they build a high-rise structure. The Empire State Building has Indiana limestone exterior walls, 8 inches thick. The floors are also 8 inches thick, consisting of 1 inch of cement over 7 inches of cinder and concrete. All the columns, girders, and floor beams are solid steel, covered with 1½ to 2 inches of brick, terra cotta, and concrete. There is virtually no opening in the floors. And, there are no HVAC air ducts penetrating fire partitions, floors, and ceilings. Each floor has its own HVAC unit. The elevators and utility shafts are masonry enclosed. And, for life safety, there is a 4-inch brick-enclosed so-called "smoke-proof stairway." This stairway is designed to allow people to leave a floor without smoke following them and filling up the stairway. The smoke-proof stairway has an intermediate vestibule that contains a vent shaft, so any smoke that seeps out the occupancy is sucked up a vent shaft.

Concrete Removal

Since the end of World War II, builders designed away most of the concrete from the modern high-rise construction. First, the concrete they eliminated was the stone exterior wall. They replaced them with the "curtain walls" of glass, sheet steel, or plastics. This curtain wall acted as a lightweight skin to enclose the structure from the outside elements. Next, the 8-inch-thick concrete floors went. They were replaced with a combination of 2 or 3 inches of concrete on top of thin, corrugated steel sheets (fig. 20–3).

Fig. 20–3. The World Trade Center towers had 3-inch concrete floors on fluted metal decks supported by bar joist steel beams.

Next, the masonry enclosure for stairs and elevators was replaced with several layers of sheet rock. Then, the masonry smoke-proof tower was eliminated in the 1968 building code (fig. 20–4). It contained too much concrete weight and took up valuable floor space. Next, the solid-steel beam was replaced by the steel truss. And, finally, the concrete and brick encasement of steel columns, girders, and floor supports was eliminated. A lightweight spray-on coating of asbestos or mineral fiber was sprayed over the steel. This coating provided fireproofing. After asbestos was discovered to be hazardous, vermiculite or a volcanic rock ash substance was used as a spray-on coating for steel. Outside of the foundation walls and a thin 2 or 3 inches of floor surface, concrete almost has been eliminated from high-rise office building construction. If you look at the WTC rubble at Ground Zero, you see very little concrete and lots of twisted steel.

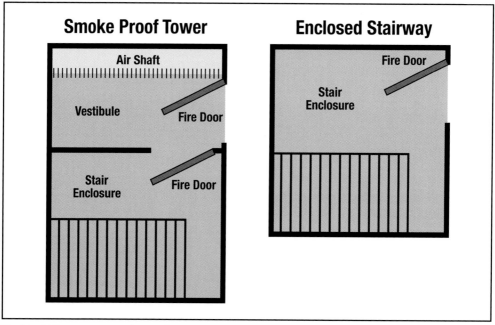

Fig. 20–4. There were no smoke proof stair towers in the World Trade Center towers.

The Performance Building Code

How did lightweight high-rise construction evolve since World War II? It evolved with the help of the *performance code*. After World War II, builders complained about building codes. They said they were too restrictive and specified every detail of construction. They called the old building codes *specification codes*. They complained that the codes specified the size and type and sometimes even the make of a product used in construction. They decried the specification code as old-fashioned. They wanted the building codes changed to what they called *performance codes*. They wanted the building codes to specify the performance requirements only and not specify the size and type of building material to use.

For example, with fire-resistive requirements, they wanted the code to state just the hours of fire resistance (one, two, three, or four hours) required by law and not to state the specific type and material used to protect structural steel and enclosures for stairways and elevators shafts. For example, a performance building code states the steel has to be protected against heat of flames for one, two, three, or four hours during a fire. It does not state what to use as a fire-resisting material. This performance code signaled the end of concrete encasement fire protection and allowed a spray-on fire protection for steel- and plasterboard-enclosed stairs and elevator shafts.

Builders hailed the New York City building code of 1968 as a good performance code. However, some fire chiefs decried it as a law that substituted frills for real construction safety. The asbestos spray-on coating of steel trusses used in the WTC towers was considered by the Chief of the New York City Fire Department, John T.

O'Hagan, to be inferior to concrete encasement of steel. Writing in his book, *High-Rise Fire and Life Safety* in 1976, he listed the following problems of spray-on fire protection of steel:

1. Failure to prepare the steel for spray-on coating adhesion. The steel should be clean and unpainted. Primer paint, rust, and dirt allowed spray-on fire retarding coating to scale and fall away from steel during construction.
2. Poor or uneven application of the spray-on fire retardant was discovered during post-fire investigations.
3. Variation of spray-on material during manufacture makes it ineffective.
4. Lack of thoroughness in covering the steel during application is a problem.
5. Failure to replace spray-on material dislodged by other tradespeople performing work around the steel during the construction of the building.
6. The spray-on fire retardant must adhere to the steel. The air flow from central air systems can blow the fire retardant off the steel.

The WTC towers built by the Port Authority of New York did not comply with the New York City building or fire codes (fig. 20–5).

Fig. 20–5. Because the Port Authority was a bi-state agency, it was exempt from complying with the New York City building or fire codes.

Lessons Learned

1. The building codes should require the fire safety directors of all high-rise office buildings in America to *conduct an evacuation drill* with all occupants leaving the building. The time it takes for total evacuation should be recorded and given to the fire department, the police department, and the building department. When responding to a terrorist incident or fire in a high-rise building, the fire and police departments should know how long it will take to evacuate a building. (In 1993, during the first terrorist attack on the World Trade Center, it took 5 hours to evacuate everyone from the two towers.)

2. The building codes should require *exits and stairway widths* in new high-rise office buildings to have sufficient capacity to allow evacuation of all occupants within the time limit of the floor's fire-resistance rating. For example, when a floor of a building has a three-hour fire-resistive rating, evacuation of all occupants should be accomplish within three hours.

3. The building codes should require steel columns, girders, and floor beams to be *encased in masonry*. The present-day use of spray-on fire retardant is ineffective. Post-fire investigations reveal the spray-on type of fire retardant has scaled off and the unprotected steel beams warp, sag, and twist and the concrete cracks. This is revealed in floors sagging and cracking during a fire.

4. The building codes should *limit the use of lightweight bar joists, truss floor* construction in high-rise buildings. Full-scale fire testing, conducted by the National Fire Protection Association, has shown unprotected bar joist steel trusses can fail after 5 or 10 minutes of fire exposure. The WTC was the only high-rise office building with long span steel bar truss floors and hollow-tube bearing walls.

5. The building codes again should require *the use of smoke-proof stairways* in high-rise office buildings. This type of stairway allows people to escape fire and a terrorist attack inside a smoke-free and non-toxic environment.

6. The building codes should require stairs and elevator *shaft ways to be enclosed in concrete*, not plasterboard. Concrete enclosures will increase the protection of occupants leaving a building during a fire or terrorist bomb.

7. The building codes should require better *insulation of electric wiring in elevators*. Water run-off from sprinkler systems and firefighters' hose streams used after a terrorist bomb explosion or fire can cause electric malfunction of elevators. (A bucket of water can make some elevators fail.)

8. The building codes should require high-rise buildings to have a Phase III elevator. This *Phase III elevator* is an elevator that could be used by firefighters for access to upper floors and, more importantly, by disabled persons for evacuation during fires and terrorist incidents. A Phase III elevator is one enclosed by walls that protect it from toxic fumes, smoke, fire, and water.

9. The building codes should *limit the ducts of heating, ventilation, and air-conditioning (HVAC) systems* in high-rise office and residence buildings to only one or two floors. A central air system serving 10 or 20 floors has shaft ways and duct systems that penetrate fire-rated floors, walls, partitions, and ceilings. As a result, during a fire or terrorist attack, smoke or toxic fumes can spread throughout large areas of high-rise buildings. Fire-resistive used to mean fire does not spread from one floor to another, barring an explosion or collapse. This definition does not apply to modern, lightweight, fire-resistive, high-rise buildings built today.

10. The building codes should *prohibit the type of structural framework used in the World Trade Center*. The structural framework (tube frame) in the World Trade Center that allowed 110 stories to collapse in 10 seconds was hollow-tube bearing wall construction, with long span bar joist truss floors and steel column support only in the center core area (fig. 20–6).

Fig. 20–6. The World Trade Center had tubular bearing walls; some only one-half-inch thick on the upper floors.

11. The building codes should require *greater thickness of concrete in floors* in high-rise buildings. The 2 or 3 inches of concrete over corrugated steel used today cracks, sags, buckles, and warps during most serious high-rise fires and must be replaced. Floor construction is the weak link in today's high-rise buildings.

12. The building codes should have *zero tolerance for non-sprinklered high-rise buildings*. Automatic sprinklers should protect all high-rise buildings. Firefighter hose teams can extinguish approximately 2,500 square feet of fire with one hoseline. Today's high-rise office buildings have 20,000 or 30,000 square feet of open-floor space. This is beyond the extinguishment capability of a team of firefighters.

13. The building codes should require *an antenna installed in all high-rise buildings* that would enhance the radio transmission of fire department radios. (McKinsey Company report, *Increasing FDNY's Preparedness*)

14. Federal buildings, state buildings, and Port Authority buildings should *voluntarily comply with New York City or local building and fire codes*.

15. The fire service should conduct pre-fire plan inspections of all high-rise buildings to determine the fire protection weakness. For example, does the building have automatic sprinklers? A smoke activation fire alarm system? 24-hour supervision? A public address system to give instructions to occupants? Stairways and doors designated with identifying letters and floor numbers so that if necessary, occupants can be directed into evacuation stairs? The fire inspector should serve violation orders to ensure that minimum required fire protection is incorporated into the building. Fire inspectors should also evaluate the building systems and determine how they will assist or hinder a fire operation. For example, will the fire department portable radio system work and transmit messages in the building from lobby to roof and cellar? Will the central air system spread smoke from floor to floor by convection currents inside the ducts, even after the system is shut down? Will the elevator system fail when wires are wet from run-off water from sprinklers or hose streams? And will the fire sprinklers and standpipe system deliver adequate water pressure and volume when used by the local firefighters?

National Institute of Standards and Technology's 2005 Final Report on the Collapse of the World Trade Center

There are 16 recommendations on the future construction of high-rise buildings that have major impact on the fire service: numbers 1, 4, 5, 6, 8, 12, 13, 16, 17, 18, 20, 21, 22, 23, 24, and 25. The entire report can be read and obtained online at *wtc.nist.gov*.

Fire chiefs, fire inspectors, and fire planners who are about to have high-rise construction in their communities should study these 16 recommendations closely and, as NIST urges, take steps to have these recommendations implemented without waiting for building and fire codes to be changed. The following are the recommendations and my comments.

1. NIST recommends progressive collapse be prevented in buildings through the development and nationwide adoption of consensus standard and code provisions and a standard methodology be developed to reliably predict the potential for complex failure in structural systems subject to multiple hazards.

Comments: (Progressive building collapse) Progressive collapse is a danger in a building with a large open-floor area that has no supporting columns or girders reinforcing the long span floor beams. Progressive collapse is a danger in lightweight high-rise buildings where steel trusses are used over large open-floor areas and steel columns and girders are eliminated to create more rentable floor space, and where the weight of masonry is removed so a framework of lightweight steel columns and girders can be used in construction.

Even before the World Trade Center towers collapsed on 9/11, the history of fire-resistive buildings in the United States had ominous implications. In the 1970s, One New York Plaza, a fire-resistive high-rise building, suffered a fire that burned two floors. This was the first time a fire spread from floor to floor in a New York City fire-resistive building. The definition of fire-resistive construction before this fire was: *a fire would not spread from floor to floor, barring an explosion or collapse.* The Los Angeles First Interstate Bank fire in the 1980s spread throughout five floors in a fire-resistive building. In the 1990s, the Meridian Plaza fire in Philadelphia burned nine floors of a fire-resistive building to a crisp. The Meridian Plaza was the first high-rise fire in the history of the American fire service where a structural engineer said the building was in danger of progressive collapse. And, it was the first fire in the history of the American fire service where the fire chief withdrew all firefighters from a burning, fire-resistive high-rise building. The high-rise fire-resistive building was deemed unsafe and demolished. Now we have the World Trade Center tower that suffered a progressive collapse from the fire that was caused by the terrorists' planes' impact.

During construction, the World Trade Center towers were considered by the New York City Fire Department to be a "lightweight" high-rise building. The towers were the lightest buildings ever constructed in New York City. They were not skeleton steel framework, using columns and girders supporting beams throughout the floors. Rather, for example, 60-foot-long open-web bar joists supported the floors. There were no girders or columns supporting these long span lightweight steel truss beams. This was a bearing wall high-rise building. These outer bearing walls were hollow tubes. The thickness of the hollow-tube bearing walls near the top of the towers was only ¼ inch thick. When the plane struck the north tower, the photo clearly shows the wings penetrated the hollow-tube wall. Contrary to some testimony by engineering firms, the twin towers were not "robust buildings with great redundancy." As the engineer in charge described them, these towers were a "leap to lightness."

4. NIST recommends evaluating and improving the technical basis for determining construction classification and fire-rating requirements and considering factors such as full evacuation of occupants, redundancy in active fire protection standpipe and sprinklers, redundancy in passive future protection features critical to structural integrity, ability to withstand maximum credible fire without collapse, need for compartmentation to retard large fire spaces.

   Comments: (Construction classifications) Today, there is no technical basis for the classification of fire-resistive construction. The term *fire-resistive construction* is meaningless. Most high-rise buildings classified as fire-resistive structures—from a firefighting perspective—are not. The goal of a fire-resistive building used to be "to confine fire to one floor, barring an explosion or collapse that would destroy part of the compartmentation." The walls, floors, and ceilings of a fire-resistive building were supposed to contain the fire. This is not true today. For example, in Nevada, 85 people died in a fire-resistive building that spread smoke through a hotel via the central air system. The term, *passive fire-resistive*, no longer exists. One of the reasons modern skyscrapers cannot stop the spread of fire and smoke is the central air-conditioning system installed in some of them. A central air system in a high-rise building may interconnect 10 to 20 floors for the purpose of heating and cooling. Ducts, shafts, and poke-through holes penetrate fire-resistive floors, walls, and ceilings. These air-conditioning openings and holes allow fire and smoke to spread throughout the 10 or 20 air-conditioned floors of a building. The high-rise hotel fire in Las Vegas, Nevada, spread smoke through the central air-conditioning system and killed 85 people, most of them in rooms on upper floors. The air system was not equipped with smoke detectors arranged to shut down the system during an emergency. In addition, the fire dampers—shutters designed to stop spread of fire in ducts and shafts of the air conditioning system—did not close properly.

Another avenue of fire spread in high-rise buildings is from window to window above; sometimes called auto-exposure. The nine-story, Philadelphia Meridian Plaza fire was primarily spread by auto-exposure—flame spread from window to window above. Auto-exposure has always been a fire spread problem with high-rise buildings; however, the flames spreading to the floor above in older buildings divided into small rooms used to be confined to a manageable area. Today's high-rise buildings have large open-floor spaces, so when fire spreads to the floor above through a window, it engulfs a large open-floor space that firefighters cannot extinguish with hose streams.

5. NIST recommends improving the century-old standard (time temperature testing fire) of fire-resistance. Future fire-resistance testing of building components, assemblies, and systems of buildings must be under realistic fire and load conditions.

    Comments: (Testing for fire resistance) The current test fire used to determine fire-resistive ratings was developed in 1918. The time temperature curve used as a test fire was based on several fires in ordinary construction and there were no plastics in the building fires; just combustibles, such as wood and paper. Another criticism of the century-old standard to determine fire-resistance is that the test fire does not represent a full-scale fire. The testing is carried out on small-scale test samples. Another problem with fire testing is that after the test to determine fire-resistance, the hose stream impact test is not administered for plaster partition walls. In the past when a masonry wall was tested for fire-resistance, after the fire test, the wall was struck with a hose stream using a $1^{1}/_{8}$-inch nozzle, discharging 250 gallons of water per minute. In order to pass the fire-resistance test, the masonry wall had to resist collapse by the hose stream impact. Today, plaster walls—not masonry walls—enclose stairs, apartments, and exit passageways and firefighters' hose streams are collapsing these plaster walls during firefighting.

6. NIST recommends developing criteria, test methods, and a standard for sprayed-on, fire-resistive materials (SFRM), commonly referred to as fireproofing or insulation.

   Comments: (spray-on fire-resistive material) The exact thickness of the SFRM used to protect steel was brought into question by NIST after the collapse of the World Trade Center towers. The thickness of the SFRM spray-on first was applied ¾ inch thick. Then, after tests in the 1980s, the thickness of the WTC SFRM insulation was increased to 1½ inches thick. The thickness of the 1½-inch fluffy spray-on was said to provide fire-retarding protection of the steel bar joists supporting the floors for two hours. During an investigation after 9/11, no fire testing documentation was found to justify the ¾- or 1½-inch thickness to provide a two-hour fire rating. Another problem at the Would Trade Center towers was the fluffy, spray-on fire retarding was difficult to adhere to the steel round bar web members of the 60-foot-long span floor beam trusses. In 2004, the New York City Fire Department prohibited the use of open-web steel joists in nonresidential high-rise buildings until appropriate fireproofing standards are developed and promulgated.

   Today, almost 40 years after being used to protect steel, there is no standard thickness for fire-retardant agreed upon by the construction industry. Only now after the WTC collapse and 2,749 deaths, is the construction industry looking for a standard use for SFRM fireproofing that is used for fire-retarding steel in high-rise buildings. Adhesion, consistency, and thickness are three criteria that the construction industry now determines are necessary for the SFRM effectiveness during a fire. Adhesion is important for steel fire retarding because the air movement by a central air system blows the fluffy fireproofing off the steel. Even if it is applied over the steel properly, the air movement and vibrations in the building shake the insulation off the steel. In the WTC towers, the SFRM was mistakenly applied to steel that first had been covered with primer paint. Tests at NIST laboratories showed the SFRM over the primer paint was only one-third to one-half of the adhesive strength. The so-called direct application of the spray-on method of fire-protecting structural steel is a cheaper and faster method than other methods, such as encasement and membrane protection. A vermiculite, volcanic rock or mineral fiber is mixed in a liquid and sprayed on the steel to provide fire-retarding. This method of spray-on fire protection of steel was known to be ineffective by the fire service since its introduction in the 1970s. After every serious fire in fire-resistive high-rise office buildings in New York City, the floors crack and sag and the SFRM is missing from sagging and warped structural steel girders and beams.

   - The steel is not prepared properly to allow the spray-on material to stick properly.
   - The spray-on slurry often is not mixed properly.
   - Workers do not apply the spray-on material evenly.
   - Other workers doing subsequent work nearby easily remove the critically important fire protection.
   - The vibrations and HVAC air movement in the plenums cause the SFRM to fall off and applications over primer painted steel reduce effectiveness by one-third to one-half.

8. NIST recommends fire resistance of structures be enhanced by developing a performance objective of an uncontrolled fire that results in a burnout without causing a partial or global total collapse.

   Comments: (Uncontrolled fires resulting in burnout) Fire testing for ratings for floor assemblies analyzes the ability of only a small sample. There is no test for determining a structure's capability during a full-scale, uncontrolled (real-world) fire. During uncontrolled high-rise fires that completely burn out a floor area, the floors above the burnout are cracking, sagging and opening up. In some instances, floors have to be shored up to prevent collapse before firefighters can enter the burnout area and perform salvage and overhaul. At the 1970 One New York Plaza high-rise fire, the major structural damage was floor collapse. Twenty thousand square feet of concrete and steel floor deck and 150 steel floor beams were replaced. The fire caused the floor above to buckle, crack, and heave upward. Floors and partition walls were slanted upward and sideways. No floor slab collapsed, but firefighters could not safely enter the floor area that was cracked and buckled upward because of the collapse potential.

   The floor of a skeleton steel-frame high-rise structure is usually corrugated steel—10-by-12-foot sheets covered with 3 or 4 inches of concrete. This combination steel and concrete floor deck is supported by steel beams, girders, and columns in a gridiron design. When the heat from a fire destroys the ceiling and heats up the underside of a floor's corrugated steel, the girders and beams warp and sag, permitting sections of concrete above to crack at the seams and then sag, warp, twist, and buckle.

   This floor-weakening effect also was in evidence at the 1980s First Interstate Bank fire in Los Angeles and the 1990s Philadelphia Meridian fire. At the Banker's Trust Company Building fire in New York City in 1993, at 49th Street and Park Avenue, an officer reported that the floor above the fire he was searching was beginning to buckle and sag. After the blaze was extinguished, the fire department shored up the floor girders, beams, and floor deck to prevent collapse. At a high-rise fire in Montreal, Canada, that burned several floors, a 10-by-12-foot section of floor collapsed to the floor below. The lesson learned from these fires is that floors are the weak link in the construction of fire-resistive high-rise buildings. Floor supports must be tested in large-scale fires simulating complete burnout of a floor.

12. NIST recommends the performance of the redundancy of active fire protection sprinklers, standpipes, hose, fire alarms, and some management systems be enhanced to accommodate greater risks associated with increasing building heights, population, open spaces, high-rise activities, fire department response limits, transient fuel loads, and higher-threat profiles.

    Comments: (Sprinkler and standpipe systems) One of the "giveaways" to the construction industry from the 1968 New York City Building Code was permission to have one single water riser supply for both a sprinkler and standpipe system. Before this 1968 code change, a standpipe system required a water supply riser separate from the sprinkler system. Redundancy of water supply was lost in New York City high-rise buildings with this compromise. The first step in getting some redundancy in high-rise building fire protection systems is to again have separate risers for standpipe and sprinkler systems and have those supply risers in separate stairways.

13. NIST recommends fire alarm and communication systems in buildings be developed to provide continuous, reliable, and accurate information on the status of life safety conditions at a level of detail sufficient to manage the evacuation process during building fire emergencies.

    Comments: (Communications for building occupants) Living and working in a high-rise building is more dangerous than working in a low-rise building. If the fire is above the reach of fire department equipment, the strategy of ladder rescue and exterior hose streams is taken away. Add this to the absence of timely information of how to react during the fire from the fire chief to the building occupants and you have the danger of large loss of life. High-rise office buildings in many states are required to have public address systems where a building manager or fire chief can give occupants on every floor instructions regarding what to do during a fire or emergency. This important communication system does not exist in high-rise residence buildings. There must be a communication system in every high-rise building to instruct occupants how and when to evacuate during a fire or emergency.

    When a fire occurs in a high-rise building, the incident commander cannot give trapped people advice when there is no communication system. If there is no public address communications system to give instructions during the fire or emergency, hundreds of people start to leave or call the fire dispatcher asking for help. People are trapped by smoke and heat in stairways and hallways. Also, the telephone calls quickly overload firefighting radio traffic.

    Comments: (Communications for firefighters) The firefighter communication problem in high-rise buildings is that their portable radios do not work, because of the steel and concrete in the structures. If you cannot communicate at a high-rise fire, there cannot be firefighting command and control or evacuation of the building. In *Increasing FDNY's Preparedness*, McKinsey recommends stationary repeater systems be installed in high-rise buildings to enhance fire department radio communications. The standard for fire department radio communication should be that radio transmissions travel from the lobby command post to the roof and from the lobby command post to the lowest below-grade area.

16. NIST recommends public agencies, non-profit organizations, building owners, and managers develop and carry out public education and training campaigns jointly to improve building occupants' preparedness for evacuations in case of building emergencies.

    Comments: (Fire safety education) The public has lost confidence in the firefighter's instructions during a fire. During the early stage of the WTC attack when only the north tower was burning, fire officials told people in the south tower to stay in the building and go back up to their offices. Those who obeyed the instructions went back and died and those who did not heed the instructions and left, survived. The order to stay in place was correct. People leaving the south tower would have been in danger from collapsing parts of the burning north tower. Also, thousands of people from the south tower in the plaza would have slowed the evacuation of the people from the north tower. No one could have imagined a second plane would have struck the south tower and trapped the people who returned to their offices. Even today, the impression of most people is that the order to return to the building was wrong.

Today, because of misinformation on how to react in a high-rise building during a fire, many people will disregard instructions from a building manager or fire chief and leave a burning high-rise building. I believe this already has led to an increase in deaths in burning high-rise buildings. People are dying in smoke-filled halls and stairs during a fire when it would be safe to stay in the office or apartment. People will immediately try to leave the building at the first sign of fire and this can be deadly. Because of this misinformation, fire officials and building mangers must launch a nationwide fire safety campaign to educate occupants of high- and low-rise buildings how to react during a fire.

17. NIST recommends tall buildings be designed to accommodate timely, full building evacuation of occupants when necessary for power outages, earthquakes, tornadoes, hurricanes, fires, explosions, and terrorist attacks. Stairwell capacity and stair discharge door width should be adequate to accommodate counter flow due to emergency access by first responders.

    Comments: (Total evacuation time of high-rise buildings) At least once a year, building management should conduct an evacuation drill where all occupants leave the high-rise office building. The time it takes for total evacuation should be recorded and given to the fire, police, and building departments. When responding to a fire, emergency, or terrorist incident in a high-rise building, the fire and police departments should know how long it would take to evacuate a building. Critical decisions—whether to fight a fire or evacuate the building—depend on this information. Defend-in-place strategy is used in a high-rise building where time necessary for total evacuation would be impractical. Defend-in-place strategy is a plan to fight a fire while people remain inside the building. To do this effectively, the incident commander must know the time it would take for full evacuation of any building.

    Comments: (Stairway capacity) Building and fire codes should require exits and stairway widths in new high-rise office buildings have sufficient capacity to allow evacuation of all occupants within the time limit of the floor's fire-resistance rating. For example, when a floor of a building has a two-hour fire-resistive rating, total evacuation of all occupants should be accomplished within two hours. During the 1993 bombing of the WTC, it took four hours to evacuate the towers, which had floors with two hours of fire resistance. There must be a safety correlation between total evacuation of a high-rise building and fire protection of its floor beams, girders, and columns. And fire chiefs must correlate the time of interior firefighter attack and the time of fire-resistive ratings of floors. Incident commanders should not continue interior firefighting attack in unoccupied burning high-rise buildings for 5 to 10 hours where columns, girders, and floor fire-resistive ratings are 2 or 3 hours.

Comments: (Smoke-proof towers) This was another giveaway in the 1968 building code. The requirement for a smoke-proof tower in every commercial building taller than 75 feet was left out of the new code. When the Port Authority left out the smoke-proof tower, they did not substitute another enclosed stair. Building and fire codes should reinstitute the requirement of smoke-proof-stairways (fire towers) in high-rise office buildings. This type stairway allows people to escape fire inside a smoke-free and nontoxic environment. The addition of this stair also would allow the counter flow of first responders.

18. NIST recommends egress systems be designed to maximize remoteness of egress components—stairs, elevators, and exits—without negatively impacting travel distances.

    Comments: (Remote exits) Exits should be remote from each other. If fire blocks one exit, there should be another exit at the other end of the floor available for escape. Pre-World War II buildings had exits on each floor remote from each other. A commercial building had an exit at one end of the floor and another exit at the other end of the floor. These were remote exits. In the modern high-rise buildings, for economic reasons, all utilities, elevators, bathrooms, stairways, and exits are gathered in the center "core" of the building. Unfortunately, there is no longer an exit at each end of the floor area, both lead to the same core area. The definition of the word *remote* in modern building codes is defined to encourage the core floor design. The distance defined as *remote* is now 15 feet. This allows all exits to be located in the center core of a floor area, so stairs near each other are still considered *remote* and the high-rise complies with building codes. This is another giveaway of the 1968 New York City building code. This code change is an example of builders, designers, and code officials living up to the letter of a law and not the spirit of a law. At the WTC, if the stairs were truly remote from each other at each end of the floor areas, there might have been an escape for the people trapped above the fire after the plane crashed into the north tower.

20. NIST recommends next generation evacuation technologies be evaluated for future use, such as protecting and hardening elevators, exterior escape devices, and stairway descent devices, which allow all occupants equal opportunity for evacuations and facilitate emergency access.

    Comments: (Hardening elevators) The building and fire codes should require new and existing high-rise buildings to have a hardened elevator (sometimes called a *Phase III elevator*). (Phase I is an elevator recapture system and Phase II is for firefighter use.) This Phase III elevator is one that is enclosed and protected from smoke, fire, and hose stream and sprinkler water, so that it can be safely used by firefighters or occupants during a fire. Firefighters can use the Phase III elevator for access to upper floors and, most importantly, to aid disabled persons in evacuation during fires and emergencies. An 8-year study of 179 major high-rise fires in New York City found elevators failed due to fire, heat, and water one-third of the time when used by firefighters.

    Comments: (Exterior escape devices) Please notice this recommendation says evacuation devices should be "evaluated." This is not the same as recommending "installation." The fire service should closely evaluate any exterior escape device. Many in the fire service, including myself, are very skeptical of exterior escape devices. The problem with such unorthodox escape systems is the building designers will want to substitute or eliminate an important safety measure, such as an enclosed stair or exits, for an exterior escape device that may never be used. The fire service must not let the design professional have people who are trapped in high-rise fires depend on outside escape or parachutes in lieu of a stairway. We need more and larger enclosed stairways, not economical exterior escape devices. The fire service must require safe, dependable stairs that are remote from each other. Our experience with innovative building designs, such as fire towers, scissor stairs, remote exits, stair widths, and elevator wiring has left us disappointed.

21. NIST recommends installation of fire-protected and structurally hardened elevators to provide timely emergency access to responders and allow evacuation to mobility-impaired building occupants.

    Comments: (Hardened elevators) England already has these hardened (Phase III) elevators. They call them "firefighter lifts." An important feature that must be included in the so-called "hardening" of elevators is improved insulation of the wiring in elevator shafts. Building codes have not considered waterproof insulation of electric wiring in elevators when installing Phase I recapture systems and Phase II firefighter service. Hopefully, water-resistant wiring will be included in the hardening of Phase III elevators. Water run-off from sprinkler systems and firefighter hose streams often cause electric malfunction of elevators. Water shorts the electric wiring in elevators and prevents firefighters from using them at fires. I have personally seen a bucket of water make an elevator fail during a fire.

22. NIST recommends installation, inspection, and testing of emergency communication systems, radio communications, and associated operating protocols to ensure fire communications in buildings with challenging environments and these systems be used to identify, locate, and track emergency responders inside and outside the building.

    Comments: (Firefighter radios) The radios used by firefighters in America are a national disgrace because:

    - Firefighter radios do not work in high-rise buildings, tunnels, or subways.
    - There is no interconnectivity with police and emergency medical personnel.
    - There is interference from nearby commercial radio signals.
    - Some radio systems have single channels for fire dispatch and fireground messages, which has contributed to firefighter deaths.
    - Tent shelter material used by firefighters at wildfires blocks radio signals.
    - Antennas being erected by broadcast cable companies interfere with firefighter radio transmissions.

After 9/11 and the World Trade Center fires and collapse, McKinsey consulting company did a study of the disaster called *Increasing FDNY Preparedness* and recommended the following solutions to firefighter radio problems:

- A short-term solution is a portable booster (repeater) radio.
- A long-term solution is a permanent booster (antennas) installed in buildings. This is called infrastructure.
- The building codes should be changed to require permanent boosters (antennas) installed in high-rise buildings, tunnels, and subways.

McKinsey also recommended the cost for FDNY radio upgrades be obtained by the local government providing tax incentives to building owners or by security grants from the Department of Homeland Security.

A major obstacle impeding effective firefighter radio transmission is the absence of a standard for fire department radio transmissions. After working in a high-rise district and experiencing poor or no fire communications during fires in high-rise buildings, tunnels, and subways, here are the standards or specifications for radio transmission:

- In a high-rise building, firefighters' radios must transmit signals from lobby to roof and from lobby to lowest sub-basement.
- In a vehicle tunnel, the radios must transmit a signal the entire tunnel length, from opening to opening.
- In a subway tunnel, a minimum radio transmission should be three station stops. This is the distance required when moving trains to the nearest station to evacuate riders.

23. NIST recommends procedures and methods for gathering, processing and delivering critical information through integration of relevant voice, video, graphical, and written data to enhance situation awareness of all emergency responders.

    Comments: (Dedicated fire person in charge) High-rise buildings are small cities and should have a person dedicated to the fire safety in the building. This person in charge is sometimes known as a fire safety director. The fire safety director assists the fire personnel upon arrival, and is separate from a security director. The fire person in charge is trained and certified in building fire safety, fire prevention, and fire evacuation of occupants. To assist a trained person in charge, the high-rise building should be equipped with a fire command system in the lobby of the building. This fire command system should have voice capability so occupants can report fires to the lobby command post and to provide evacuation instructions from a fire safety director at the lobby command post to occupants on all floors. Also at the lobby command post, there should be a video display computer terminal to visually track smoke (smoke detectors) and fire spread, and generate graphical readout reports for the incident commander and sector officers upon their arrival. All information generated by the computer should be saved and printed out on hard copy and retained for post-fire analysis.

24. NIST recommends the establishment and implementation of codes and protocols for ensuring effective, uninterrupted operation of command and control systems for large-scale emergencies. State, local, and federal jurisdictions should implement NIMS (National Incident Management System). The jurisdiction should work with the Department of Homeland Security to review, test, evaluate, and implement an effective, unified command and control system.

    Comments: (Unified command and the National Incident Management System) In addition to this NIST recommendation, the McKinsey consultant report, *Increasing FDNY Preparedness*, recommended that New York City adopt the NIMS that is used by fire and police throughout the country. New York City does not use NIMS. They adopted a management system called Coordinated Incident Management System (CIMS), which has police in command of fires, emergency medical incidents, hazardous material incidents, building collapses, search-and-rescue, and mass casualty when terrorism is suspected. The Chief of the New York City Fire Department, Peter Hayden, stated at the 2006 national Fire Department Instructors Conference (FDIC) that the city's adoption of CIMS in place of NIMS is unsafe for residents and the firefighters of New York City. Chief Hayden stated these incidents that police command are fire department core competencies, not the police department's core competencies. Core competencies—fires, emergency medical incidents, hazardous material incidents, building collapses, search-and-rescue, and mass casualty incidents—are incidents firefighters train for and do every day and to which they are committed. Under CIMS, the police will have command of these incidents that they do not train for, do not do every day, and to which they are not committed. Police officers are committed to law enforcement. CIMS is doomed to fail, says Chief Hayden. He has stated there were no lessons learned from 9/11 and there will be a repeat of the same mistakes at the next terror attack in New York City.

25. NIST recommends nongovernmental and quasi-governmental authorities that build, own, or lease buildings and are not subject to local building and fire codes be encouraged to provide a level of fire and building safety that equals or exceeds the level of safety that would be provided by strict compliance with the code requirements of the local government.

    Comments: (Port Authority compliance with New York City building and fire codes) The Port Authority of New York and New Jersey is rebuilding on Ground Zero again. The Port Authority of New York and New Jersey built the World Trade Center towers that collapsed in ten seconds and eight seconds and where 2,749 people died. The Port Authority of New York and New Jersey still does not comply with the building and fire codes of New York. The Port Authority continually states they will equal or exceed the requirements of New York City's buildings and fire codes when they rebuild. But according to the *Final Report on the Collapse of the World Trade Center Towers*, records show they did not equal or exceed the New York City building and fire codes when constructing the towers in the 1970s. In fact, they repeatedly chose less building and fire safety requirements when designing the original twin towers. For example, most of the design for the twin towers was done in the late 1960s when New York City's 1938 building code was being rewritten and changed in 1968. The 1938 building code contained more restrictive fire safety requirements than the newly adopted New York City building code of 1968. When the Port Authority was given the choice to use the restrictive 1938 code or the less restrictive 1968 code, they chose the 1968 code standards. In addition, when recommendations of designers or contractors called for safer standards or further testing, the Port Authority chose to ignore the suggestions. The Port Authority continually chose the faster, more economical design for the towers. The following examples of the NIST report, pages 55-58, 69-71, show: "(1) when choosing between construction class 1 A (a more resistive fire-resistive building) and 1 B, the Port Authority chose 1 B. (2) Port Authority chose fire partitions separating tenant spaces that only extended to the underside of the ceiling instead of extending to the underside of the floor above. Fire and smoke could spread above the ceilings. Only after a warning from the general contractor, fire partitions separating tenants were extended to the floor above during later floor alterations. (3) The Port Authority did not install automatic sprinklers in the 110-story towers as recommended by the Fire Department. Only after a fire in 1975 did a slow, 25-year installation of sprinklers take place. (4) After sprinklers were installed, fire partitions were removed to create a large floor area as an allowance for the sprinklers. (5) When designing exits, doors and stairs, the Port Authority chose the 1968 code requirements which eliminated a fire tower (smoke-proof) stairway and they did not substitute another stair and so the number of stairways were reduced from six to three; they also reduced the width of the doors leading to the three stairs, from 44 inches wide to 36 inches wide. (6) When the architect recommended 2 inches of spray-on fire-retarding for fire-retarding floor steel web members and 1 inch for the top and bottom steel chords, he also recommended the spray method of protecting steel could not be determined without testing. The Port Authority disregarded this advice and directed a ½-inch-thick coating of spray-on fire retarding for trusses. During this NIST investigation, the Port Authority confirmed there was no record of fire endurance testing of this innovative (spray-on) thermally protected floor system used in the towers. This section of the NIST report confirms the families of the 9/11 victims and Sally Regenhard's Skyscraper Safety Committee fears that the Port Authority of New York and New Jersey cannot be trusted to equal or exceed the building and fire code requirements as they claim.

Ironically, today, New York City is rewriting a new building and fire code and the Port Authority once again is rebuilding towers at Ground Zero. And the New York City Building Commissioner's prior employer used to be the Port Authority of New York and New Jersey. The New York City Building and Fire Departments better closely supervise the rebuilding of Ground Zero to ensure there is not a repeat of this tragedy. In my opinion, the Port Authority should not be trusted again to comply with or exceed the New York City building and fire codes.

Lessons Learned

NIST states these recommendations are both realistic and achievable. Building owners and public officials should evaluate these recommendations and, where warranted, take steps to implement these fire safety suggestions without waiting for changes to occur in building and fire codes, standards, and construction practices.

Note: For more information on why the World Trade Center towers collapsed, visit *wtc.nist.gov*.

21

HIGH-RISE BUILDING COLLAPSE

The most important lessons learned from 9/11 were from the collapse of World Trade Center Building 7. Building 7 collapsed solely because of fire destruction. Towers 1 and 2 collapsed because of plane impact and jet fuel fires. Not so, for Building 7 of the World Trade Center complex. Uncontrolled fire, fueled by ordinary office furnishings, caused a "global collapse" of this 47-story, fire resistive, skeleton steel high-rise office building (fig. 21–1). Fire in Building 7 was started by burning parts of collapsing Tower 1 crashing into it. Flames spread because automatic sprinklers failed, and because the FDNY decided not to fight the high-rise blaze. The National Institute of Standards and Technology (NIST) states, "This was the first instance of fire causing the total collapse of a tall building." (NIST, "Final Report of Collapse of World Trade Center Building 7," August 2008.)

On August 21, 2008, the National Institute of Standards and Technology (NIST) published the final report on the collapse of Building 7. The report ruled out several suspected causes of the collapse, such as the unusual truss construction in the building spanning over a Con Edison generating plant; or burning diesel fuel from tanks stored in the building; or a controlled terrorist demolition. The NIST report concludes that fire caused the collapse. And that the fire was fueled solely by office furnishings. The same type of furnishings found in high-rise office buildings throughout the nation.

A burning 47-story high-rise office building crumbled down covering the streets and rooftops, on all sides with tons of steel, masonry, and plaster dust. This collapse has major implications for the fire service. The collapse of Building 7 must be considered a benchmark in the fire service. Consider for example:

1. Structural stability of a burning, modern, lightweight high-rise, skeleton steel building must be questioned.
2. A high-rise can collapse after a 7 hour free burn time.
3. Ten floors of fire brought down the high-rise building.
4. A 47-story skeleton building can totally collapse in 13 seconds.
5. Passive fire protection provided by construction did not stop fire spread.

It must be stated the collapse of World Trade Center 7 was a rare event. However the fire service mission is to plan for rare events. Now if firefighters cannot extinguish a fire in a high-rise and the blaze spreads from floor to floor, and they find themselves huddled in stairways hoping for fire-resistive construction to contain the blaze, while all the office furnishings are consumed by flame, (called "controlled burning"); then collapse of the building must be considered. This happened to us in New York City, during the One New York Plaza high-rise fire in 1970; and to firefighters in Los Angeles at the First Interstate fire in 1988; and to Philadelphia firefighters at the One Meridian Plaza fire in 1991. From this day forward, when an uncontrolled high-rise fire happens, the fire commander must plan for the rare event of a global collapse of the high-rise burning building.

Fig. 21–1. World Trade Center Building 7 burning before building collapse

The Construction

World Trade Center Building 7 was an irregular shape (trapezoid) 329×144 feet 47 stories high, type 1-C; skeleton steel, fire-resistive office building. Columns, girder, and floor beams supported a composite concrete and steel deck floor. The concrete deck was 2.5 inches thick. The core and exterior columns carried loads to the foundation; however the structure was built over a Con Edison electric generating plant. To do this, some columns transferred loads across girders and trusses between the fifth and seventh floors over the smaller structure. Connections between beams, girders, and columns were simple shear, seated connections. The NIST report states these connections did not contain the stamp of a professional engineer. World Trade Center Building 7 collapse started, as do most building failures at the connections. Specifically, thermal expansion from the heat of fire on the 13th floor caused the fluted steel floor deck to expand more than the concrete above and the connections of the composite concrete and steel Q decking failed. The floor beams expanded and pushed out a girder; the girder pulled away from a column; this unconnected column buckled; and a global collapse of Building 7 started.

The NIST experts describe this event as "progressive collapse." Progressive collapse occurs when an initial load failure spreads from the structural element to structural element, resulting in the collapse of an entire structure. Global collapse is total collapse. These two terms: *progressive collapse* and *global collapse* are now associated with high-rise fire-resistive construction for the first time. The fire service must now acknowledge this possibility first described by engineering experts from the NIST study of the collapse of the World Trade Center buildings.

Fire Protection of Steel

The spray-on fire retarding material (SFRM) covering steel in Building 7 was assumed sufficient for a Type 1-C fire-resistive building. The Type 1-C building requires that the fluffy spray-on protect columns for 2 hours and beams for one and a half hours. The thickness of spray-on fire retarding (estimated by NIST) was less than 1 inch thick throughout the entire steel structure, including 875 covered columns and girders, 5-inch covered beams, and 0.375 covered floor deck undersides. The steel is supposed to be unpainted for good adhesion of the spray on fire-retarding material, however evidence suggested steel was covered with primer paint. The structure had a fully installed automatic sprinkler, which did not function on the lower floors where the main body of fire grew. The sprinklers on the lower floors of WTC 7 were supplied by water mains and the upper floors were supplied by gravity supply tanks. The collapse of Towers 1 and 2 damaged the water mains. Figure 21–2 depicts the World Trade Center complex layout.

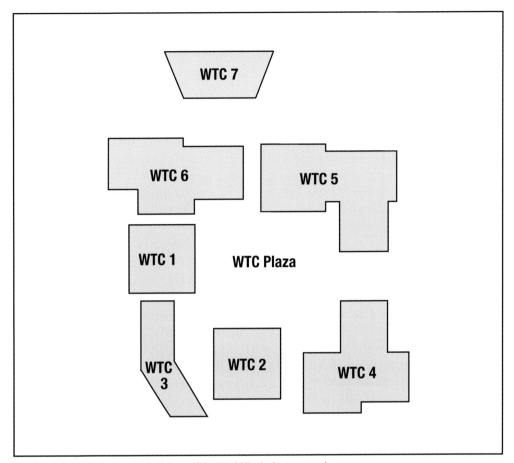

Fig. 21–2. Layout of the seven buildings of the World Trade Center complex

The Fire

Fire spread into Building 7 from burning debris spewed outward by the collapse of World Trade Center Tower 1. Burning desks, computers, chairs, partitions, and papers crashed through the windows of Building 7 and ignited small fires on several floors. This burning debris, technically named *flying brands* are usually seen blowing in the air over a lumberyard fire, or spreading ahead of a raging wild fire. These were *urban flying brands*—flaming pieces of office furnishings, along with the steel structure, flying across streets crashing through windows, landing on roofs, starting "spot fires" ahead of the main conflagration. Burning debris was expelled outward each time a tower collapsed; creating a rainstorm of flying brands onto firefighters, bystanders, fire trucks, automobiles, roof tops, building setbacks, and through windows of adjoining buildings. This burning flying debris mixed with the dust flows that choked and blinded everyone for 20 minutes each time a building collapsed. When the dust cleared, hundreds of cars, fire trucks, and broken furnishings in the street were ablaze. The facades of adjoining buildings were ripped open and fires burned everywhere, including multiple floors of Building 7.

Flames were first observed in Building 7 on the south east corner of at the 22nd floor around 12:10 p.m. (Tower 2 collapsed at 9:58 a.m.; Tower 1 at 10:28 a.m.; Building 7 at 5:20 p.m.). Flames were recorded by video on floors 19, 22, 29, 30, but those fires did not spread. The fires that caused the global failure of Building 7 were first observed on the west side of the building; first on floor 12 and then on floors 7, 8, 9, (not 10) 11, and 13 with a small fire reported shortly before the collapse on floor 14.

Fire Spread

Horizontal fire spread occurred on several floors unimpeded by partitions. Flame spread across large open floor areas and plenum areas, the space above the ceilings, and progressed along the west side of Building 7, and then turned the north corner and spread east across the north side. The wind was coming from the north at 11 to 22 miles per hour. There was very little smoke visible from the north side of the building. Only flames could be seen in many burned-out windows. Wind blowing through these open north side windows drove flame and large amounts of smoke out of the south side of Building 7. Heavy smoke billowed out many floors on the south side.

Vertical inside fire spread inside the burning structure was hidden from view of NIST investigators, but experience tells us that flames probably spread up through the usual openings: open stairs called "access" or unprotected convenience stairs, central air conditioning shafts and ducts, poke-through holes in floors, elevator shafts, and utility closets, and cracks in concrete floors.

Vertical fire spreading outside Building 7 was observed by investigators on the west and north side. This was a limited amount of auto exposure—the spread of flame from window to window. However, window breakage from the heat of flames, which allows auto exposure to take place, was documented. Window breakage and falling glass from the metal frame is due to the unequal expansion rate when heated by flame. The exact failure of large sections of window glass breaking due to the heat of fire was observed on videos. First, the center section of window glass, heated by fire, would break and fall outward and this would leave small jagged pieces of glass surrounding the opening still fixed in the metal frame. Next, the remaining jagged pieces of glass and sometimes the metal frame would fall away later leaving the entire opening unobstructed, allowing flame to blow out of the window opening, and in some instance spread upward to heat a window above—thus, auto exposure.

The Collapse

According to the National Institute of Standards and Technology, Building 7 did *not* collapse because of unusual construction of a transfer truss beam spanning over a Con Edison generating plant or because of diesel fuel stored in the building or because of structural damage caused by falling Tower 1 or because of bar joist truss construction or because of controlled demolition by terrorist's explosives. NIST states fire destruction alone caused the collapse of Building 7. This cascading progressive building collapsed after 7 hours of uncontrolled burning, if you determine that the first sign of fire was on floors 19, 22, 29, and 30 at 10: 28 a.m.—the time of collapse of Tower 1. However, if you consider the first sign of fire was on the west side of the building on floors 12, 7, 8, 9, 11, 13, and 14 starting at 12:10 p.m., then the global collapse of Building 7 occurred after only 5 hours and 10 minutes of uncontrolled burning.

The exact cause of building failure was thermal expansion, causing a progressive collapse of a column on the 13th floor. The floor beams pushed out a girder; the girder between columns 44 and 79 lost its connection with column 79 that provided support for a long floor span on the east side of the building; the column buckled and a cascading progressive collapse of Building 7 started. The first detectable indicator of WTC 7 collapse was the downward movement of the eastern side of the roof line, progressing to the west side and the total collapse of the high rise structure. Figure 21–3 depicts a bird's-eye view of Building 7.

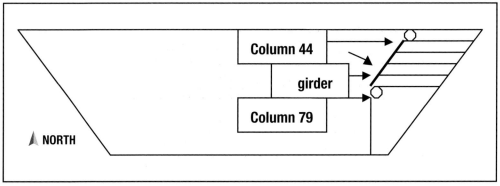

Fig. 21–3. World Trade Center Building 7 from above

Progressive Collapse

The NIST study of Building 7 highlights the nature of a progressive collapse of building elements. Progressive collapse, also known as disproportionate collapse, is defined as the spread of local damage from an initiating event, from element to element, eventually resulting in collapse of an entire structure or a disproportionately large part of it. The term *progressive collapse* was first introduced to the fire service in 1972 by the post-fire analysis of the Boston Vendome Hotel collapse, which killed nine firefighters. At this fire, a progressive collapse of a 100-year-old heavy timber building was started when a bearing wall failure caused columns, girders, floor beams, and enclosing walls to collapse. The progression of failure at the Boston fire was from the top of the structural hierarchy to the bottom, it went from a bearing wall to column to girder to beam. In the World Trade Center Building 7 study, the progression of collapse was reversed. Progress of failure in Building 7 was from the bottom of the structural hierarchy to the top, it went from a beam to girder to column. In both cases the result was a global collapse of the entire structure.

Content Fuel Load

The fire that heated the steel in the 47-story high-rise building was fueled by office furnishing, such as desks, chairs, wood or laminate partitions, tons of paper, computers, carpeting, cubicle work stations, book shelves, high-density file storage, and miles of combustible insulation on computer cables. An important factor in the fire spread of combustible content in Building 7 was the open floor design. Only the core was divided by a demising partition wall (a partition that extends from floor slab to floor slab above). In addition to open floors there were also "dwarf partitions" separating some areas. These walls extend only to the underside of the ceiling and do not stop or limit fire spread above the ceiling. Flames spread in the plenum area feeding on combustible cable insulation above the ceiling. An open floor design as in Building 7 creates 10 or 20 thousand square feet of open floor area and allows fire to spread beyond the extinguishing ability of firefighters hose streams. Building 7 had a sprinkler system that did not work. Some fire protection engineers and code officials erroneously believe if a building has automatic sprinkler system protection, there is no need for walls and partitions to limit spread of fire. The fire service appreciates sprinklers, but also recommends compartmentation—fire partitions to subdivide a large floor area and limit fire spread. Compartmentation, along with automatic sprinklers is the required fire protection for a high rise building.

20th-Century Warning

A warning of what was to come on 9/11—global collapse of a high-rise skeleton steel burning buildings—occurred in Philadelphia on February 23, 1991, when the 38-story Meridian Plaza building burned for hours. During the fire, the building was declared structurally unsound by an engineer. Chiefs inside notified the commander of collapse indicators during the fire: walls in all three stairs were cracking and moving. In one stair what had been a 2-inch crack had grown to a fist-sized opening. Floors had moved as much as 3 feet. I-beam flange connections were cracked. Fire-resistive material on the beam in the stairway had fallen off and the unprotected steel structural members were twisting, moving, and starting to elongate. The structural engineer called to the scene to analyze the building after inspecting floors 22, 23 and 24, stated structural problems were substantial. He believed that particularly with the free-burning fire on floors 22, 23, and 24, now spreading to floors 25 and 26, internal collapse was possible. Based on this report of structural instability and after several more hours of firefighting proved futile and after retrieving the bodies of three dead firefighters, Philadelphia Fire Chief Roger Ulshafer made a major decision and abandoned firefighting efforts. He withdrew all firefighters from the 38-story free-burning high-rise building.

This was the first instance in the history of the fire service where firefighters were withdrawn from a burning high-rise building. Hose lines inside were abandoned. Only hoselines directed from buildings across a 60-foot-wide street and supplying the sprinkler partial system that only covered floors 11, 15, 30, 31, 34, and 35 continued supplying water. The fire totally burned out floors 22 to 30. When the flames reached the thirtieth floor, nine sprinkler heads extinguished fire. The fire was declared under control after 19 hours, so each floor burned for approximately 2 hours. If the top eight floors had also burned, the fire would have burned 16 additional hours, for a total of 35 hours. This fire could have burned for a day and a half.

Long-Duration High-Rise Fires

Building 7 burned for 7 hours and collapsed. This fire did not progress from floor to floor as a typical high-rise fire will. Multiple fires started on several floors. One expert in the NIST report states it can take on average 1½ hours of burn time per floor. All combustibles on a floor are consumed during this time while firefighters attempt to hold the stair enclosures with hoselines. Another example of a long duration fire is the One New York Plaza high-rise fire on August 5, 1970. It burned two floors and took five hours. It was a good thing firefighters extinguished the blaze because this 50-story building had 16 additional floors above. If they had burned at the same rate, the fire chief would have stood at the command post for an additional 38 hours, a total of 43 hours. The chief would have been in the street almost two days gazing up at flame spreading from floor to floor. On May 4, 1988, the First Interstate fire in Los Angeles burned five floors in approximately 4 hours. There were 46 more floors above for a total of 62 stories. If the LA firefighters had not stopped this fire at floor 16, the chief would have been giving orders for 32 more hours total of 36 hours; hopefully after the occupants and firefighters had been withdrawn. In the future, there will probably be a high-rise fire spewing smoke and flame into the sky for several days and in the future, unfortunately, there will probably be another global collapse of a burning high-rise building.

We now have evidence that high-rise, fire-resistive buildings can collapse from fire. We also know high-rise fire-resistive steel frames are only protected against heat for a maximum of four hours. This information poses the question to fire chiefs, *How long should firefighters remain inside a burning high rise building that may collapse?* This question will be answered by tomorrow's fire chiefs.

A Future High-Rise Fire

A fire chief responds at 2 a.m. to a second alarm high-rise fire in the center city, downtown area. A radio progress report states: "Fire building is 30-story, office building, 100 by 200 feet, and irregular shape, Class I, fire-resistive construction. Fire fully involves the 20th floor. All occupants removed. Exposure 1 is a street, and similar building; exposure 2 a street, and similar; exposure 3 a street, unknown; exposure 4 a street, and similar. By order of Battalion 1, transmit a third alarm".

At the command post, the fire chief questions the incident commander. The commanders responds, "Chief, we have two lines operating in A stairway. They cannot advance out of the stair enclosure, too much fire. B stair is the evacuation stair. I have second battalion as the operation's chief on the floor below. Everyone's out. All cleaning personnel are accounted for. Got a tower ladder 1 on the exposure 2 side, but the fire is above the reach of the hose stream. Looks bad; it's up to the building now." He turns and adds, "Here is the night manager."

"What can you tell us about the fire floor?" the chief asks.

The man responds, "The sprinklers are out of service, don't know why. A law firm rents the 20th to 24th floors, there are unprotected (open) convenience stairs connecting all four floors. The floors are one large open area. And the HVAC system serves the twentieth to thirtieth floors, I shut it off."

The chief orders a fourth alarm transmitted and actives mutual aid response for reinforcements. Having just finished reading the NIST report about the collapse of World Trade Center Building 7; he sizes up his fire and thinks:

- If fire spreads to the 10 floors above, at 90 minutes average floor burn time there could be a 15-hour-long fire.
- World Trade Center Building 7 collapsed when 10 floors burned for seven hours.
- The skeleton steel framework is protected from heat by probably fluffy spray-on retarding and that protects steel columns for only 4 hours.
- This building is irregular (diamond) shape. Building 7 was irregular—trapezoid.
- This building was constructed in 1987 around the same time as WTC Building 7 and is lightweight steel construction.
- The glass windows heated by fire in this building are crumbling down—auto-exposure vertical fire spread is beginning.
- This building contains the same type office furnishings as Building 7—carpet, desks, paper, etc.
- The floors in this building are large open design, 100×200 feet, approximately 20,000 square feet, beyond the capability of 2½-inch attack hoselines.
- A global, cascading, progressive collapse of this 30-story high-rise office building is possible if the firefighters cannot extinguish this blaze soon.
- The FDNY firefighters were ordered to withdraw four blocks (approximately 800 feet away) from burning Building 7; command was set up at Chambers and West Street.

A firefighter interrupts, "Chief we just got a report fire spread to the floors 21 and 22."

Lessons Learned

Before the WCT 9/11 NIST report, collapse of a modern high-rise building was not in a fire chief's vision of imaginable events. We saw plenty of local collapse after a fire. Post-fire investigations revealed concrete floor sagging, façade collapse, buckled steel beams, and girders pushed aside, but not cascading progressive collapse (global) from fire destruction. Since 9/11, the fire service must plan for total global collapse of a high-rise building during fire. This NIST study must be used by the fire service to think the unthinkable.

Note: For more information on the World Trade Center Building 7 collapse, search the Web for: *NIST.wtc7*.

22

POST-FIRE ANALYSIS

When on the morning of August 2, 1978, I heard that six firefighters had died in a timber truss roof collapse at a burning Brooklyn supermarket, I asked myself some very disturbing questions: *What did I know about truss roofs?* Not enough to have ordered the men off the roof before it collapsed. *Would those six men have died if I were the chief-in-command of that fire?* Yes, they would have. After the incident, I remember that I looked through two books on firefighting strategy and noted that both mentioned truss roof construction and its collapse danger very briefly, with only a few lines of warning. (Francis Brannigan's book, *Building Construction For The Fire Service*, covered the subject more thoroughly.) Then I tried to recall my experience with timber truss roofs in my 20 years of firefighting.

My earliest recollection was in the late 1950s, when our company was inspecting an old auto parking garage, and our captain pointed to the open under side of the roof and said that the roof was dangerous. I remember seeing crossed wood web members at the underside of a very high roof. Then, when I was a backstep firefighter in the early 1960s, a fire occurred in an old vacant movie theater on my day off. Responding firefighters talked about the collapse of its timber truss roof the next day. Ten years later, a classmate in a college night course told me about a timber truss roof collapse that had occurred the day before, showing me a photograph of a building with a large opening in the roof. In my years of firefighting and academic and promotional study of fire protection, those few instances were my only contact with the timber truss roof—one of the most dangerous structures that exists, from a firefighting point of view. It would not have been enough to influence my strategy if I had commanded at the supermarket fire.

Today I know quite a lot about truss roofs, but at the time I asked myself how I could, in good conscience, command an operation and be responsible for the lives of other firefighters if I had such scanty knowledge of collapse dangers. After the Brooklyn supermarket truss roof collapse, one of the charges brought against the City of New York and the New York Fire Department was that none of the department's written training material mentioned the subject of truss roof collapse. The lawyers stated that this lack was a failure of the municipality to train its employees properly. Looking back, I find it hard to believe that, with the enormous number of bulletins, circulars, and directives issued to the department by the division of training, a subject as important as truss roofs was never covered. But it was not, nor were many other collapse dangers that firefighters confront. The fire department training division now supplies a considerable amount of printed information on truss roofs and other collapse dangers.

The city and the FDNY were also criticized for not providing formal training to chief officers. In my case, when I entered the fire department in 1957 and served as a probationary firefighter, I received three months of training. As a newly promoted company officer in 1963, I received six weeks of instruction. But when I was promoted to chief officer in 1973, I received no training. This pattern is typical for chiefs throughout most of the United States. Training is almost nonexistent, primarily because local governments cannot justify the cost of training small groups of chiefs, and very few chief officers are qualified and willing to teach fire strategy and tactics.

Today, the city provides training to FDNY chiefs through a chief officer's development program. Chiefs attend conferences where they are encouraged to exchange information about firefighting strategy, tactics, and fireground safety and to discuss the latest life safety and fire protection innovations. The National Fire Academy at Emmitsburg, Maryland, assists New York with the costly burden of training chiefs through its two-week course on fireground strategy and tactics, "Command and Control of Fire Department Major Operations." Designed for battalion chiefs and tour commanders, the course provides chiefs with a general, nationwide view of the role of the fireground commander and of operations at fires and emergencies. The course, held at the National Emergency Training Center, prevents the danger of "inbreeding" and a "know-it-all" attitude that can develop among firefighters in large or small fire departments, where members receive only limited in-service training.

Fig. 22–1. Waldbaum's Supermarket timber truss roof collapse killed six FDNY firefighters in 1978.

The final report of the fire department's investigation into the supermarket roof collapse stated that the firefighters had died partly because "the extent, the severity, and to some extent, the location of the fire had not been clearly defined prior to the collapse." One of the most important size-up duties of first-in chiefs and company officers is locating the fire and determining its severity. This information lays the foundation for the entire operation. First, it determines the number of firefighters and amount of equipment needed to control the blaze. Second, until the location and extent of the fire are known, firefighters cannot determine the overall life hazard, the most effective point of fire attack, and the most efficient method of venting heat and smoke. Sometimes, however, the location and severity of a fire cannot be assessed by the chiefs and company officers first on the scene. This inability is often caused by alterations or unusual construction of the fire building. In a fire that killed 12 FDNY firefighters in 1966, the exact location of the fire was not discovered because of alterations. Similarly, the severity and extent of the Brooklyn supermarket fire was not "clearly defined" because of three factors: an unusual alteration called a "rain roof," the fire-retarding compartmentation of the trusses, and the large heat collection space created by the bowstring design of the structure (fig. 22–1).

Rain Roof

Because the supermarket's bowstring roof was sagging and rainwater was accumulating at its depressed center, a second roof, or "rain roof," had been built over its center (fig. 22–2). The rain roof recreated the curve at the highest point of the roof and enabled water to roll to the sides of the trusses, where scuppers and drains removed it to the street. The rain roof was not simply another layer of tar paper placed on top of the original roof, nor was it a so-called "raised" or "inverted" roof, where the roof deck is built up on a flat roof by placing 2×4-inch wood members on top of roof beams and a single deck atop the wood members. Instead, the rain roof at the Brooklyn supermarket consisted of wood beams and a roof deck laid on top of the old tar roof. There were two roof decks, then, which acted as an insulating barrier between the fire and the men on the roof and also delayed roof venting. When firefighters cut the top rain roof with a power saw and pulled up the plywood deck with hooks and Halligan tools, they discovered another roof deck beneath the top one. To accomplish ventilation, they had to cut another, smaller opening within the original opening. The rain roof had been laid down the center or highest point of the curved roof, where the heat and fire below were greatest.

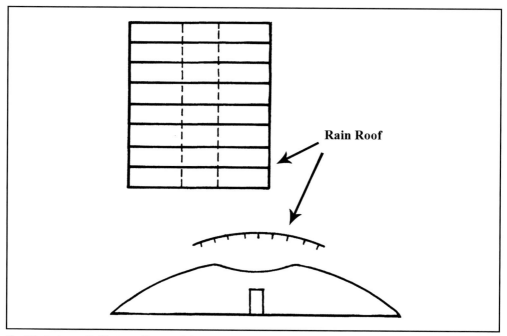

Fig. 22–2. A so-called "rain roof" covered the center one-third of the supermarket roof.

Fire-Retarding Compartmentation

Inside the large, enclosed roof space of the supermarket were seven bowstring timber trusses, spaced 20 feet apart. A door opening was located at the center of each bowstring truss, with a walkway providing access through each truss from the front to the rear of the large roof space. Truss numbers 2, 4, and 6 were fire-retarded and covered on one side with one-hour plaster board. Held open by a fusible link, the automatic fire doors would close automatically when the link was melted by the heat of the fire. Trusses 1, 3, 5, and 7 were exposed, unprotected wood trusses. When the fire extended from the mezzanine floor into the large roof space, it spread between trusses 4, 5, and 6. The automatic fire doors on trusses 4 and 6 closed, confining the flames between those trusses (fig. 22–3). The fire grew on both sides of truss 5, which eventually collapsed. A 4,000-square-foot section of roof deck fell with it, the roof deck split apart, and fire roared out into the 40×100-foot crevice.

Twelve firefighters tumbled into this fire-filled trench; miraculously, six fell through to the floor of the supermarket and were able to make their way to safety. Six others were burned to death; the bodies of several of them were found on top of the display shelves. Because the fire was confined between trusses 4, 5, and 6, there was little or no indication of fire when firefighters pulled down the ceiling beyond this area. When they looked up into the roof space, they saw no sign of fire, so they had to open the ceiling between trusses 4 and 6. The same confusing effect occurred when roof vent openings were cut. When firefighters made the cut between trusses 3 and 4, they saw little evidence of heat or fire. When they made the next cut, several feet away, between trusses 4 and 6, a raging fire exploded through the vent opening.

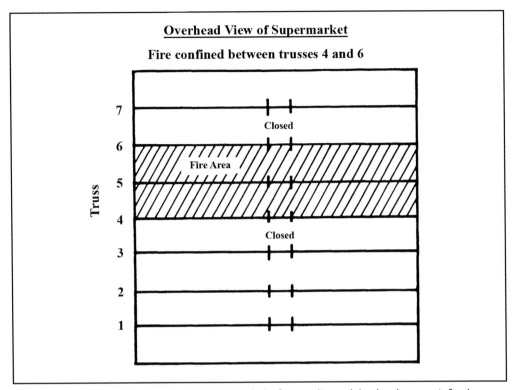

Fig. 22–3. The fire was confined between trusses 4 and 6 by fire retarding and the closed automatic fire doors. The first truss to fail was truss 5.

Bowstring Truss Roof Design

The height of the truss roof space above the ceiling in the supermarket was 10 feet at the center, and the distance from the floor to the ceiling inside the store was 16 feet. Thus, the total distance from floor to underside of the truss roof was 26 feet. The fire started in the mezzanine floor, burned upward past a double ceiling, and spread out into a large roof space. Ordinarily when a firefighter enters a burning room with an 8 or 10 foot high ceiling, the distance that the heat and smoke are banked down below the ceiling is one of the factors used in determining the severity of the fire. The lower the firefighter is forced to crouch to escape the heat, the more serious the fire. Conversely, if a firefighter can walk upright, the fire is considered small or not serious. In an occupancy with a high ceiling or a bowstring truss roof space that can collect a great deal of heat and fire, firefighters can be misled about the severity of the fire. There may be two conflicting statements about the fire's size and severity: Firefighters on the roof, after venting the fire, will report a serious fire; meanwhile, a firefighter walking upright underneath the heat and flame may report a fire of little severity. In a fire department with a history of interior firefighting, like New York City's, the report from inside the fire building is often considered the most accurate report of fire severity. At a truss roof building or a structure with a high ceiling where crew members give conflicting reports, however, the roof report of a serious fire may be the most accurate assessment.

Conclusion

The supermarket's roof was not the first bowstring truss roof that collapsed in the 140-year history of the New York City Fire Department. After the Brooklyn supermarket collapse, chiefs, company officers, and firefighters came forward and related many past incidents of truss roof collapses. Somehow, though, these fires had not been documented or made known to chiefs at any training conferences. Only the chiefs and officers assigned to areas that included older industrial and manufacturing buildings (such as the Hunt's Point section of the Bronx and the Greenpoint section of Brooklyn) were aware of the dangers of truss roofs. The Brooklyn supermarket, on the other hand, had been built in 1952, in a predominantly residential area.

Over the years, the fire service has expended little effort documenting and recording its firefighting experience. Though we have developed extensive pre-planning procedures to prepare for all types of fires and emergencies, we have not adequately explored the advantages of *post-fire analysis*—examining a fire operation after it occurs from the perspectives of strategy, tactics, and fireground survival. For years, insurance companies have conducted investigations of fires after they occurred, but always from a property protection point of view. The fire service should consider analyzing significant fires by documenting them on fact sheets, diagramming the fire building, and photographing the fire scene after the operation. This information could then be distributed throughout the fire department or to the entire fire service, if it is of great significance. If the knowledge and experience of the earlier truss roof collapses had been communicated to all chief and company officers in the city, perhaps the Brooklyn supermarket collapse would not have been fatal.

What is a post-fire analysis? The best way to explain it is first to explain what it is not. A post-fire analysis is *not*

- A cause-and-origin investigation from the point of view of an arson investigator
- A fatal fire investigation from the point of view of a fire education officer
- A loss control investigation from the point of view of an insurance investigator

A post-fire analysis is a strategy, tactics, and safety investigation from the point of view of a fireground commander. A post-fire analysis of strategy and tactics should be conducted after a major fire or an ordinary fire where a significant event occurs. Some specific examples of occasions when a fireground analysis should be conducted are:

- When a firefighter or fire officer is killed
- When an explosion, backdraft, or boiling liquid expanding vapor explosion (BLEVE) occurs
- When a structural failure occurs at which a firefighter is injured
- When a large loss caused by fire occurs

When any one of these events occurs, as soon as the fire has been extinguished, steps should be taken by the officer in command of the fire to conduct a post-fire analysis. The officer in command or another chief officer should perform the fire analysis to ensure that it receives the point of view of a fireground commander. There are some major or unusual fires that occur only once or twice in the career of a firefighter.

These fires should be thoroughly documented, analyzed, and used as case studies to train future fireground commanders. Realistic training programs simulating strategy, tactics, and safety procedures can only be developed from information obtained from post-fire analysis. When an analysis of a fire is done properly, all firefighters, officers, and chiefs in a department can gain valuable experience—not merely those who fought the blaze. For years after the fire, training can occur as a result of a post-fire analysis.

There are many benefits in conducting a post-fire analysis. The greatest benefit occurs to the chief officer doing the analysis. This officer learns about firefighting strategy, and his skills of fireground investigation and analysis are developed. The ability to analyze a fire scene is an extremely valuable skill, which unfortunately, in the past, has been surrendered by chiefs to insurance investigators, arson investigators, and fire protection engineers. There are three parts to a post-fireground analysis. Part one is a printed form or fact sheet that is filled out and includes questions about the fire, the structure, strategy, and tactics. Part two is a fireground diagram—an overhead line drawing showing the fire building, exposures, apparatus, hose and ladder placement, point of origin, collapse, and explosion areas. Part three is photographic documentation of significant factors identified in parts one and two of the post-fireground analysis.

The Fire Fact Sheet: Part One of the Post-Fire Analysis

To complete part one of a fireground analysis, the chief officer must revisit the site of the fire and examine the burned-out structure, interview firefighters who were at the fire, and study information collected for the official fire record or report. A post-fire analysis may have to begin immediately after the fire has been extinguished. In high-value districts, unless there is a court order not to disturb evidence of a fire suspected to be caused by arson, the scene will be completely cleaned up the day after the fire. In such a situation, the third part of the investigation (the photo documentation) must be completed first, before the scene is altered. The diagram would be the next step, and part one, the form, can be the last phase of the post-fire analysis to be completed.

Two important facts of the fire analysis fact sheet are the point of fire origin and the cause of the fire. The room of origin is most often the area with the deepest charring of burned wood. To find the exact point of origin within the fire room, look for the lowest point of deep char. There are exceptions, but at 99% of fires, the point of origin can easily be determined by a close examination of the fire area. After the point of origin is determined, the cause of the fire should be determined. If the point of origin is near a mattress or an upholstered chair, this can be considered the first material ignited and smoking carelessness suspected as a cause. Occupants often voluntarily offer information about the cause of the fire. Fires suspected to be arson must be referred to the fire investigation specialist; however, the officer in command is responsible for determining the point of origin and the cause of the fire for the official fire record. This skill of analyzing a fire scene can only be learned by doing it.

The Structure

Fire spread beyond the point of origin is primarily determined by a structure's design and construction. After an officer has conducted several post-fire analyses, patterns of fire spread within similar types of buildings quickly become evident. Similar vertical and horizontal avenues of flame spread can be identified when analyzing fires in similar types of buildings. Experienced fire officers know the reoccurring fire spread dangers associated with each type of construction in their communities and take immediate actions to protect these areas with their strategy and tactics. Fire spread design problems associated with various construction types are discovered quickly when documented in post-fire analysis. There are five standard types of building construction found throughout the nation:

1. Fire-resistive construction
2. Noncombustible/limited combustible construction
3. Ordinary construction
4. Heavy timber construction
5. Wood-frame construction

Known reoccurring fire spread weakness associated with each one of the standard construction types are

- Central air-conditioning duct systems allow the spread of smoke, heat, and flame throughout a fire-resistive building.
- A combustible roof deck allows the spread of flame and smoke throughout a noncombustible building.
- Concealed voids and spaces allow fire spread throughout ordinary constructed buildings.
- Large amounts of structural fire loading, in combination with unprotected window openings, create tremendous radiated heat waves which spread fire to exposed buildings when heavy timber constructed buildings burn.
- Combustible exterior walls with small horizontal separations between buildings allow rapid fire spread between wood-frame buildings.

Although these are the known fire spread weaknesses associated with standard construction types, today, many variations on the five types have been built in communities throughout the country. New avenues of fire spread in addition to the ones mentioned will be learned quickly by post-fire analysis documentation.

Burning building collapse is another important fireground event which should be recorded in a post-fire analysis. Just as there are certain fire spread weaknesses associated with construction types, there are particular collapse weaknesses associated with construction types. The manner in which a building fails during destruction by fire is often repeated. The same kind of brick walls located at the same portion of a burning building collapse in similar ways. Walls of burning wood-frame buildings break apart and fall down in reoccurring, recognizable collapse patterns. Steel columns, girders, and beams in similar types of buildings buckle and fail at the same general point when heated to their yield point. Some general collapse characteristics of burning standard building construction types are:

- Fire-resistive concrete buildings suffer localized ceiling collapse by spalling chunks of concrete.

- Secondary steel floor support beams of fire-resistive steel skeleton buildings separate from the main girder at the connection.
- Noncombustible buildings suffer early roof collapse.
- Parapet walls of ordinary brick-and-joist construction buildings often collapse near the roof level or at setbacks.
- Fully involved burning heavy timber constructed buildings suffer floor collapse followed by masonry wall collapse.
- Wood-frame buildings suffer bearing wall collapse with instantaneous floor collapse.

When recording a burning building collapse on a post-fire analysis fact sheet, a fire department investigator should state the first structure to fail. Many times, the most dramatic collapse or the collapse structure that kills a firefighter is not the first structure to fail. For example, a roof failure could cause a wall failure. This is defined as a "progressive collapse." This collapse occurs when the initial structural failure spreads from structural element to structural elements, resulting in the collapse of an entire structure or a disproportionately large part of it. At one fire, a firefighter was injured when a utility pole collapsed on a fire truck; however, the pole collapsed because it was struck by a falling masonry wall. In fact, the wall collapse was caused because the roof collapsed first. It is more valuable to a safety officer to identify on the post-fire analysis fact sheet the first structure to fail, in addition to any resulting secondary or even tertiary subsequent progressive collapse caused by the initial structure failure.

In addition to identifying the first failed structure, the officer should determine the cause of collapse. If the evidence is destroyed by fire and the exact cause cannot be discovered, the investigating officer should state a subjective opinion rather than state it as unknown. This subjective opinion can be based on interviews with officers or firefighters near the collapse at the time of failure. If a subjective opinion is stated instead of a documented cause, a statement that the cause given is a subjective opinion should be recorded. The most frequent causes of burning building collapse are: fire destruction explosion, backdraft, BLEVE, impact of hose streams or equipment striking a structure, water accumulations trapped on upper floors or roofs, and ice formation from hose streams. In many instances, there are two or more causes of a burning building collapse. All suspected causes should be stated.

Another important fact about a collapse which should be recorded is the stage of fire growth in which structural failure occurs. There are three identifiable stages of a building fire: growth stage, fully developed stage, and decay stage (fig. 22–4). During the growth stage of a fire, the structural framework of a building is not yet involved with flames. In most instances, the fire is still contained by plaster walls and confined to the building content or furnishings. If the fire is not extinguished at this stage and continues to grow, flashover occurs; the fire may destroy the confines of the plaster walls and spread to the structural supports of the building. The building's structural framework is often affected by fire during the second and third stages and collapse may occur. The second and third stages of a fire are the periods when most burning building collapses happen.

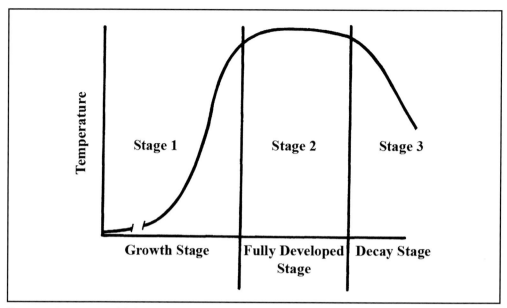

Fig. 22–4. Most structural failures occur during the second and third stages of a fire.

Strategy

The dictionary defines strategy as *the planning and directing of large-scale military actions*. The fire service defines strategy as *the planning and directing of the actions of large numbers of firefighters, apparatus, and equipment*. A fireground commander must have thorough knowledge of, and the ability to put into use, many different types of strategies or plans. The most important one is a firefighting strategy or plan. It may be a general, step-by-step plan such as: locating the fire; protection of life (civilians and firefighters); prevention of fire extension; confinement of the fire on all sides; advancing and extinguishing the fire; salvaging property and overhauling to prevent rekindling; and securing and safeguarding the premises before leaving the scene. Or it may be simply: locate, confine, and extinguish. After deciding the firefighting strategy, the fireground commander then must have the ability to direct fire companies to perform any number or any type of firefighting tactics necessary to accomplish the strategy.

In addition to the fire strategy, many other plans have to be at the fingertips of the fireground commander, and the plans may have to be put into operation at a moment's notice. The following are a sample of some strategic plans which a fire commander must be able to use effectively at a large-scale fire: assuming command, giving situation radio reports, and expanding an incident command system. The commander must designate fire officers for specific pre-planned divisions (geographic areas) and groups (functions) to command at fires, and establish several plans including:

- A fireground designation plan to define exposures surrounding the outside of the fire building (A, B, C, D)
- A fire building designation plan defining areas inside a large-area building or a row of stores inside a large-area building

- A fire communication plan that can divide and safely monitor portable radio messages into two or more separate channels—a command channel and a tactical channel
- A plan to establish the location and identification of a command post at large-scale operations where the transfer of responsibility can take place smoothly between arriving chiefs and to which incoming fire companies report for orders and assignments
- A plan for establishing alternate water supply and pressure sources when the primary water supply fails, or is inadequate or nonexistent

All of these plans may not be needed at minor fires; however, the next fire you respond to may be a major fire, and all the plans may have to be put into use at once. Every fireground commander must be able to function effectively as the commander or division or group officer within each one of the plans. When the post-fire analysis fact sheet is filled out, the officer, answering whether any of the strategic plans were used at the fire, should also be asking himself, *am I thoroughly familiar with these plans? Could I put these plans into operation effectively? Could I be an effective sector officer if my officer in command ordered these plans into operation?*

Tactics

Fire department tactics are the operations of a fire company performed at a fire. The company's tactics are procedures that achieve the fireground strategy. For example, a strategy of protection of life at a fire in a private house will be achieved by companies performing tactics such as hose stretching, ladder raising, forcible entry, window venting, knot tying, roof cutting, team searching, ladder climbing, and radio communications. The post-fire analysis fact sheet requires answers about company tactics. This information can assist fireground commanders and can help firefighters to learn and evaluate the strategy used at a fire. When the fire service began to describe how it extinguished fires, the first explanations of firefighting strategy came from chiefs who studied successful company tactics. Tactics came before strategies. The firefighting strategies we use today were developed from analyzing successful, time-proven company tactics. The groups of tactics called strategies have been given priorities, and some, such as life safety, are considered more important than others, such as fire extinguishment. By analyzing the company tactics recorded on a post-fire analysis fact sheet, a present-day fireground commander can learn strategy.

The Fire Diagram: Part Two of the Post-Fire Analysis

Part two of a post-fire analysis is a diagram of the fire building. The officer preparing the fire analysis diagram should identify on it the point of fire origin, the area of fire damage, exact position of fatalities, hose line and ladder placement, and collapse and explosion destruction. These are critical details of a fire which cannot be captured by a photograph or a fact sheet; the relationship between these important factors can only be revealed through a diagram or drawing.

When a diagram of a post fire analysis is drawn, there are certain guidelines which should be followed. For example, standard fire symbols should be used to enable the diagram to be interpreted by fireground commanders of other communities and departments (fig. 22-5); the diagram should be an overhead line drawing with exposure number I at the bottom of the page; all writing and symbols on the diagram should be readable from left to right when exposure number I is at the bottom of the page; the date and time of the fire, the alarm box number, and the address of the fire building should be noted at the top of the page; the drawing of the building outline need not be to scale, but the relative size of the structure and areas inside the structure should be discernible; the construction type, height in stories, width, and length should be indicated in the lower left hand corner of the building diagram; occupancies should be indicated in the drawing (fig. 22-6).

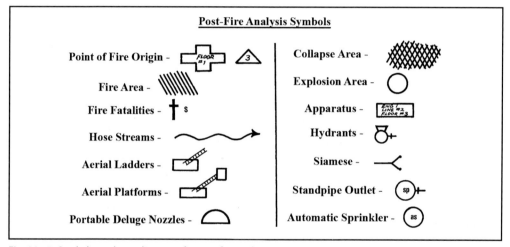

Fig. 22–5. Symbols used on a diagram of a post-fire analysis

The area of fire origin inside a fire building drawing is pinpointed by a Greek cross (one with arms of equal length). If the building drawn is multistory, the floor of fire origin is indicated with the symbol for fire origin. The area of fire damage within the structure is designated by diagonal lines. This marking should show only the area of active flaming, not of smoke or heat damage. The exact position of a fatality is indicated by a cross; the smaller portion of the main stem of the cross represents the victim's head, and the horizontal bar represents the arms. If the victim is discovered face up, the letter *S* is written near the cross, indicating *supine*; if the victim is face down, a *P* next to the cross indicates a *prone* position.

Another important fact shown on a diagram of a post-fire analysis is the position of all the attack hose, aerial, and ground-based master streams. Hose streams leading from pumpers are shown by a line with an arrow for the nozzle. The pumper is shown as a rectangle. An aerial ladder stream can be indicated by a small ladder leading from a rectangle. An aerial platform can be shown as an aerial ladder with a box at the tip. A portable deluge ground nozzle position can be identified with a semicircle, the round portion facing the fire building indicating horizontal range. Apparatus placement can be shown with a rectangle. Inside the apparatus rectangle, indicate the company number. Also, if a hose or master stream device was operated by this apparatus, the sequence number showing when the stream was placed in operation should be written.

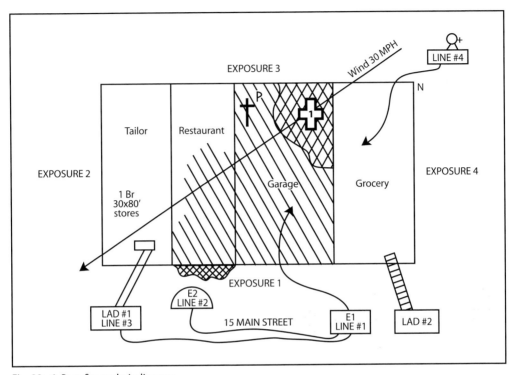

Fig. 22–6. Post-fire analysis diagram

If the building has several stories or levels, there should be a floor number indicating the floor of stream operation inside the apparatus rectangle, along with unit number and hose stream number. A collapse or explosion will sometimes occur inside a fire building, and this area should be shown on the post-fire analysis diagram. A portion of floor or roof collapse within the fire area can be shown by crossed diagonal lines. The crossed or double diagonal lines will distinguish a small collapse area inside a large fire area, which is shown by the single diagonal lines. If a masonry exterior wall collapses at a fire, the area of rubble that covers the sidewalk or collapse zone outside the building line should also be shown by crossed diagonal lines. If a backdraft or explosion occurs, this fact can be indicated within a fire area by a circle showing the area of shock waves or blast.

An important strategy consideration that must be shown on the post-fire analysis diagram is the wind direction; a line with an arrowhead indicating the wind direction should be drawn across the page. In addition, to show the relative geographic position of the fire building, north is indicated by another arrow with the letter N written across it. The street name or number in front of the fire building near exposure A should be written at this point. There are standard symbols which can be used for a post-fire analysis diagram.

Fire Photographic Documentation: Part Three of the Post-Fire Analysis

A photographic documentation of a fire site is the third and most important part of a post-fire analysis. A post-fire scene photograph is an extremely valuable visual aid which can be used over and over again to train new fireground commanders. A photograph taken after the fire can sometimes show more about strategy and tactics than a photograph taken during the fire operation. The photographic results of a fire operation will furnish proof that a firefighting strategy is effective or ineffective. A picture of a collapse area, even after the event, is highly dramatic and has a great effect. The purpose of part three of a post-fire analysis, the photographic documentation, is to prepare a photographic training session to supplement the information and facts compiled in the fire analysis fact sheet and the diagram. A color slide of a fire scene projected on a large screen in a classroom can reveal the lessons learned at a fire to a hundred fire officers with much more realism than a lecturing instructor with charts and word pictures.

Compiling the photographic documentation of a fire scene, however, is more difficult than filling out the fact sheet of part one and drawing the diagram for part two of the post-fire analysis. Photographic documenting requires two skilled individuals: a photographer who knows his equipment and how to use it and a fireground commander who knows what pictures to take. Actually there are many good photographers, but there are very few chiefs who know how to direct the photographing of a fire scene. A photographer called to the scene of a fire will expect the fire officer to decide what photographs to take. The following are some guidelines as to how to photograph a fire scene after the fire has been extinguished. The photographer should be requested to take a sufficient number of 35-millimeter color shots to create a 25-picture presentation. (This task will require many more than 25 photographs.) Two minutes of classroom instruction associated with 25 color slides projected on the screen will create a 50-minute lecture. Photographs of important details such as collapse area and avenues of fire spread should be taken from two viewpoints: a close-up view and one from a distance, which shows the important detail in relation to the rest of the structure. When preparing the slide lecture photographs, the lecturer should attempt to vary the pictures between close-up views and distant, full views. This technique provides variety of scenes and helps to maintain audience attention. One of the photographs should be a close-up of the diagram prepared for part two of the post-fire analysis. The apparatus placement and hose and ladder positions are revealed in the diagram. Another photograph should be taken from a high point, looking down into the roof of the fire building; a higher adjoining building's rooftop or a raised tower ladder or aerial can be used to obtain this photograph. If possible, the photographer should show the entire fire building and the four exposure sides from this bird's-eye view.

Exposures

Individual photographs of the four sides of the fire building should be taken. An analysis of why a certain exposure was protected first can be shown by these photographs. The distance between the fire building and an exposed structure should be shown if it was a factor in the strategy. Place an object whose size is known in the photograph for distance reference. A discussion of priorities used to protect exposures from radiant, convection, and conduction heat can be assisted by photographs of exposed buildings.

Avenues of fire spread

How a fire spreads from the point of origin should be photographed. Flames often extend from the fire area in both a vertical and a horizontal manner. The point where the flames enter the avenue of fire spread and exit the concealed space or opening should be photographed.

Point of fire origin

Working backward from the exposures, through avenues of fire spread, the chief should direct that several photographs be taken of the point of fire origin; these photographs should be designed to explain the technique of determining the point of fire origin. Burn patterns, wood charring, first material ignited, and the heat source of the fire should be photographed. Here portable floodlights powered by a portable generator will be required to light the area sufficiently for it to be photographed.

Explosions

Explosions are occurring at an increasing number of fires. The exact cause of this dangerous event should be determined in the post-fire analysis. The explosion could have been caused by a backdraft, ignition of accumulated natural gas, a broken gas pipe or meter, a boiling liquid expanding vapor explosion (BLEVE), the ignition of excess flammable vapors of an arsonist's accelerant used to start the fire, and the ignition of flammable gases created by burning insulation of electrical wires—another increasing cause of explosions at fires. If the explosion was caused by a broken gas pipe or fixture, this defect should be photographed. If the arson investigator determines that the detonation was caused by the vapor of a flammable liquid, the container holding the combustible liquid should be the subject of a picture. A photograph of the ruptured cylinder causing a boiling liquid expanding vapor explosion should be obtained if that was the cause of the explosion. A photograph of a burned-out electrical supply panel housing the suspected insulation which generated the explosive gas should also be taken. After the cause of the explosion has been recorded on film, the structural damage caused by the shock wave must be photographed. Any fallen walls, collapsed floors, roofs, or ceiling, or the area of glass fragments of blown-out windows must be documented. This material provides very important fire-ground survival information.

Collapse

When structural failure occurs at a fire, this result should be photographed. The area covered by a falling wall should be recorded. The collapse danger zone should be measured and photographed—how far stones or bricks rolled beyond the fallen wall, in addition to how far the wall fell outward. Voids or pockets of safety in which a firefighter could be trapped after a floor, roof, or ceiling collapse are important facts to record. A picture of the first structure to fail along with the second or third structure to collapse in progression should be taken. The type of connection used to fasten structural members together is a factor in many collapses; in some instances, the connection is the point of failure. How a roof beam is held up by a brick wall, how a steel beam is bolted to a girder, and how web members of a lightweight wood truss are fastened to the chords, how the web section of a wood I-beam burns and fails, how a bar joist truss distorts,are all important connections which influence collapse and should be photographed, even if they did not fail.

Fatality: civilian or firefighter

If a firefighter dies or is seriously injured, the exact position of the body should be recorded. An outline of the body should be drawn in chalk and photographed. After a close-up of this spot, a wide-angle photograph of the area should be taken, showing the firefighter's location, together with the nearest escape route such as a window or door. The point of firefighter entry should be retraced and photographed also. The cause of death or injury should be recorded; it may be a collapsed structure, an area in which a fall occurred, a crushed or damaged fire truck, or furnishings which caused flame spread or toxic smoke generation.

Water supply

The most important strategy and tactical consideration at a fire is water supply and pressure. The water source, the pump pressure, and any built-in fire protection water system used in the fire operation should be photographed. Defective or broken hydrants or the absence of a water main system, which may cancel out the best fire strategy must also be recorded on film. The amount of water supplied by the main, tanker, or lake should be on the post-fire analysis fact sheet.

Lesson Learned

Today, post fire analyses are conducted by National Institute for Occupation Safety and Health (NIOSH) and by National Institute of Standards and Technology (NIST) and the National Fire Administration (NFA). However this is usually only when there is a firefighter fatality. These investigations concentrate on safety and survival at a fire. They do not address firefighting strategy and tactics. Many fire departments still do not conduct in-house post-fire investigations of major fires and so there is little record of valuable firefighting experiences. Because of this deficiency, a tremendous amount of knowledge and experience is being lost and is not available to improve strategy and tactics for future fires. A post-fire analysis, consisting of a record of the fire strategy and tactics used at a fire; a detailed diagram of inside and outside the fire building and surrounding exposures; and finally a photographic document of the scene is part of the history of a fire department. This information can provide future firefighters with important, accurate, and documented lessons learned. Noted philosopher Jorge Santayana once said, "Those who can not remember the past are condemned to repeat it." Post-fire investigations help you remember the past.

23

Early Floor Collapse

An engine company is advancing the first attack hoseline 10 feet inside the first floor of a burning two and half-story wood frame dwelling. The firefighter directs a stream "over his head and all around," breaking up heat waves and driving flames back. Moving forward, smoke is banked down. The floor feels hot. There is zero visibility. The nozzle firefighter backed up by a captain and second firefighter move forward. Directing the hose stream, the firefighter feels a strong backwards tug on the hose, a cracking noise, and a shout behind him. Turning; there is no one behind him. A section of floor collapsed. The officer and backup are gone—they fell through the floor. Smoke and flame rise up through the floor opening. Leaping backwards over the opening and then getting up and scrambling out to the front porch, the firefighter shouts "Chief, the floor collapsed! The company is in the cellar!"

This *"through the door through the floor"* story is being repeated throughout the country. A veteran fire officer said to me, "Floor collapse seems to be happening during the early stages of a fire, too often, and at too many fires. Some one has to sound an alarm." For example, in the FDNY, Howard Carpluk, Mike Reilly, and John Clancy. Across the river in New Jersey, East Franklin Fire Department, Kevin Appuzzio. Across the nation, Brant Chesney, Forsythe County, Georgia; John Ginocchetti, Tim Lynch, Manlius County, New York; Steve Smith, Wea Township, Indiana; Arnie Wolfe, Green Bay, Wisconsin. All of them went "through the door through the floor" with collapsing first floors (fig. 23–1).

Sometimes the cause is old construction, other times bad renovations, but most of the time, it's rapid failure of preengineered floor construction or lightweight truss construction. Preengineered wood materials have long been identified by the fire service as the cause of early collapse incidents at post-fire investigations. Finally the scientific community is catching up. Underwriters Laboratories (UL) has an online presentation summarizing a research study of structural stability of engineered lumber and fire conditions (*www.ul.com/fire/structural.html*). They examine the hazards and assess risk for the life safety of building occupants and firefighters in the program. The study was funded by a Department of Homeland Security grant. The fire service now has scientific proof of claims of early collapse of structures built with lightweight, preengineered wood materials, specifically, wood trusses and wood I-beams.

Fig. 23–1. Steve Smith, Wea Township FD, was the first firefighter to die in a wood I-beam floor collapse.

A type of floor collapse that seems to occur over and over again is the first floor of a private dwelling collapsing into a cellar fire. Firefighters fall into the basement after the floor collapses and are killed by products of combustion, flame, heat, smoke, or gas.

- Question 1. What causes a first floor to collapse?
- Question 2. Why don't the firefighters size up the fire as a cellar and stay off the first floor?
- Question 3. Why is the floor collapse increasing?

The following are some answers to these questions.

Cause of Floor Collapse

Cellars are often unoccupied spaces. They contain large amounts of storage and heating units. If a fire occurs, it can quickly grow unnoticed. Cellar fires are often delayed alarm fires. There will be a delayed notification to the local fire department. Undetected fire progresses through three stages: growth stage, fully developed stage, and decay stage. An undetected cellar fire may progress from the growth stage to the fully developed stage before it is discovered and before the arrival of first responders. Flashover—full room involvement—can happen in the transition from the growth stage to the fully developed stage. Once the blaze is burning freely, it starts to attack the structure and the collapse danger begins. The collapse danger of a cellar fire is the collapse of the first floor. A long-burning cellar fire can feed on the underside of the first floor. Upon arrival of the first responders, smoke and fire may involve the upper floors of a dwelling, hiding the real origin of the blaze down in the basement or cellar.

Most basements are unfinished. There is no cellar ceiling. The underside of the first floor is open joist construction. In a finished basement with a ceiling, the ceiling barrier can give floor supports 20 or more minutes of fire protection. Without a ceiling, any fire will quickly weaken the underside of the exposed floor joists and the first floor quickly becomes a collapse danger during a cellar fire. A first floor will collapse faster than second floor level or even attic floor if it is without ceiling protection. Even if there is a ceiling protecting the underside of a first floor, there may be panel missing or the light fixtures may not have fire-rated frames or some of the ceiling panels not be properly set in the frame, allowing cracks for fire to spread to the first floor construction. Often, unoccupied cellar areas get less maintenance than upper floors.

Why Don't Firefighters Size Up a Cellar Fire?

Why don't the first-arriving firefighters locate the origin of the fire in a cellar and use defensive firefighting tactics? The answer is the fire in a cellar may be overlooked during the initial size-up, especially when heavy smoke is pushing out the upper floors and attic. The large open stairway in a private dwelling often extends from the cellar to the second-floor bedrooms. Smoke from the cellar rises up the open stairs to the upper floor bedrooms. A private dwelling of balloon construction has concealed spaces that extend from the cellar to the attic. In balloon framing construction, unlike post and girt, or platform wood construction, the exterior wall studding has 16-inch spaces that extend from the cellar foundation sill directly up to the attic space. If you have a cellar fire in a balloon constructed building, you can have smoke and fire coming out of the attic when you actually have a cellar fire. Fire and smoke from the tightly sealed up burning cellar can enter the exterior walls spaces at the foundation sill and travel up the walls spaces of balloon construction to the attic.

The increase of "through the door, through the floor" collapse, makes initial size-up upon arrival critical for first responders. In addition to the size-up of smoke and flame from outside the building, first responders must then ask themselves "Is this a cellar fire?" When responding to a dwelling fire and smoke involves one of the upper floors, the cellar must be checked to see the blaze has not originated there. The small basement windows or double cellar doors at the side or back of the house should be checked for signs of cellar fire. At strip store fires, tragic experience over the years has taught incident commanders to order someone to check the cellar for fire origin. This is because of the danger of floor collapse. In New York City, 12 firefighters were killed when a first floor collapsed into a burning cellar. Have someone check the cellar for fire. This size-up is even more critical if you know the building is lightweight wood truss or I-beam construction (fig. 23–2).

Fig 23–2. Lightweight wood I-beams have a wood web member and a 2×2 inch top and bottom flange.

If smoke is coming from a chimney and there is a fireplace in the finished basement it could be a cellar fire. If smoke is coming from the attic in a balloon frame wood building it could be a cellar fire. If fire is coming from the first or second floor and there is an open cellar stair it could be a cellar fire.

After a lecture on lightweight wood truss collapse and a slide show that names 21 firefighters who have died, I am sometimes asked, "Chief do you know anyone who has died from wood I-beam floor collapse?" My answer was no until I went to Lafayette, Indiana. The firefighters from Wea Township, Indiana, told me about Chief Steve Smith. He died in a floor collapse during a cellar fire on June 26, 2006 (fig. 23-3). Chief Smith is the first firefighter to die by the floor collapse of a wood I-beam floor failure.

The Lafayette, Indiana, building was one-story wood frame 1,200-square-foot private dwelling built in 2004. The fire started in the unfinished basement that had no ceiling. A lightning strike hit the cable TV roof antenna. The TV cable ran down through the building into the cellar and through the laminated glue-and-wood shaving web section of several wood I-beams supporting the first floor. The hot antenna wire ignited the cellar wood I-beam. Smoke showed from the first floor of the building. Chief Smith, first on the scene, entered the first floor to make a quick size-up before the first engine arrived. He entered the front door, made a left turn, and the floor collapsed. The first floor caved in less than 10 feet from the front door, and he fell into the burning basement, where he was trapped.

Chapter 23 | **Early Floor Collapse**

Fig. 23–3. The wood I-beam floor collapsed and killed Chief Steve Smith.

A post-fire investigation (NIOSH F 2006-26) and photo analysis revealed that the cellar was unfinished and there was no ceiling. The first floor was supported by wood I-beams. During the post-fire investigation, the open joist wood I-beams showed exactly how the fire burned and the floor collapsed: First the hot electric cable wiring extended through holes drilled through the web section ignited several I-beams simultaneously. Flames spread outward from the wire, consuming the web sections constructed of wood shavings and glue. When the web sections burned away completely, the top flanges of the I-beams (2×2-inch wood pieces) bowed downward at the center. The top flange compressed downward to the bottom flange and then the bottom flange broke and fell away from the burned web section and top flange (fig. 23-4). There was no beam support for a section of the floor deck. A section of floor deck and several burned wood I-beams collapsed under the weight of Chief Smith as he entered the first floor to locate the fire's origin.

Fig. 23–4. The destruction of wood I-beam floor by fire

In 1984, James Pressnall was the first firefighter to die by collapsing lightweight wood truss construction using sheet metal surface fasteners, and in 2006, Chief Steve Smith was the first firefighter to die by collapse of wood "I" joist. Because of an increase in "through the door through the floor" collapse incidents, firefighter survival tactics and rapid intervention team (RIT) rescue tactics have been developed. These tactics have been successfully used by firefighters surviving a through-the-door, through-the-floor collapse, and by firefighters in rapid intervention teams who have rescued firefighters after a collapse.

Survival Tactics for Trapped Firefighters

Transmit a Mayday. State your location to the best of your ability, state the floor, room, area or division—A, B, C, D side. You must give rapid intervention team firefighters as much information about your location as possible. Stay in radio contact with command and rescuers. Manually activate your PASS alarm and maximize sound control—this can alert nearby firefighters to your location and they can come to your aid before a rapid intervention team. Remain in place unless heat and fire forces you to move. If you have to move due to fire or smoke, search for a window or door if possible. Notify command of your movement to escape fire and indicate the direction of your movement from collapse area if possible.

Use a tool to tap metal or a hard surface to make noise for rescuers and tell rescuers of any special tools necessary to get to you. Point your flashlight beam to the ceiling to assist rescuers, and stay low. Conserve air supply in your mask by trying to control your breathing, while communicating with command and rescuers. Do not remove your facemask. If your mask runs out of air, stay low, breath into your turnout coat, or crawl to a floor pipe recess or floor corner and try obtain air from a concealed space.

Admittedly these are difficult survival procedures to accomplish in a life-and-death situation when you're caught and trapped after a floor collapse, however survivors have used these life-saving actions and recommend their use.

Rescue Tactics for Rapid Intervention Teams

RITs standing by at a cellar fire should preplan a first floor collapse with trapped firefighters in a cellar. If a first floor collapses and a firefighter falls into a cellar, consider the following rescue procedures.

Have ground ladder placed down the collapse floor opening. A firefighter can sometimes climb up a ladder through the floor collapse opening. Divert a charge hoseline and quickly direct it down through the collapse floor opening. This will cool some fire and burning material around a trapped firefighter after a collapse into a burning cellar, and immediately lets the trapped firefighter know you recognize his location and the life-threatening situation. Extra self-contained breathing masks could be dropped down the floor opening to trapped firefighters for use if their masks run out, and extra flashlights could also be dropped down a floor collapse opening to trapped firefighters.

Rescuers must use caution after a floor collapse. If the floor deck around the floor collapse opening has sloped and tilted downward, there could be a "slide effect." The edges of a floor deck tilting downward could cause rescuers to slide into the cellar though the floor collapse opening. A ground ladder over the opening or supported by firefighter at one end could be used to support rescuers approaching a floor collapse area. Immediately after a floor collapse trapping a firefighter into a cellar, simultaneous rescue should be attempted from adjacent spaces such as adjoining buildings or cellar rooms or any horizontal opening, in addition to rescue from above.

Lessons Learned

- Building using preengineered wood trusses or I-beams should be identified and preplans for defensive firefighter drawn up.
- First responders should know upon dispatch if a building has preengineered truss or wood I-beam floor or roof construction. This can be accomplished by dispatcher's relaying this information to all first responders by radio during the response. The incident commander should have detailed written preplans of any building with preengineered construction material in the auto and ready for use at a fire. Defensive

firefighting procedures should be known by all chiefs, company officer and firefighters and recorded in the preplans.

- For example a pre plan could recommend a defensive firefighting standard operating procedures (SOP) such as the following:
    - Content fire: standard firefighting procedures
    - Structure fire: remove occupants and conduct exterior firefighting procedures
- A floor deck may appear intact even after supporting wood I-beams have burned away and failed. The floor deck may then suddenly collapse without warning.
- Upon arrival at a house fire in a building with truss or wood I-beam construction, first size-up a fire and determine if the fire origin is the cellar. Check side cellar foundation windows for signs of smoke or fire. Open any cellar doors at the side or rear to check for fire or smoke.

Note: For more information on floor collapse, search the Web for: *Firefighter Fatality Investigations: NIOSH F 2006-24; F 2005-09; F 2004-05; F 2002-11; F 2001-16; F2001-26; F 97-04.*

Epilogue

Are Architects, Engineers, and Code-Writing Officials Friends of the Firefighter?

After finishing a lecture on why the World Trade Center collapsed, a chief, who is also an engineer and was in attendance at the FDNY Bureau of Training, came to me and said, "Chief, you were rough on architects and engineers."

That night, I thought about what he said. I realized that there is an edge in my voice when I talk about architects and engineers and even code officials since the World Trade Center tragedy. Perhaps he was right.

But then I started thinking about architects, engineers, and building code officials and building construction changes I have witnessed during the past three decades. I then asked myself, *are architects, engineers and building code writers friends of the firefighter?* I had no answer. I started asking that question of groups of firefighters during lectures I gave around the country. Quite often, the answer I received was *no, architects, engineers and code officials are not friends of the firefighter*. Some firefighters would say, *I don't know*. Very rarely would someone defend these design professionals.

Based on this very limited sampling, it appears there is a small rift between the architectural, engineering, and code-writing communities on one side and the fire service community on the other side. This sentiment seems to be related to building construction methods and material and building code changes incorporated into buildings today. This is unusual, because across the country, engineering schools (not architects or building code officials) often team up with their local fire departments.

For example, for many years, Polytechnic University in New York City, an engineering school, has teamed up with the FDNY and created a Center for Fire Safety Engineering. In the 1970s, Polytechnic helped the FDNY conduct full-scale tests of high-rise buildings and row houses. Polytechnic engineers worked with FDNY when writing building code changes for high-rise office buildings and, today, they are helping the department prepare for 21st-century firefighting. Additionally, fire protection engineers Glenn Corbett and Charles Jennings teach at John Jay College—the Fire Science Institute—and are leading the way for an investigation into the World Trade Center collapse.

Twenty years ago, I was invited by Manhattan College's civil engineering department to teach a course on fire protection design. This was in response to the MGM-Grand Hotel fire in Nevada. Worcester Polytechnic Institute in Massachusetts is the premier fire protection engineering school in the nation and is working with the National Institute of Standards and Technology (NIST) conducting the landmark 2010 "Crew Size Study" with the International Association of Firefighters (IAFF). And the National Fire Protection Association (NFPA), whose membership consists largely of architects and engineers, provides most of the fire protection literature and technical information to the fire service.

So why, with all this interaction, is there a misunderstanding between the firefighter and the architects, engineers, and code officials of this country? What could possibly be bothering the fire service? Why is there an edge in my voice when I speak about safety and survival on the fireground? I offer some reasons:

- **Steel bar joist truss construction.** The lightweight steel bar joist was used to support floors in the World Trade Center. This floor support is another form of lightweight floor and roof construction used throughout the country that has the fire service alarmed and mistakenly is blamed on architects, engineers, and code officials. When unprotected, lightweight bar joist beams can fail within 5 to 10 minutes of fire exposure. To my knowledge, the World Trade Center—constructed by the Port Authority—was the only high-rise office building in New York City to use lightweight bar joist construction.

- **Fire protection of steel.** Since the 1960s, builders have used sprayed-on fire protection covering steel. Instead of the heavy concrete encasement used in pre-World War II fire-resistive buildings, a lightweight mineral fiber is sprayed on steel to protect it from fire. The change to spray-on fire protection of steel has been fought by the fire service since its introduction in the New York City building codes. In 1976, Chief John O'Hagan outlined the problems in his book, *High-Rise Fire and Life Safety*:
    1. The spray-on slurry often is not mixed properly.
    2. The steel is not prepared properly to allow the spray-on material to stick properly.
    3. Workers do not apply the spray-on material evenly.
    4. Other workers doing subsequent work nearby easily remove the critically important fire protection.

The (NIST) National Insititute of Standards and Technology investigation of the World Trade Center Towers collapse found there is no basis for thermal protection of steel in floors. And the movement of air in plenum areas above a ceiling blow the fire retarding material off the steel. And it must be replaced to be effective.

Again, I believe the architects, engineering, and code-writing community is being blamed unfairly for this inadequate fire protection. Other building code changes that have firefighters questioning the fire safety concerns of design professionals are:

- **Fire-resistive construction.** The concept of *passive fire-resistive* has been allowed to slip by. At one time, a fire-resistive building was a structure that—barring a collapse or explosion—would confine a fire to one floor. This is no longer true. In the 1970s, New York had a two-floor fire in One New York Plaza. In the 1980s, Los Angeles had a five-floor fire in the First Interstate Bank building. And, in the 1990s, the One Meridian Plaza building in Philadelphia suffered a nine-floor fire. Today, there is no fire-resistive building. If sprinklers or firefighters do not extinguish the fire, the building will not confine it. Today NIST defines the a fire-resistive building as one that will "burn and not collapse." No mention of restricting fire spread.

- **Evacuation of occupants.** Stair design and capacity still are based on the fact that a building is fire-resistive and fire will be confined to one floor.

Therefore, the reasoning is stairs need not have the capacity to hold all the people of the building. Stairs are designed to allow only a limited number of people to leave a building. The rest must stay in place during the fire. This is a so-called "defend-in-place" firefighting strategy, and it is based on the building being fire-resistive, which many in the fire service (including me) no longer believe is true.

- **Large open-floor-area design.** Client-driven architects and engineers have constructed buildings of a size that is beyond the control of firefighters using hose streams. These buildings contain 30,000 to 40,000 square feet of open floor space. Designers did not know or care that a typical fire company can extinguish only about 2,500 square feet of fire. If these buildings are not protected with automatic sprinklers, firefighters cannot extinguish a fire inside these large-area structures.

- **Floor construction.** The use of a 2- or 3-inch concrete floor over corrugated steel I-beams has failed at every multiple-alarm fire in New York City. Steel floor beam supports sag, warp, and twist. The 4-inch concrete floor above sags when the steel cracks and heaves. Smoke and flames spread to the floor above. Floor beams and concrete floor surface must be replaced after every serious fire. This started at the 1970s fire in One New York Plaza, where 130 steel floor beams were replaced and 20,000 square feet of concrete floor was removed. It continues to happen today. For example, in a fire in 1993 at the Bankers Trust building on Park Avenue, floors were seriously damaged and had to be shored up before firefighters could enter, perform salvage, and overhaul the smoldering offices.

- **Scissor stairs.** The scissor stair is another innovative design recently incorporated into the building codes. Enclosing two stairs in one enclosure is a cost-saving item that the firefighter is concerned about, especially since the stair enclosures now can be constructed of two layers of sheetrock instead of masonry. Engineers now say they want to "harden" the construction of high-rise buildings since the second terrorist attack on the World Trade Center. That is strange. It seems to me someone has been "softening" buildings for the past 50 years.

    In its investigation, the Federal Emergency Management Agency (FEMA) stated that the stairways of the World Trade Center buildings were clustered together in the core area. In pre-World War II buildings, stairways were required by law to be located on remote portions of a large floor area. This way, if fire blocked one exit, occupants could go to another remote area of the floor and find an exit. Exit stairways used to be at each end of a floor area. The New York City building code written in 1968 defined the meaning of the term "remote" when it applied to exit stairs. The building code defined exits as "remote" if they were more than 15 feet away from each other.

- **Controlled inspections.** Another innovation in the 1968 building code was the "controlled inspection." Building or fire officials need not always go to a construction site to inspect a process or material. Rather, a so-called "controlled inspection" is allowed. This is accomplished when an architect or engineer sends a written affidavit to the Buildings Department that the material or process has been inspected and meets the building code requirements. The "controlled inspection" cuts down on the chance of bribery, but some say it is akin to the fox guarding the hen house. The architect that signed off on the renovation of a store in the Bronx where Howard Carpluk and Mike Reilly were killed in a floor collapse has been indicted by the District Attorney.

After thinking more about whether architects, engineers (design professionals), and code officials are friends of the fire service, I realize this is an unrealistic question. Of course, they are not friends of the fire service. Neither is the fire service a friend of the architect, engineer, and code official. We are all professionals, not friends. Architects are responsible for designing great buildings. Architects are also responsible for fire safety in the design. However, fire safety is a small part of their overall responsibility. Engineers are responsible for designing the structural framework of the architect's plans at the lowest cost to the owner. Engineers are not responsible for fire safety. Firefighters are responsible for fire safety.

The fire service should not depend on design professionals for any special consideration. We must perform our mission of fire prevention and fire extinguishment. We must conduct building pre-plan inspections and approvals. We must recommend changes for legislation that will make buildings safer, based on knowledge gained during fire inspections and fire tragedies. And, when fire safety laws are not enacted by code officials, we must continue to save lives no matter how unsafe the design or construction of buildings.

Index

A

abandoned derelict buildings
    burning building collapse and, 3–4
    as target hazards, 3
    weather eroding/ravaging, 4
access stair, 20
active fire protection, 13
aerial ladder, 73
aerial platform large-caliber master stream, 226–227
    design changes for, 227
    search-and-rescue and, 241
    sounds of, 230
age of building, burning building collapse and, 2–3
Aldridge, Todd, 161
Alfred P. Murrah Federal building, 238
Appuzzio, Kevin, 319
apse, 124
arch, 13
architects/engineers/code-writing officials
    controlled inspections and, 329–330
    firefighter endangered by, 327–330
    floor construction and, 329
    large open-floor-area design by, 273, 329
    scissor stairs and, 329
area wall, 33
asbestos removal, 222
asphalt roof, noncombustible/limited combustible construction and, 42
attic space, quick access to, 114
automatic sprinkler protection, 42, 46
axial load, 26

B

balloon construction, 13, 14, 204, 321
bar joist collapse, 4
beam(s), 19
    C-, 143–145, 148–149, 150, 328
    cantilever beam support, 13
    carriage, 181, 185
    cold-formed steel, 15
    continuous beam support, 13, 81
    fire cut, 22, 135, 136
    floor, 80, 81, 85
    header, 23
    I-, 24, 25, 143, 322–323
    laminated, 24
    restrained end of, 29
    simple supported, 13, 81
    tail, 23
    trimmer, 23, 32
    unrestrained end of, 29, 33
bearing wall, 33
    fire destruction of, 205
Beddia, Robert, 222
Beetle, Joseph, 173
Benge, Mark, 161
BLEVE. See boiling liquid expanding vapor explosion
boiling liquid expanding vapor explosion (BLEVE), 308, 317
Boswell, Joseph, 117, 161
bowstring timber truss roof collapse, 72, 307
    evacuation before, 115
    inward/outward collapse, 58
    truss position influencing, 115
bowstring truss
    common buildings with, 112
    strength/stability of, 109
Brace, Charles, 118
braced-frame construction, 14, 15
    methods for, 202–204
Brackenridge, Pennsylvania floor collapse, 1, 2, 77–78
Brady, Michael, 93
brick cavity roof beam connection, 134
brick-and-joist construction, ordinary, 40
    collapse hazards of, 51
    concealed spaces in, 43
    fire problems of, 43–45
    fire spreading in, 44
bridge truss, 15
broken glass
    commercial v. residential, 7
    as falling object, 7
    from venting, 7

Brooklyn, New York, 104
Bucca, Ron, 189
Building 7, of World Trade Center
    aerial view of, 298
    collapse of, 293–301
    construction of, 295
    content fuel load of, 299
    fire in, 296–297
    horizontal fire spread of, 297
    lessons learned from, 301
    progressive collapse of, 298
    7 hours of uncontrolled burning in, 297–298
    spray-on fire retarding material and, 295
    thermal expansion of, 298
    vertical fire spread of, 297
building codes, for high-rise buildings, 278–280
    antenna installation and, 280
    column/girders/floor beams in masonry, 278
    compliance of, 290–291
    concrete floor thickness and, 279
    concrete shafts and, 278
    evacuation drill and, 278
    exit/stairway widths and, 278
    fire protection weakness inspection and, 280
    heating, ventilation, and air-conditioning system duct limited by, 279
    lightweight bar joist/truss floor limited by, 278
    local building/fire code compliance and, 280
    non-sprinkler zero tolerance and, 280
    Phase III elevator and, 278, 287
    smoke-proof stairway and, 278, 287
    structural framework prohibited by, 279
building construction. *See also* abandoned derelict buildings
    brick-and-joist, ordinary, 40, 43–45, 51
    collapse hazards for types of, 48–52, 310–311
    firefighting problems/structural hazards and, 39–52, 310
    fire-resistive, 40–42
    five standard types of, 40–48, 310
    of floors, 79

    heavy timber, 40, 45–46, 51–52
    noncombustible/limited combustible, 40, 42–43
    wood-frame, 40, 47–48, 52, 203–204
building design, fire service knowledge of, 11–13, 130
building element, collapse hierarchy of, 24
building under construction
    burning embers and, 220–221
    collapse hazards of, 211–224
    fire protection systems and, 222–224
    firefighter experience of, 211–212
    formwork fires and, 218–219
    lessons learned from, 224
    safety/stability of, 213–215
    scaffolding fires and, 219–220
    site hoist collapse and, 221
    worker shanty fires and, 218
burning building collapse
    abandoned derelict buildings and, 3–4
    building age and, 2–3
    case study on, 7–9
    construction method/lightweight materials and, 4
    data on, 5–9
    documentation lack on, 9
    factors for, 2
    firefighting exit strategy for, xiv
    general information on, 1–10
    illegal renovations and, 4–5
    information lack regarding, 6
    investigations on, 6, 9
    as leading cause of firefighter deaths, xiii, 1
    post-fire analysis for, 10
burning embers, 220–221
Burns, Michael Cielicki, 77
buttress, 15, 16, 124

C

Cahill, Paul, 172
Callan, Joseph, 266
canopy, 60
    collapse of, 62–63
    skylights and, 62
cantilever beam support, 13
capstone, 20
Cardinal bowling alley timber truss collapse, 104–106

Index

Carpluk, Howard, 5, 80, 319, 330
carriage beams, 181, 185
cast-in-place concrete structure, 216
C-beam roof construction, 143–145, 328
    hazards of, 148–149
    I-beam replaced by, 143
    ventilation tactics for, 150
ceiling. *See also* suspended ceiling
    attachment of, 167
    membrane, 146
    overload of, 176
    suspended, 167–174
    timber truss roof concealed by, 112–113
    timber truss roof not concealed by, 113
    types of, 166–174
ceiling collapse, 163–177
    concrete, 48, 217
    danger of, 166, 175
    firefighter deaths and, 173
    firefighter experience of, 163–166
    firefighter rescue from, 175
    firefighter trapped by, 169
    lessons learned from, 175–177
    master stream operations and, 225–226, 233–234
    of place of worship, 123
    preparations/cautions for, 175–177
cellar fires
    danger of, 320–321
    size-up errors of, 321–324
Center for Fire Safety Engineering, 327
chancel, 124
Chesney, Brant, 161, 319
church/temple fire, defensive operations at, 119–122
Clancy, John, 79, 319
Cliffside Park, New Jersey, 104
cockloft, 31
code-writing officials. *See* architects/engineers/code-writing officials
cold-formed steel beam, 15
collapse. *See also* ceiling collapse; floor collapse; roof collapse; World Trade Center collapse
    bar joist, 4
    of Building 7, of World Trade Center, 293–301
    building element hierarchy of, 24
    burning building, xiii, 1–10
    canopy, 62–63
    of construction site hoist, 221
    construction type hazards of, 48–52
    cornice, 63
    crane, 215
    curtain-fall wall, 16, 57–58
    early floor, 319–326
    of flat roof, 127–140
    floor beam, 80, 81, 85
    floor deck, 79, 80, 85
    global, 22, 293, 295
    of high-rise building, xiii, 213–214, 273, 293–302
    of I-beams, 322–323
    illegal renovations and, xiii, 4–5
    inward/outward, 16, 58–60, 200, 202, 205–208
    lean-over, 16, 200, 201
    lean-to floor, 16, 17, 83
    of lightweight steel roof/floor, 141–154
    of lightweight wood truss, 155–162
    of marquee, 62, 70
    of masonry wall, 53–64
    master stream operations and, 225–236
    multilevel floor, 28, 82
    of multilevel floor, 28, 80–82, 85–86, 88
    of 90-degree-angle, 16, 55–56, 200, 201
    Oklahoma City, 239
    during overhaul/salvage, 234–235
    pancake floor, 17
    of parapet walls, 51, 61, 65–74
    of peak roof, 91–102
    photographic documentation for, 317–318
    in place of worship, 122–123, 124
    progressive, 29, 280–281, 295, 298
    safety precautions prior to, 255–270
    search-and-rescue for, 242
    secondary, 18, 60, 249–250
    of slate/tile shingles, 97–99
    of spalling concrete, 216–217
    of stairway, 179–187
    standard definition of, 6
    of Strand Theater, 1
    structural, 6, 7, 74, 93–94
    tent floor, 18
    "through the door and through the floor," xiii, 205, 319, 321, 322, 324
    of Vendome Hotel, 1, 5, 78, 80
    V-shape floor, 18, 19, 84
    of wall, 16, 35, 52–60, 123, 200–202, 205–208
wall hierarchy of, 35

of wood-frame building, 199–209
collapse hazards
    of brick-and-joist construction, ordinary, 51
    of building construction types, 48–52, 310–311
    of building under construction, 211–224
    of fire-resistive construction, 48–50
    of heavy timber construction, 51–52
    of noncombustible/limited combustible construction, 50–51
    of suspended ceiling, 169
    of wood-frame construction, 52
collapse site
    police and, 239
    rubble removal, 251
    securing of, 238–240
collapse zone, 60
    establishment of, 262–264
    firefighters entering, 263
    horizontal/vertical, 36–37
    for master stream operations, 230
    miscalculations of, 264
    positioning apparatus and, 232
Collins, Brian, 118, 161
column, 18, 19
communication/fire alarm, National Institute of Standards and Technology on, 285
compression stress, 30
computer-designed high-rise building, 274
concentrated load, 26
concrete
    case-in-place structure of, 216
    ceiling collapse of, 48, 217
    encasement of, 145–146
    as fire resistive construction, 48
    floor collapse of, 49–50, 216
    floor thickness of, 279
    removal from high-rise building, 275
    shafts of, 278
    spalling, 216–217
concrete ceiling collapse, 48, 217
concrete encasement
    danger of, 146
    as fire protection method, 145–146
concrete floor collapse, 49–50, 216
construction
    brick-and-joist, ordinary, 40, 43–45, 51
    C-beam roof, 143–145, 148–149, 150, 328

of Empire State Building, 274–275
fire-resistive, 40–42, 48–50, 281–282, 328
of floor, 79, 328
frame tube, 22
heavy timber, 40, 45–46, 51–52
of high-rise building, 213–214, 280–291
lightweight floor, 79
lightweight wood truss, 156–160, 162, 205
noncombustible/limited combustible, 40, 42–43, 50–51
platform wood-frame, 28–29, 204–205
"post-and-grit," 204
of roof, 96, 108, 130–132, 134–138
of stairs, 184–185
steel open bar joist, 51
of timber truss roof, 108
Type I resistive, 147
Type II noncombustible steel, xiii, 133, 147
Type III, 120
wood-frame, 40, 47–48, 52, 203–204
of World Trade Center, 272–274
construction methods. *See also* building construction
    balloon, 13, 14, 204, 321
    braced-frame, 14, 15, 202–204
    burning building collapse and, 4
    classification of, 281–282
    frame tube, 22
    platform wood-frame, 28–29
    for roofs, 131–132
construction site hoist collapse, 221
continuous beam support, 13, 81
convenience stair, 20
Coordinated Incident Management System (CIMS), 290
coping stone, 20
    of parapet wall, 69
corbel, 20
    as roof connection, 135
Corbett, Glenn, 327
corner safe areas, 21, 209
corner wood-frame building, 202–204
cornice, 60
    collapse of, 63
    of parapet wall, 69
crane collapse, 215
curtain-fall collapse, 16
    of masonry wall, 57–58

D

danger(s)
- of ceiling collapse, 166, 175
- of cellar fires, 320–321
- of concrete encasement, 146
- of fire escape, 189–198
- of membrane ceiling, 146
- of parapet walls, 67
- of place of worship collapse, 116, 122–123
- reporting of, 74, 256–257, 258
- of sheet metal surface fastener, 158
- of slate/tile shingle collapse, 99
- sloping roof, 99, 102
- of spray-on fire retarding material, 146, 283
- of stream destruction, 227
- of suspended ceiling, 167
- of timber truss roof, 109
- of venting, 7
- zone of, 36, 37

danger area
- collapse zone establishment of, 262–264
- cordoning off of, 262
- for falling objects, 7
- firefighter evacuation of, 264
- increased supervision of, 260
- investigation of, 259–260
- lighting up of, 258–259
- reporting of, 74, 256–257, 258

dead load, 26
Dean, Phillip, 118, 161
deck, 21
- fluted metal (steel), 22, 49

deck collapse
- floor, 79, 80, 85
- roof, 95–96, 97

defensive overhauling, 89, 235
deflection, 21
demising wall, 33
Department of Homeland Security, 289, 319
Deutsche Bank building demolition, 213, 222
Dolney, Robert, 264
dome, 125
Dupee, Joseph, 107

E

Earl, Thomas, 173
early floor collapse, 319–326
- cause of, 319, 320–321
- firefighter experience of, 319–320
- lessons learned from, 325–326
- rescue tactics for, 325
- survival tactics for, 324–325

Ebenezer Baptist church collapse, 121
eccentric load, 26–27
egress systems, National Institute of Standards and Technology recommendation for, 287

elevator
- hardened, 288
- in high-rise building, 278, 280, 287–288
- Phase III, 278, 287

Emanuelson, David, 77
emergency exit evacuation
- assembly point for, 268–269
- evacuation signal for, 268
- of firefighters, 266–270
- timing of, 269
- tools/equipment for, 268
- training for, 269–270

Empire State Building
- airplane crash into, 274
- construction of, 274–275
- HVAC system of, 274
- World Trade Center v., 273–274

engineers. *See* architects/engineers/code-writing officials
Environmental Protection Agency (EPA), 222
evacuation drill, for high-rise buildings, 278
exit strategy, xiv
- emergency, 266–270
- fire resistive rating and, 87
- training for, 266–267

explosions/backdrafts, parapet wall collapse from, 72
explosive spalling, 217

F

facade, 21, 60
falling objects
- broken glass as, 7
- danger zones for, 7

definition of, 6
examples of, 6–7
responsibility of, 8
structural collapse v., 7
Federal Emergency Management Agency (FEMA), 329
Federici, Francis, 94
Final Report on the Collapse of the World Trade Center Towers, 290
fire attack
   hoseline destination of, 124
   time limits for, 87–88
"fire breeders," 88
fire cut beam, 22, 135, 136
Fire Department Instructors Conference (FDIC), 290
fire diagram
   apparatus/supply positions and, 314–315
   for post-fire analysis, 313–315
   sample of, 315
   symbols for, 314
   victim location and, 314
fire escape danger, 189–198
   counterbalance stairways and, 196
   drop ladders and, 194–195
   exterior screened stairway and, 190–192
   firefighter experience with, 189
   firefighter use and, 190
   hazards of, 190–196
   ladders v., 195, 196
   lessons learned from, 196–198
   party balcony fire escape and, 192–193
   standard fire escape and, 193–194
   step collapse and, 196
fire extinguishment
   search-and-rescue and, 240–241
   suspended ceiling and, 172–173
fire fact sheet
   collapse cause and, 311
   fire stages and, 311
   fire strategy and, 312
   first structure failure and, 311
   of post-fire analysis, 309–313
   structure and, 309–310
   tactics and, 313
fire load, 22
   size-up error and, 111
fire protection
   active, 13
   in high-rise building, 284–285

sprinklers/standpipe connection as, 284–285
of steel, 295–296, 328
Fire Protection Handbook (National Fire Protection Association), 73
fire protection methods
   concrete encasement as, 145–146
   for lightweight steel roof/floor collapse, 145–148
   membrane ceiling as, 146
   spray-on fire retarding material as, 146–148
fire protection systems, 222–224
fire service
   building design knowledge of, 11–13, 130
   emergency exit strategy needed for, 267
   exterior escape device evaluated by, 288
   fire resistance rating understanding by, 86
   high-rise building communication and, 285
   pre-fire plan inspection for, 280
   priorities of, 87–88
   radios of, 288–289
   tactics of, 313
   truss identification and, 160
fire spreading
   in brick-and-joist construction, ordinary, 44
   exterior, 124
   in fire-restrictive construction, 41
   in heavy timber construction, 46
   horizontal, 297
   in lightweight wood truss construction, 159–160
   in noncombustible/limited combustible construction, 43
   photographic documentation of avenues for, 317
   suspended ceiling and, 172
   vertical, 297
   weaknesses caused by, 310
fire stages, 312
   fire fact sheet and, 311
   floor collapse and, 88–89
fire strategy
   for attic space quick access, 114
   fire escape v. ladders and, 195, 196
   fire escapes and, 196–198
   fire fact sheet and, 312

fire resistive ratings and, 87–88
of incident commander, 312
from interior to exterior, 235–236
lessons learned from, 113
for lightweight wood truss collapse, 160
post-fire analysis of, 312
priorities of, 87–88
for roof venting, 114
for timber truss roofs, 112–114
for timber trusses concealed by ceiling, 112–113
for timber trusses not concealed by ceiling, 113
from World Trade Center Building 7 lessons, 301
fire wall, 33, 264
firefighter, trapped, survival tactics for, 324–325
firefighter deaths
    burning building collapse as leading cause of, xiii, 1
    ceiling collapse and, 173
    column failure and, 2
    floor collapse and, 5, 77
    from lightweight wood truss collapse, 161
    National Fire Protection Association on, 1–2
    from parapet wall collapse, 70
    protective clothing influencing, 140
    structural collapse v. falling objects and, 7
firefighter evacuation
    complete withdrawal of, 265–266
    of danger zone, 264
    emergency exit of, 267–270
    during One Meridian fire, 87–88
    partial withdrawal of, 264–265
    training for, 266–267
    World Trade Center collapse and, 266–267
firefighter withdrawal
    complete, 265–266
    factors influencing, 265
    from high-rise building, 299–300
    partial, 264–265
fire-resistance rating, 22
    builder's understanding of, 86–87
    exit strategy and, 87
    factors influencing, 86
    fire service understanding of, 86

fire strategy and, 87–88
noncombustible v., 154
time limits for, 87–88
fire-resistive construction, 281–282, 328
    automatic sprinkler protection and, 42
    collapse hazard of, 48–50
    fire problem of, 40–41
    fire spreading in, 41
    heat-activated fire damper and, 41
    reinforced concrete buildings as, 48
    steel skeleton construction as, 48
fire-resistive rating, 47
fire-resistive testing, 282
fire-retarding compartmentation, 306
fire-retarding insulation
    for steel bar joist system, 132
    in World Trade Center, 132
firestops/fire barriers, 44
First Interstate building fire, 87, 284, 294
"flash fire," 153
flat roof
    joist spaces of, 130
    structural elements of, 95
    venting operation for, 128
flat roof collapse, 127–140
    connections and, 134–138
    evacuation and, 140
    firefighter experience of, 127–129
    lessons learned from, 140
    National Fire Protection Association 10-year study on, 129
    risk factors for, 131
    size-up errors and, 130
    wood joists and, 133
floor beam collapse, 80, 81
    risk management for, 85
floor collapse
    beam, 80, 81, 85
    causes of, 88, 320–321
    concrete, 49–50, 216
    deck, 79, 80, 85
    fire attack time limits for, 87–88
    fire protection methods for, 145–148
    fire stages and, 88–89
    firefighter deaths and, 5, 77
    lean-to, 16, 17, 83
    lessons learned from, 89
    maintenance lack contributing to, 88
    multilevel, 28, 80–82, 85–86, 88
    during overhaul/salvage, 89
    overloading contributing to, 88

pancake, 17, 85
possible reasons for, 48
renovation contributing to, 88
risk management for, 85–86
structural hierarchy of, 78
tent, 18, 84
testing for, 86
types of, 79–81
V-shape, 18, 19, 84
floor construction, 79, 328
floor deck collapse, 79, 80
risk management for, 85
fluted metal (steel) deck, 22, 49
flying brands, 296
force, 22
Ford auto dealership timber truss collapse, 104, 106–107
formwork fires, 218–219
frame tube construction, 22
Frantz, Rick, 77
freestanding parapet wall, 67–68
freestanding wall, 33
fuel load, 11
Fyfe, Cyril, 161

G

gable roof, 92
structural elements of, 95
gambrel roof, 92
gang nails, 156
Ginocchetti, John, 161, 162, 319
girder, 19, 22, 136
global collapse, 22, 293, 295
gothic, 125
Graffagino, Joseph, 222
gravity load, 22
gusset plate, 22–23, 131, 156

H

Hackensack, New Jersey, 104, 113
hardened elevators, 288
hat truss, 23
Hayden, Peter, 290
header beam, 23
heat-activated fire damper, 41
heating, ventilation, and air conditioning system (HVAC), 40
duct limit for, 279
in Empire State Building, 274
smoke detector in ducts of, 42, 282
heavy timber construction, 40
automatic sprinkler protection in, 46
collapse hazards of, 51–52
fire problems of, 45–46
fire spreading in, 46
fuel load of, 45
high-rise building
antenna for, 280
codes for, 278–280, 290–291
communication/fire alarm in, 285
computer designed of, 274
concrete removal from, 275
elevators and, 278, 280, 287–288
evacuation drills for, 278, 328–329
fire protection in, 284–285
firefighter withdrawal from, 299–300
floors as weak link of, 284
long-duration fires in, 300
National Institute of Standards and Technology recommendation for construction of, 280–291
remote exits in, 287
smoke-proof stairways in, 278, 287
sprinklers needed in, 280, 284–285
stairway capacity and, 286, 328
standpipe connection in, 284–285
total evacuation time of, 286
high-rise building collapse, xiii, 293–302
under construction/demolition, 213–214
lessons learned from, 302
lightweight material and, 273
High-Rise Fire and Life Safety (O'Hagan), 276–277, 328
Hill, James, 117, 161
hip roof, 92
firefighter caught in collapse of, 93–94
structural elements of, 95
hollow-tube bearing walls, 278, 279, 281
horizontal collapse zone, 36, 37
Hudgins, John, 161
hydraulic overhauling, 89

I

I-beam, 24, 25
C-beam replaced by, 143

collapse of, 322–323
illegal renovations
    burning building collapse and, 4–5
    collapse problems created by, xiii
    Vendome Hotel collapse and, 5
    Wonder Drug Store floor collapse and, 5
impact load, 27
incident command system, 240
    safety precautions of, 257
incident commander
    fire strategy of, 312
    safety precautions by, 259–260
Increasing FDNY's Preparedness, 285, 289
International Association of Firefighters (IAFF), 327
interstitial space, 24, 25
inverted roof, 137–138
inward/outward collapse, 16
    bowstring timber truss roof and, 58
    causes of, 205–208
    of masonry wall, 58–60
    of wood-frame building, 200, 202, 205–208

J

Jennings, Charles, 327
John Jay College, 327
joist, 24
    open-web steel bar, 28, 148–149, 272

K

Kane, Kevin, 169, 173
kip, 24
Korwatch, Harry, 96

L

ladder
    aerial, 73
    drop, 194–195
    fire escape v., 195, 196
    roof, 101
laminated beam, 24
lateral load, 27
lean-over collapse, 16

    of wood-frame building, 200, 201
lean-to floor collapse, 16, 17
    supported/unsupported, 83
lean-to voids, supported/unsupported, 247
lightweight floor construction, 79
lightweight materials
    benefits/hazards of, 4
    burning building collapse and, 4
    National Fire Protection Association on, 4
lightweight steel roof and floor collapse, 141–154
    fire protection methods for, 145–148
    firefighter experience of, 141–143
    horizontal ventilation and, 142
    lesson learned from, 154
    ventilation tactics for, 150
lightweight wood truss collapse, 155–162
    construction and, 156–158
    fire strategy for, 160
    firefighter death from, 161
    firefighter experience of, 155–156
    lessons learned from, 160–162
lightweight wood truss construction, 205
    collapse and, 156–158
    fire spread and, 159–160
    identification of, 160
    increased use of, 156, 162
    sheet metal surface fastener and, 157–158
lintel, 24, 65–66
    steel, 68
    support of, 67
live load, 27
load(s)
    axial, 26
    concentrated, 26
    dead, 26
    eccentric, 26–27
    fire, 22, 111
    fuel, 11, 45, 299
    gravity, 22
    impact, 27
    lateral, 27
    live, 27
    over-, 88, 176, 208
    static, 27
    stress of, 27, 152–153
    torsional, 27
    wind, 27
load stress, 27, 152–153

London Fire Brigade, 242
L-shaped stairs, 182, 183
Lynch, Tim, 161, 162, 319

M

mansard roof, 92
Marquard, Earnest, 173
marquee, 60
    collapse of, 62, 70
    water accumulation and, 229
masonry smoke-proof tower, 275
masonry wall collapse, 53–64
    curtain-fall type of, 57–58
    factors contributing to, 56
    firefighter experience of, 53–54
    flanking protecting positions for, 64
    inward/outward type of, 58–60
    lessons learned from, 63
    90-degree-angle type of, 55–56
    secondary collapse and, 60
    types of, 55–60
    wall attachment and, 60–63
master stream operations
    aerial platform large-caliber, 226–227, 230, 241
    ceiling collapse and, 233–234
    collapse caused by, 225–236
    collapse zone for, 230
    firefighter experience of, 225–226
    flanking building and, 232
    gallons of water delivered by, 227, 228
    lessons learned from, 235–236
    overhauling and, 234–235
    positioning of, 232, 233
    renovated buildings and, 232
    sounds of, 230
    stone/brick veneer wall collapse and, 230
    strategies for, 234
    stream destruction and, 227–229
    stream direction and, 229
    water accumulations and, 229
    in windows, 231–233
Mayday signal, 268
Mayo, Lewis, 161, 171
McKenna, John, 135
McLaughlin, Peter, 173
membrane ceiling
    dangers of, 146
    as fire protection method, 146
Michelson, Alan, 161
Moran, Michael, 117
mortise, 27, 28
mortise-and-tenon connection, 205, 207
movie marquees/canopies, parapet wall collapse and, 70
multilevel floor collapse, 28, 80–82
    realization of, 88
    risk management for, 85–86

N

National Emergency Training Center, 304
National Fire Academy, 304
National Fire Administration (NFA), 318
National Fire Protection Association (NFPA), 1–2, 9, 42, 73, 144, 327
    flat roof collapse 10-year study by, 129
    on lightweight materials, 4
    Standard 200 classifying method of, 47
National Incident Management System (NIMS), 290
National Institute for Occupational Safety and Health (NIOSH), 6, 9, 318
National Institute of Standards and Technology (NIST), xiii, 6, 9, 85–86, 132, 318, 327
    on building code compliance, 290–291
    on construction classification, 281–282
    egress systems recommended by, 287
    on evacuation time, 286
    on fire alarm/communication system, 285
    on fire protection, 284–285
    on fire safety education, 285–286
    on firefighter radios, 288–289
    on fire-resistive testing, 282
    hardened elevators recommended by, 288
    high-rise building construction recommendation by, 280–291
    on information methods/procedures, 289
    on person in charge, 289
    on progressive collapse prevention, 280–281
    recommendations from, 292
    on smoke-proof stairways, 287
    on spray-on fire retarding material, 283

on stairway capacity, 286
on uncontrolled fires/burnout, 284
on unified command systems, 290
on World Trade Center collapse, 280–291, 293
on World Trade Center collapse, Building 7, 295, 297
nave, 124
New York City Building Code, 51, 173
"giveaway" of, 284
as performance code, 276
rewriting of, 291
New York City Fire Department, xiii, 70
90-degree-angle collapse
of masonry wall, 55–56
of wall, 16
of wood-frame building, 200, 201
99-cent store floor collapse, 5
noncombustible/limited combustible construction, 40
asphalt roof and, 42
collapse hazards of, 50–51
fire problem of, 42–43
fire spreading in, 43
types of, 50
Nutter, Strawn, 161

O

O'Connell, Robert, 98
O'Hagan, John T., 276–277, 328
Oklahoma City collapse, site securing at, 239
Olson, Kevin, 161
One Meridian fire, Philadelphia, 87, 281, 294
firefighter evacuation during, 87–88
1 New York Plaza fire, 87, 281, 294
open-web steel bar joist, 28
collapse of, 148–149
in World Trade Center, 272
overhaul/salvage
collapse during, 234–235
floor collapse during, 89
master stream operations and, 234–235
responsibility during, 8
safe procedure during, 8–9

P–Q

pancake floor collapse, 17, 85
pancake void, 248
parapet wall, 21, 34, 60
building code on, 66
collapse of, 51, 61, 65–74
coping stones/cornices of, 69
danger of, 67
environmental exposure and, 69
freestanding, 67–68
snorkels/aerial platforms/aerial ladders and, 73
steel reinforcement rods embedded in, 68–69
unfinished brick of, 69–70
water accumulation and, 229
parapet wall collapse, 65–74
from explosions/backdrafts, 72
firefighter deaths from, 70
firefighter experience of, 65–66
lessons learned from, 73–74
movie marquees/canopies and, 70
planning for, 69
safety myths regarding, 73
from steel expansion, 70–71
timber truss roof collapse and, 71–72
party balcony fire escape, 192–193
party wall, 34
pass-along method, for tunneling/trenching operation, 251
passive fire protection, 28, 328
passive fire-resistive, 282
Payne, Warren, 172
peak roof collapse, 91–102
contributing factors for, 94
of structural frame, 93–94
performance building code, 276–277
creation of, 276
New York City Building Code as, 276
specification building code v., 209, 276
for wood-frame building, 208
Petronas Towers, 274
Phase III elevator, 278, 287
photographic documentation
of collapse, 317–318
difficulty of, 316
of explosions, 317
of exposures, 316–317
of fatality, 318
of fire origin point, 317

of fire spread avenues, 317
importance of, 316
for post-fire analysis, 316–318
of water supply, 318
pilaster, 28
place of worship
    aggressive interior attack on, 118–119
    bell tower collapse and, 118
    ceiling collapse of, 123
    collapse danger of, 122–123, 124
    defensive operations at, 119–122
    flames spreading in, 116–117
    imitation stone veneer surface of, 121
    interior fire spread of, 120–122
    offensive v. defensive strategy for, 120–121
    renovations and, 116
    timber truss roof collapse and, 116–119
    as Type III collapse danger, 116
    unique design/structure of, 118
    unstable elements of, 118, 122–123
    venting of, 119–120
    wall collapse of, 123
    wax coating and, 116–117
plank-and-beam roof, 96
platform wood-frame construction, 28–29, 204–205
Polytechnic University, 327
Port Authority of New York
    building code noncompliance of, 277, 290–291
    corners cut by, 290–291
    World Trade Center built by, 277, 291
"post-and-grit" construction, 204
post-fire analysis, 303–318
    benefits of, 309
    for burning building collapse, 10
    criticized during, 304
    fire diagram for, 313–315
    fire fact sheet of, 309–313
    fire service tactics and, 313
    of fire strategy, 312
    lessons learned from, 318
    photographic documentation as, 316–318
    of "through the door and through the floor" collapse, 323
    of Waldbaum's supermarket timber truss collapse, 304–309
Precious Faith Temple church collapse, 118
Pressnall, James, 161, 324

primary structural member, 29
primary venting, 124
progressive collapse, 29, 280–281, 295
    National Institute of Standards and Technology on prevention of, 280–281
    World Trade Center Building 7 as, 298
    World Trade Center collapse as, 281
purlin, 29, 71, 95

R

rain roof, 137–138, 305, 306
raised roof, 137–138
Ramos, Edward, 161
rapid intervention team (RLT), 324
    alerting of, 261–262
    collapse preplan by, 325
    rescue tactics for, 325
Reilly, Michael (Mike), 5, 80, 319, 330
reinforced concrete building
    concrete spalling of, 49
    as fire-resistive construction, 48
rescue surface, 244–249
restrained beam end, 29
ridgepole, 29
roof, 30
    asphalt, 42
    bowstring timber truss, 58
    construction, C-beam, 143–145, 148–149, 150, 328
    flat, 95, 128, 130
    gable, 92, 95
    gambrel, 92
    hip, 92–95
    inverted/raised/rain, 137–138
    ladder, 101
    mansard, 92
    peaked, 93–94
    peaked v. sloped, 92
    plank and beam, 96
    rain, 137–138, 305, 306
    raised, 137–138
    rotting of, 97
    sloping, 92, 96–97, 99–102
    space, in row buildings, 206
    support systems for, 132–133
    timber truss, 108–115
    venting of, 114
roof collapse

bowstring timber truss, 58, 72, 115, 307
flat, 127–140
sloping peak, 91–102
timber truss, 71–72, 103–124
roof connections
  brick cavity as, 134
  corbel shelf as, 135
  fire cut beam and, 22, 136
  girder support and, 136
  for inverted/raised/rain roof, 137–138
roof construction
  connections and, 134–138
  methods for, 131–132
  modern method of, 130
  for sloping roof, 96
  of timber trusses, 108
roof deck collapse, 95–96
  stability before, 96, 97
roof hazards
  disorientation and, 139
  repairs as, 138–139
  scuttle covers as, 139
  shafts as, 139–140
roof ladder, 101
roof support systems, 132–133
  identification of, 132
  steel bar joist system as, 132, 143–145
rose window, 125
  unbreakable plastic covering for, 119
  venting of, 119
rotting, of roof, 97
row buildings
  fire spread and, 205–206
  roof space in, 206

S

safety factor, 30, 274
safety precautions
  danger area cordoned off and, 262
  danger area investigation and, 259–260
  danger area light up as, 258–259
  danger area supervision increase as, 260
  danger area/collapse zone
    establishment and, 262–264
  danger report acknowledgement and, 258
  danger reporting and, 256–257
  firefighter experience of, 255–256
  incident commander and, 259–260
  lessons learned from, 270
  monitor assignment and, 262
  prior to collapse, 255–270
  public education and, 285–286
  rapid intervention team alert and, 261–262
Saint Francis Church, 116
Sanders, Gary, 118, 161
scaffolding fire, 219–220
scissor stairs, 329
scuttle cover
  as roof hazard, 139
  skylight v., 139
  venting and, 114
search-and-rescue, 237–253
  aerial platform master stream and, 241
  for collapse, 242
  control/organization needed for, 238
  eight-step procedure for, 242
  fire extinguishment and, 240–241
  firefighter experience of, 237–238
  firefighter roll call and, 250
  incident command system for, 240
  lessons learned from, 252–253
  plan for, 242, 252
  risk management and, 242
  safety first during, 252
  secondary collapse and, 249–250
  site size-up, 243
  support personnel needed for, 250–251
  time-out stage of, 245–249, 253
  transit used during, 249
  tunneling/trenching operation and, 248, 251, 252
  utility shut-off and, 244
  victim-tracking officer and, 246
secondary collapse, 18
  masonry wall collapse and, 60
  search-and-rescue and, 249–250
self-contained breathing apparatus (SCBA), 47
shear stress, 30
sheet metal surface fastener, 157–158
  danger of, 158
simple supported beam, 13, 81
site safety officer, 224
size-up errors, 150
  of cellar fires, 321–324
  fire load and, 111
  flat roof collapse and, 130
  by interior/exterior operating forces, 111

with timber truss roofs, 110–112
upward flow delays and, 111
Skinner, Stanley, 173
skylight, 62
venting of, 114, 139
slate/tile shingle collapse, 97–99
causes of, 98–99
dangers of, 99
firefighter injured by, 98, 100
sloping peak roof collapse, 91–102
contributing factors for, 94
firefighter experience of, 91–92
lessons learned from, 101–102
roof rotting and, 97
sloping roof
construction of, 96
dangers of, 99, 102
firefighter balance needed on, 99
ladders needed for, 101
pitches of, 99, 100
rotting of, 97
structural elements of, 95
types of, 92
"slow-burning" fire, 46
Smith, Kimberly, 161, 171
Smith, Steven, 79, 319, 320, 321
smoke-proof stairway, 274, 276
in high-rise buildings, 278, 287
snorkels/aerial platforms/aerial ladders, parapet walls and, 73
spalling, 49, 145
of concrete, 216–217
explosive, 217
spalling concrete collapse, 216–217
spandrel wall, 35
special occupancy structures, 215–217
concrete floor collapse as, 216
crane collapse as, 215
specification building code, performance building code v., 209, 276
split-ring metal connector, 108
spray-on fire retarding material (SFRM), 31, 145, 275, 328
Building 7 and, 295
criteria for, 148
dangers of, 146, 283
failure of, 147
as fire protection method, 146–148
National Institute of Standards and Technology on, 283
problems of, 277
testing/standard for, 283
thickness needed of, 147–148
World Trade Center and, 132, 147–148, 283
sprinkler protection, 42, 46, 280, 284–285
stairs
building codes and, 182
construction of, 184–185
L-shaped, 182, 183
scissors, 329
smoke-proof, 274, 276, 278, 287
straight-run, 182, 183, 184
types of, 182–183
U-returned, 182, 183, 184–185
stairway collapse, 179–187
building codes impacting, 182
firefighter experience of, 179–181
of intermediate landing, 185
lessons learned from, 186–187
warning signs/safety actions for, 186, 187
standard operating procedures (SOPs), 102, 263, 326
standpipe connection, 223
high-rise buildings and, 284–285
static load, 27
steel
expansion of, 70–71, 143
failure characteristics of, 143–145, 148, 151–153
fire protection of, 295–296, 328
lintel, 68
in renovated buildings, 148–151
shaping of, 143–144
temperature failure of, 143–145, 151–152
thermal protection of, 328
steel bar joist and C-beam roof construction, 143–145, 328
hazards of, 148–149
ventilation tactics for, 150
steel bar joist system
fire-retarding insulation for, 132
as roof support system, 132, 143–145
spacing and, 151
venting of, 133
wood joist v., 133
steel expansion
from fire/heat, 143
parapet walls collapse from, 70–71
steel failure, 151–154

characteristics of, 148
fire size and, 153
steel thickness and, 153
stress load and, 152–153
temperature and, 143–145, 151–152
steel open bar joist construction, New York City Building code prohibiting, 51
steel reinforcement rods, parapet wall embedded with, 68–69
steel skeleton construction, 49
as fire-restrictive construction, 48
in World Trade Center, 273–274
Stefanakis, Richard, 118
straight-run stairs, 182, 183, 184
Strand Theater collapse, 1
stream destruction
dangers of, 227
master stream operation and, 227–229
upper building danger from, 228
water weight and, 228
stream direction, 229
stress
common types of, 31
compression, 30
shear, 30
tension, 30
structural collapse, 6
falling objects v., 7
of peaked roof, 93–94
reporting responsibility for, 74
structural hierarchy, 78
supported lean-to floor collapse, 83
survival tactics, for trapped firefighters, 324
suspended ceiling, 31, 167–174
collapse hazard of, 169
concealed spaces and, 172
danger of, 167
fire entrance and, 171
fire extinguishment and, 172–173
fire spread and, 172
grid work/framing and, 168
identification of, 175
metal grid type, 170
metal hangers/framework holding up, 172–174
New York City Building Code and, 173
New York City Hall on, 173–174
removable panel type, 170, 171
types of, 170–171
weight of, 171
wood grid type, 170

T

tail beam, 23
target hazards, abandoned derelict buildings as, 3
Tate, John, 117
temperature failure
of steel, 143–145, 151–152
structural fire v. test fire, 145
tenon, 27, 28, 31
tension stress, 30
tent floor collapse, 18, 84
tent void, 248
terrazzo floor, 31–32
collapse of, 32
"through the door and through the floor" collapse, xiii, 205, 319
firefighter deaths from, 322, 324
firefighter experience of, 322
increase of, 321
timber, 32, 108
timber truss roof
attic space quick access and, 114
ceiling concealment of, 112–113
ceiling not concealing, 113
construction of, 108
dangers of, 109
defensive strategy needed for, 110
fire strategy for, 112–114
identification of, 112
lessons learned from, 115
roof venting of, 114
shape variety of, 108
size-up errors with, 110–112
timber truss roof collapse, 103–124
Cardinal bowling alley, 104–106
death history of, 107
defensive strategy needed for, 110
firefighter experience of, 103–104
Ford auto dealership, 104, 106–107
front wall collapse from, 109, 110
lessons learned from, 124
parapet wall collapse and, 71–72
place of worship and, 116–119
"tragic, timber truss trilogy" of, xiv, 104–106, 107
at Valley Stream, New York, 117
time-out stage, 253
of search-and-rescue, 245–249
victims heard during, 245
torsional load, 27

"tragic, timber truss trilogy," xiv
    Cardinal bowling alley as, 104–106
    firefighter death and, 107
    Ford auto dealership as, 104, 106, 107
    Waldbaum's supermarket as, 104–106
transept, 125
transit, 249, 262
triforium, 125
trimmer beam, 23, 32
truss, 33, 108
    bowstring, 58, 72, 109, 112, 115
    bridge, 15
    failure of, 115
    hat, 23
    identification of, 110, 112, 114, 160
    strength/stability of, 109
tunneling/trenching operation, 248, 252
    collapse site rubble removal and, 251
    pass-along method for, 251
Type I resistive construction, 147
Type II noncombustible steel construction, xiii, 133, 147
Type III collapse danger, place of worship as, 116
Type III construction, 120
Type IV heavy timber mill building, 120

U

Ulshafer, Roger, 299
Underwriters Testing Laboratory, 205
Uniformed Fire Officer's Association, 172
unrestrained beam end, 29, 33
unsupported lean-to floor collapse, 83
urban flying brands, 296
U-returned stairs, 182, 183
    construction of, 184–185
utility shut-off, for search-and-rescue, 244

V

Valley Stream, New York timber truss roof collapse, 117
Vendome Hotel collapse, 1, 78, 80
    illegal renovations and, 5
veneer wall, 35, 36
    collapse of, 230
venting
    broken glass from, 7
    C-beam roof construction and, 150
    danger of, 7
    for flat roof, 128
    horizontal, 142
    place of worship, 119–120
    primary, 124
    reasons for, 120
    of roof, 114
    of rose window, 119
    skylights/scuttle covers and, 114
    of steel bar joist system, 133
    truss identification during, 114
Veri, Frank Jr., 77
vertical collapse zone, 36, 37
victims, trapped in voids, 244–249
    calling out for, 245
    in pancake void, 248
    in supported lean-to voids, 247
    in tent void, 248
    in unsupported lean-to voids, 247
    V-shaped voids and, 246
victim-tracking officer, 256
    duty of, 248
    planning team and, 246
    research by, 248
voids
    pancake, 248
    supported lean-to, 247
    tent, 248
    unsupported lean-to, 247
    V-shaped, 246
V-shape floor collapse, 18, 19, 84
V-shaped voids, 246

W–X

Waldbaum's supermarket timber truss collapse, 104–106
    bowstring truss roof design of, 307
    conclusion on, 308–309
    fire-retarding compartmentation and, 306
    overhead view of, 307
    post-fire analysis of, 304–309
    rain roof at, 138
wall(s). *See also* parapet wall
    area, 33
    bearing, 33
    collapse hierarchy of, 35
    column, 15

demising, 33
freestanding, 33
freestanding parapet, 67–68
overload, 208
party, 34
spandrel, 35
veneer, 35
wall collapse, 35
curtain-fall, 16, 52–58
inward/outward, 16, 58–60, 200, 202, 205–208
lean-over, 16, 200, 201
ninety-degree-angle, 16
of place of worship, 123
water accumulations
marquees and, 229
parapet walls and, 229
wind load, 27
wire entrapment, 174
Wire Lathers Union Local 46, 173
Wolfe, Arnie, 161, 319
Wonder Drug Store floor collapse, 80
illegal renovations and, 5
wood floor collapse, 75–89
firefighter experience of, 75–77
wood joists
flat roof collapse and, 133
steel bar joist system v., 133
wood-frame building collapse, 199–209
exterior wall overload and, 208
firefighter experience of, 199
inward/outward, 200, 202, 205–208
lean-over type of, 200, 201
lessons learned from, 208–209
mortise-and-tenon connection and, 205, 207
90-degree angle type of, 200, 201
performance v. specification building codes and, 209, 276
three-story v. one- or two-story, 203
warning signs of, 200, 202
wood-frame construction, 40
collapse hazards of, 52
fire problems of, 47–48
methods of, 203–204
worker shanty fires, 218
World Trade Center, 238. *See also* Building 7, of World Trade Center
bearing walls/open-floor design in, 273
build by Port Authority, 277, 290
building complexes of, 296
construction of, 272–274
conventional skyscrapers v., 272
Empire State Building v., 273–274
fire resistive rating of, 87
hollow-tube bearing walls in, 278, 279, 281
as lightweight construction, 132, 281
lightweight materials in, 273
60-foot open-web bar joists in, 272
spray-on fire retarding material and, 132, 147–148, 283
stairways of, 276
steel skeleton framing in, 273–274
structural framework of, 279
World Trade Center collapse, xiii, 1, 9, 80, 271–292
bar joist collapse and, 4
of Building 7, 293–301
deaths from, 271
evacuation order in, 285–286
firefighter evacuation and, 266–267
lecture on, 327
lessons learned from, 278–280
as progressive collapse, 281
rubble of, 275
site unsecured at, 240
worship. *See* place of worship

Y

yoke control value (OSY), 223
Young, Frank, 161

Z

zone of danger
for falling objects, 7
firefighter evacuation and, 264
horizontal/vertical collapse, 36, 37
safety precautions and, 262–264